INSTRU[illegible]
SOLUTIONS MANUAL

MW01411750

APPLIED MULTIVARIATE STATISTICAL ANALYSIS

FOURTH EDITION

RICHARD A. JOHNSON
DEAN W. WICHERN

PRENTICE HALL, Upper Saddle River, NJ 07458

Executive Editor: Ann Heath
Supplement Editor: Mindy McClard
Special Projects Manager: Barbara A. Murray
Production Editor: Barbara A. Till
Supplement Cover Manager: Paul Gourhan
Supplement Cover Designer: PM Workshop Inc.
Manufacturing Buyer: Alan Fischer

© 1998 by Prentice-Hall, Inc.
Simon & Schuster / A Viacom Company
Upper Saddle River, NJ 07458

All rights reserved. No part of this book may be reproduced, in any form or by any means, without permission in writing from the publisher

Printed in the United States of America

10 9 8 7 6 5 4 3 2 1

ISBN 0-13-834202-4

Prentice-Hall International (UK) Limited, *London*
Prentice-Hall of Australia Pty. Limited, *Sydney*
Prentice-Hall Canada, Inc., *London*
Prentice-Hall Hispanoamericana, S.A., *Mexico*
Prentice-Hall of India Private Limited, *New Delhi*
Prentice-Hall of Japan, Inc., *Tokyo*
Simon & Schuster Asia Pte. Ltd., *Singapore*
Editora Prentice-Hall do Brazil, Ltda., *Rio de Janeiro*

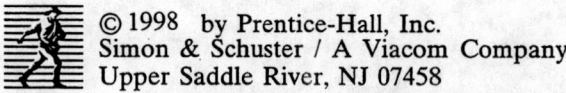

Contents

Preface

This solutions manual was prepared as an aid for instructors who will benefit by having solutions available. In addition to providing detailed answers for most of the problems in the book, this manual can help the instructor determine which of the problems are most appropriate for the class.

Most of the problems have been solved with the help of available computer software. A few of the problems have been solved with hand calculators. The reader should keep in mind that round-off errors can occur—particularly in those problems involving long chains of arithmetic operations.

We would like to take this opportunity to acknowledge the contribution of our many students, whose homework formed the basis for many of the solutions. In particular, we would like to thank Jorge Achcar, Sebastiao Amorim, W. K. Cheang, S. S. Cho, S. G. Chow, Charles Fleming, Stu Janis, Richard Jones, Tim Kramer, Dennis Murphy, Rich Raubertas, David Steinberg, T. J. Tien, Steve Verrill, Paul Whitney and Mike Wincek. Professor Sam Kotz and his students at the University of Maryland have pointed out errors and suggested improvements in previous editions of this manual, as well as the book. Deborah Smith compiled most of the material needed to make this current solutions manual consistent with the fourth edition of the text. We are grateful for her help.

The solutions are numbered in the same manner as the exercises in the book. Thus, for example, 9.6 refers to the 6th exercise of chapter 9.

We hope this manual is a useful aid for adopters of our Applied Multivariate Statistical Analysis, 4th edition, text. Corrections, comments and suggestions are always welcome.

Richard A. Johnson
Dean W. Wichern

Chapter 1

1.1 $\bar{x}_1 = 4.29$ $\bar{x}_2 = 6.29$

$s_{11} = 4.20$ $s_{22} = 3.56$ $s_{12} = 3.70$

1.2 a)

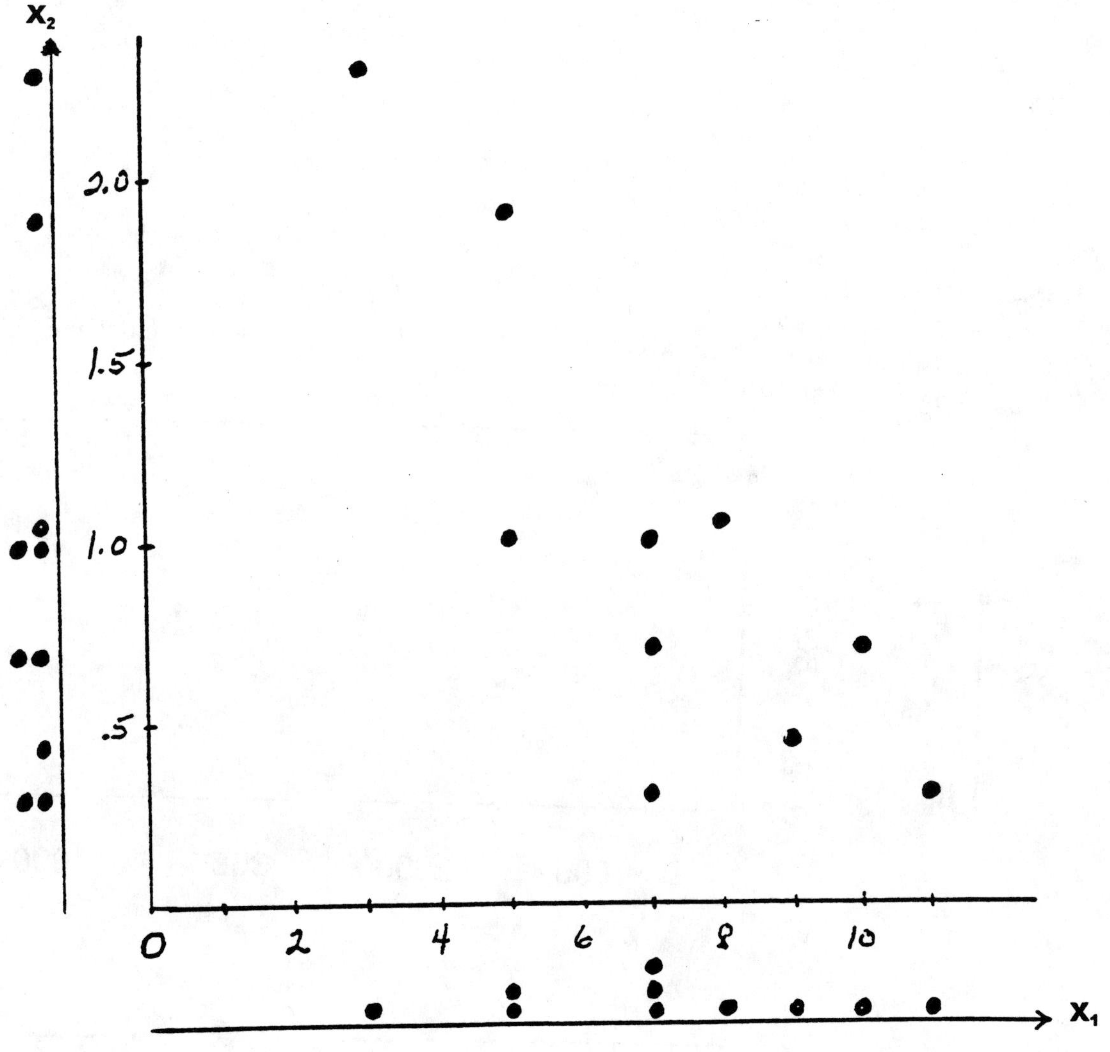

b) s_{12} negative

c) $\bar{x}_1 = 7.2$ $\bar{x}_2 = .97$ $s_{11} = 5.360$ $s_{22} = .396$

$s_{12} = -1.169$ $r_{12} = -.80$. Large x_1 occurs with small x_2 and vice versa.

d) $\underset{\sim}{\bar{x}} = \begin{bmatrix} 7.2 \\ .97 \end{bmatrix}$, $S_n = \begin{bmatrix} 5.360 & -1.169 \\ -1.169 & .396 \end{bmatrix}$, $R = \begin{bmatrix} 1 & -.80 \\ -.80 & 1 \end{bmatrix}$

1.3

$$\bar{\mathbf{x}} = \begin{bmatrix} 6 \\ 8 \\ 2 \end{bmatrix} \qquad S_n = \begin{bmatrix} 6 & 4 & -1.4 \\ & 8 & 1.2 \\ (\text{symmetric}) & & 2 \end{bmatrix} \qquad R = \begin{bmatrix} 1 & .577 & -.404 \\ & 1 & .300 \\ (\text{symmetric}) & & 1 \end{bmatrix}$$

1.4

(a). There is a positive correlation between X_1 and X_2. X_1 has long left hand tails, but X_2 has a long right hand tail.

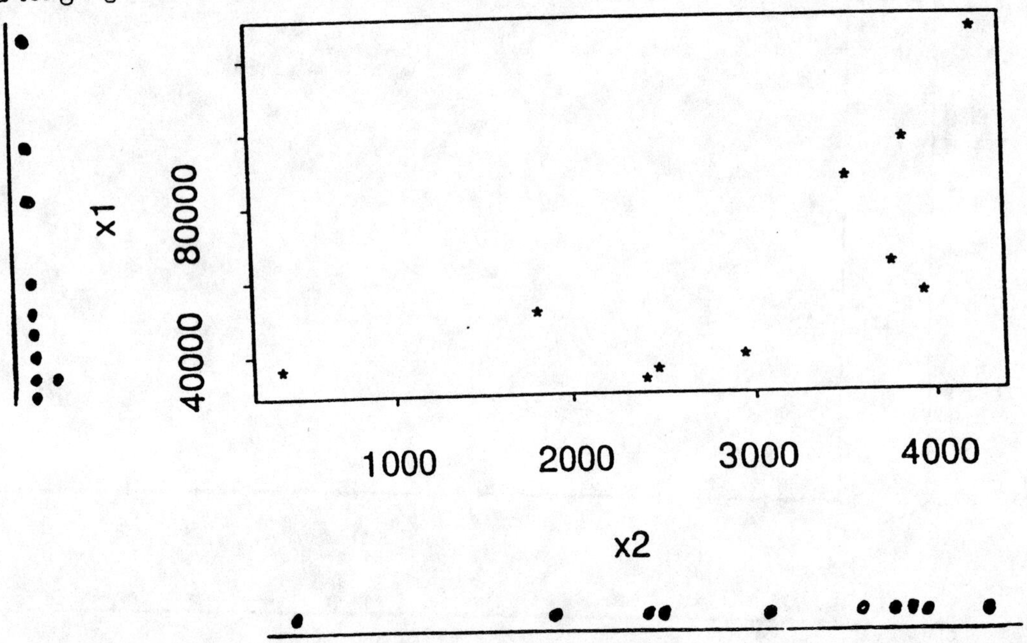

(b). $\bar{x}_1 = 62309.1$, $\bar{x}_2 = 2927.3$. $s_{11} = 900458202$, $s_{12} = 23018040$, $s_{22} = 1287018$. $r_{12} = 0.67615$.

1.5

(a). There are positive correlations among (X_1, X_2, X_3). X_1 and X_3 have long left hand tails, but X_2 has a long right hand tail.

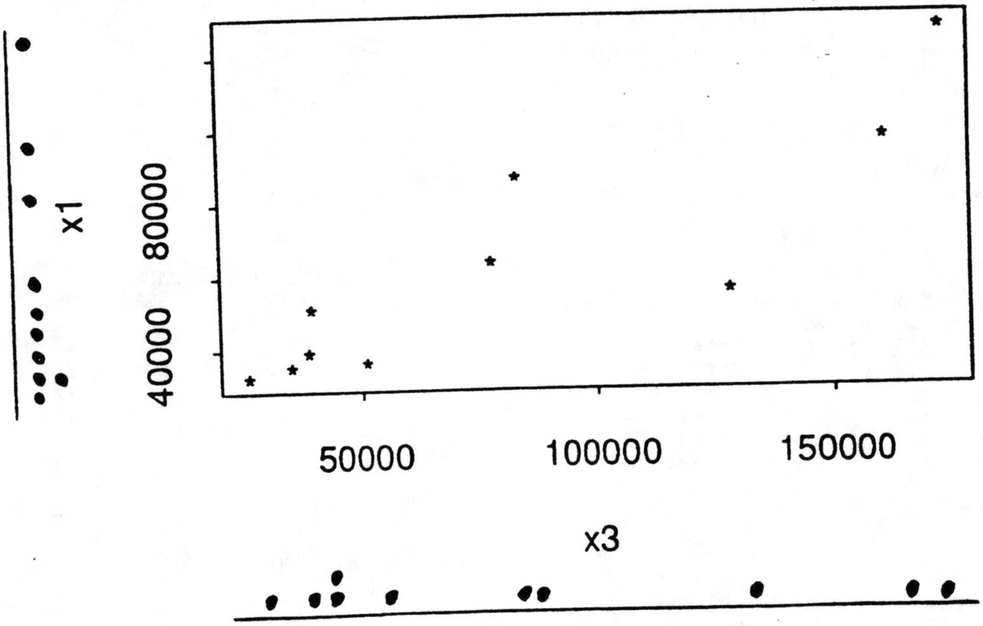

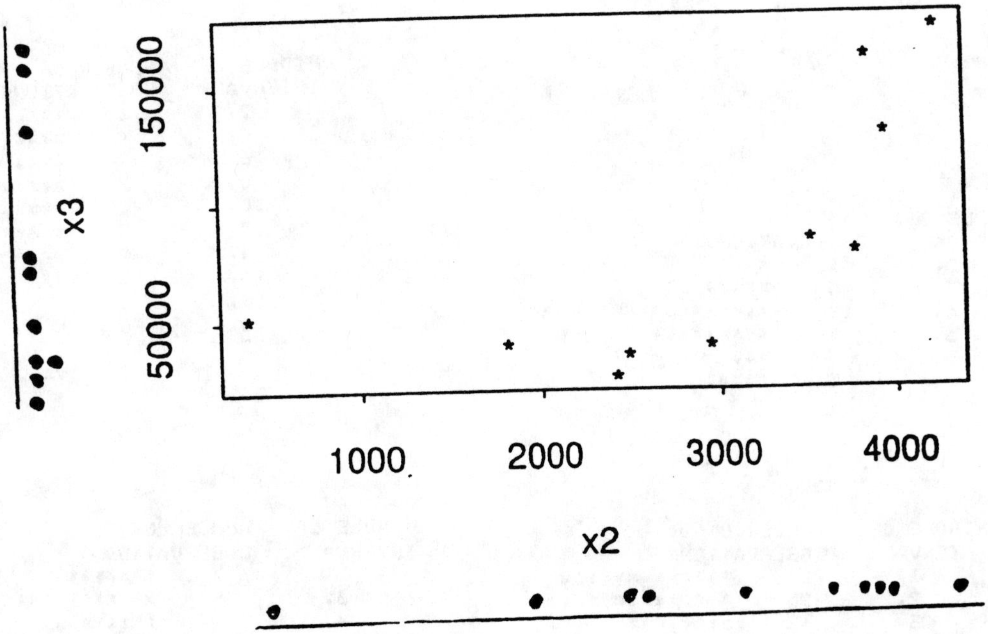

1.5 (b).

$$\bar{\mathbf{x}} = \begin{pmatrix} 62309.1 \\ 2927.3 \\ 81248.4 \end{pmatrix}, \quad \mathbf{S}_n = \begin{pmatrix} 900458202 & 23018040 & 1360644507 \\ & 1287018 & 41089157 \\ & & 2682440803 \end{pmatrix},$$

$$\mathbf{R} = \begin{pmatrix} 1 & 0.67615 & 0.87548 \\ & 1 & 0.69931 \\ & & 1 \end{pmatrix}$$

1.6 a) Histograms

X_1

MIDDLE OF INTERVAL	NUMBER OF OBSERVATIONS	
5.	5	*****
6.	8	********
7.	7	*******
8.	11	***********
9.	5	*****
10.	6	******

X_2

MIDDLE OF INTERVAL	NUMBER OF OBSERVATIONS	
30.	1	*
40.	3	***
50.	2	**
60.	3	***
70.	10	**********
80.	12	************
90.	8	********
100.	2	**
110.	1	*

X_3

MIDDLE OF INTERVAL	NUMBER OF OBSERVATIONS	
2.	1	*
3.	5	*****
4.	19	*******************
5.	9	*********
6.	3	***
7.	5	*****

X_4

MIDDLE OF INTERVAL	NUMBER OF OBSERVATIONS	
1.	13	*************
2.	15	***************
3.	8	********
4.	5	*****
5.	1	*

MTB>

X_5

MIDDLE OF INTERVAL	NUMBER OF OBSERVATIONS	
5.	2	**
6.	3	***
7.	5	*****
8.	5	*****
9.	6	******
10.	4	****
11.	4	****
12.	5	*****
13.	4	****
14.	1	*
15.	0	
16.	1	*
17.	0	
18.	1	*
19.	0	
20.	0	
21.	1	*

X_6

MIDDLE OF INTERVAL	NUMBER OF OBSERVATIONS	
2.	3	***
4.	4	****
6.	7	*******
8.	7	*******
10.	8	********
12.	5	*****
14.	2	**
16.	2	**
18.	1	*
20.	0	
22.	0	
24.	2	**
26.	1	*

X_7

MIDDLE OF INTERVAL	NUMBER OF OBSERVATIONS	
2.	7	*******
3.	25	*************************
4.	9	*********
5.	1	*

1.6 b)

$$\bar{\mathbf{x}} = \begin{bmatrix} 7.5 \\ 73.857 \\ 4.548 \\ 2.191 \\ 10.048 \\ 9.405 \\ 3.095 \end{bmatrix} \quad S_n = \begin{bmatrix} 2.440 & -2.714 & -.369 & -.452 & -.571 & -2.179 & .167 \\ & 293.360 & 3.816 & -1.354 & 6.602 & 30.058 & .609 \\ & & 1.486 & .658 & 2.260 & 2.755 & .138 \\ & & & 1.154 & 1.062 & -.791 & .172 \\ & & & & 11.093 & 3.052 & 1.019 \\ & & & & & 30.241 & .580 \\ \text{(symmetric)} & & & & & & .467 \end{bmatrix}$$

$$R = \begin{bmatrix} 1 & -.101 & -.194 & -.270 & -.110 & -.254 & .156 \\ & 1 & .183 & -.074 & .116 & .319 & .052 \\ & & 1 & .502 & .557 & .411 & .166 \\ & & & 1 & .297 & -.134 & .235 \\ & & & & 1 & .167 & .448 \\ & & & & & 1 & .154 \\ \text{(symmetric)} & & & & & & 1 \end{bmatrix}$$

The pair x_3, x_4 exhibits a small to moderate positive correlation and so does the pair x_3, x_5. Most of the entries are small.

1.7

a)

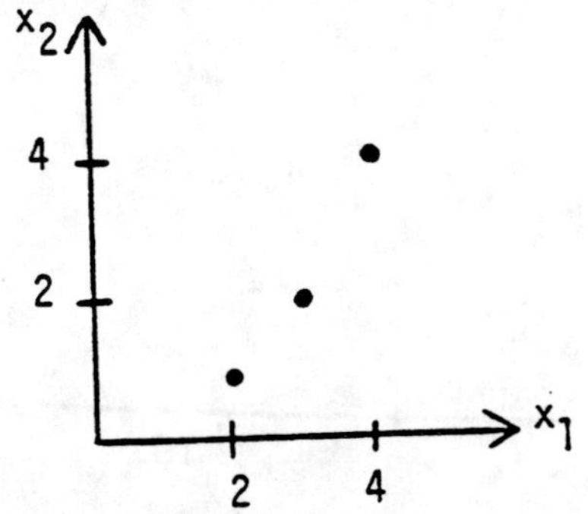

Scatter plot
(variable space)

b)

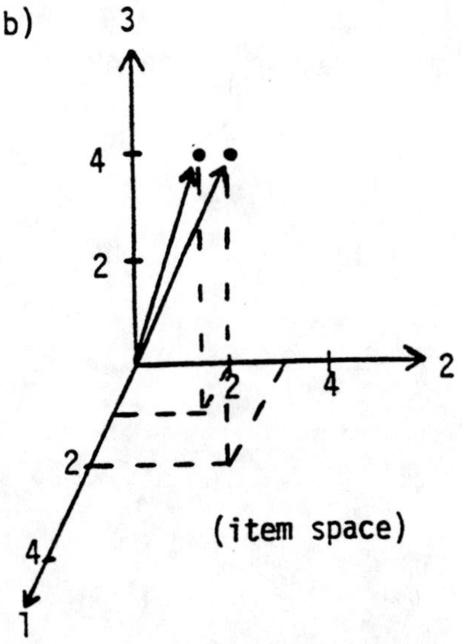

(item space)

1.8 Using (1-12) $d(P,Q) = \sqrt{(-1-1)^2+(-1-0)^2} = \sqrt{5} = 2.236$

Using (1-20) $d(P,Q) = \sqrt{\frac{1}{3}(-1-1)^2+2(\frac{1}{9})(-1-1)(-1-0)+\frac{4}{27}(-1-0)^2} = \sqrt{\frac{52}{27}} = 1.388$

Using (1-20) the locus of points a constant squared distance 1 from Q = (1,0) is given by the expression $\frac{1}{3}(x_1-1)^2+\frac{2}{9}(x_1-1)x_2+\frac{4}{27}x_2^2 = 1$. To sketch the locus of points defined by this equation, we first obtain the coordinates of some points satisfying the equation:

(-1,1.5), (0,-1.5), (0,3), (1,-2.6), (1,2.6), (2,-3), (2,1.5), (3,-1.5)

The resulting ellipse is:

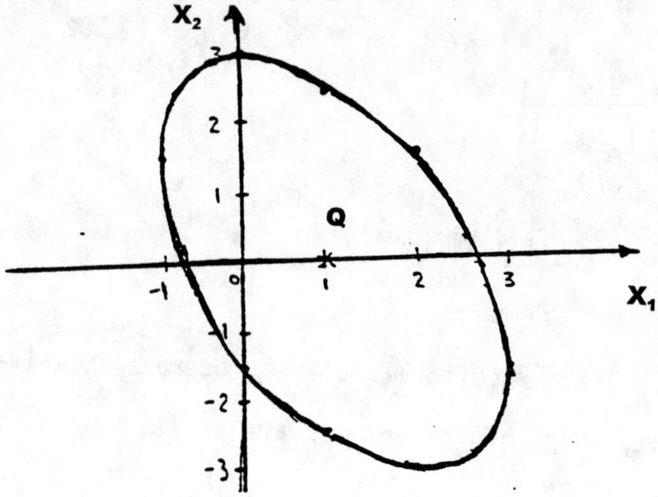

1.9 a) $s_{11} = 20.48$ $s_{22} = 6.19$ $s_{12} = 9.09$

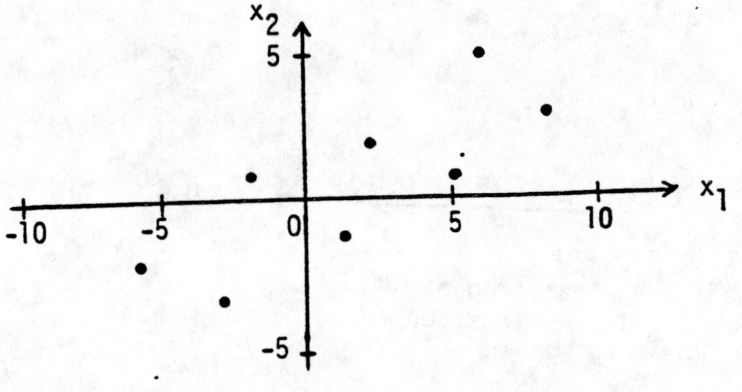

1.9 b)

$\tilde{x}_1$	-6.20	-4.10	-1.23	.37	2.73	4.83	7.70	8.43
$\tilde{x}_2$	1.27	-1.10	1.87	-1.37	.73	-1.63	1.33	-1.40

c)

$\tilde{s}_{11} = 24.90$ $\tilde{s}_{22} = 1.77$ (Note $\tilde{s}_{12} = .00$)

d)

$(\tilde{x}_1, \tilde{x}_2) = (2.72, -3.55)$

$d(0,P) = 2.72$ using (1-17).

e) $d(0,P) = 2.72$ using (1-19).

1.10 a) This equation is of the form (1-19) with $a_{11} = 1$, $a_{12} = \frac{1}{2}$ and $a_{22} = 4$. Therefore this is a distance for correlated variables if it is non-negative for all values of x_1, x_2. But this follows easily if we write $x_1^2 + 4x_2^2 + x_1x_2 = (x_1 + \frac{1}{2}x_2)^2 + \frac{15}{4}x_2^2 \geq 0$.

b) In order for this expression to be a distance it has to be non-negative for all values x_1, x_2. Since, for $(x_1, x_2) = (0,1)$ we have $x_1^2 - 2x_2^2 = -2$, we conclude that this is not a valid distance function.

1.11

$$d(P,Q) = \sqrt{4(x_1-y_1)^2 + 2(-1)(x_1-y_1)(x_2-y_2) + (x_2-y_2)^2}$$

$$= \sqrt{4(y_1-x_1)^2 + 2(-1)(y_1-x_1)(y_2-x_2) + (x_2-y_2)^2} = d(Q,P)$$

Next, $4(x_1-y_1)^2 - 2(x_1-y_1)(x_2-y_2) + (x_2-y_2)^2 =$

$$= (x_1-y_1-x_2+y_2)^2 + 3(x_1-y_1)^2 \geq 0 \text{ so } d(P,Q) \geq 0.$$

The second term is zero in this last expression only if $x_1 = y_1$ and then the first is zero only if $x_2 = y_2$.

1.12 a) If $P = (-3,4)$ then $d(0,P) = \max(|-3|,|4|) = 4$

b) The locus of points whose squared distance from (0,0) is 1 is

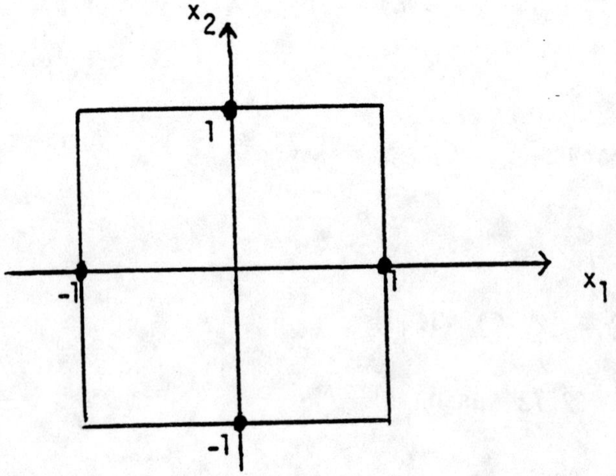

c) The generalization to p-dimensions is given by $d(0,P) = \max(|x_1|,|x_2|,\ldots,|x_p|)$.

1.13 Place the facility at C-3.

1.14 a)

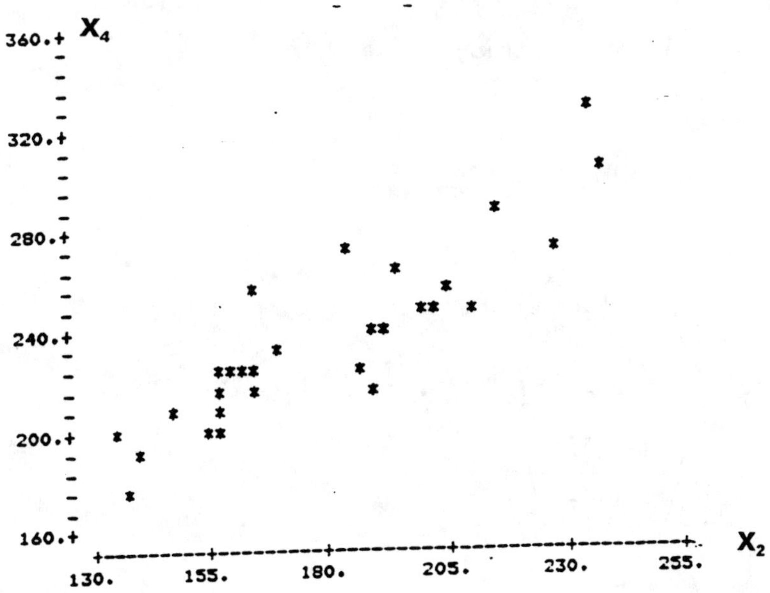

Strong positive correlation. No obvious "unusual" observations.

b) <u>Multiple-sclerosis group</u>.

$$\bar{\mathbf{x}} = \begin{pmatrix} 42.07 \\ 179.64 \\ 12.31 \\ 236.62 \\ 13.16 \end{pmatrix}$$

$$S_n = \begin{pmatrix} 116.91 & 61.78 & -20.10 & 61.13 & -27.65 \\ & 812.72 & 218.35 & 865.32 & 90.48 \\ & & 305.94 & 221.93 & 286.60 \\ & & & 1146.38 & 82.53 \\ \text{(symmetric)} & & & & 337.80 \end{pmatrix}$$

$$R = \begin{pmatrix} 1 & .200 & -.106 & .167 & -.139 \\ & 1 & .438 & .896 & .173 \\ & & 1 & .375 & .892 \\ & & & 1 & .133 \\ & \text{(symmetric)} & & & 1 \end{pmatrix}$$

<u>Non multiple-sclerosis group.</u>

$$\bar{\underset{\sim}{x}} = \begin{pmatrix} 37.99 \\ 147.21 \\ 1.56 \\ 195.57 \\ 1.62 \end{pmatrix}$$

$$S_n = \begin{pmatrix} 273.61 & 95.08 & 5.28 & 101.67 & 3.20 \\ & 110.13 & 1.84 & 103.28 & 2.15 \\ & & 1.78 & 2.22 & .49 \\ & & & 183.04 & 2.35 \\ & \text{(symmetric)} & & & 2.32 \end{pmatrix}$$

$$R = \begin{pmatrix} 1 & .548 & .239 & .454 & .127 \\ & 1 & .132 & .727 & .134 \\ & & 1 & .123 & .244 \\ & & & 1 & .114 \\ & \text{(symmetric)} & & & 1 \end{pmatrix}$$

1.15 a) Scatterplot of x_2 and x_3.

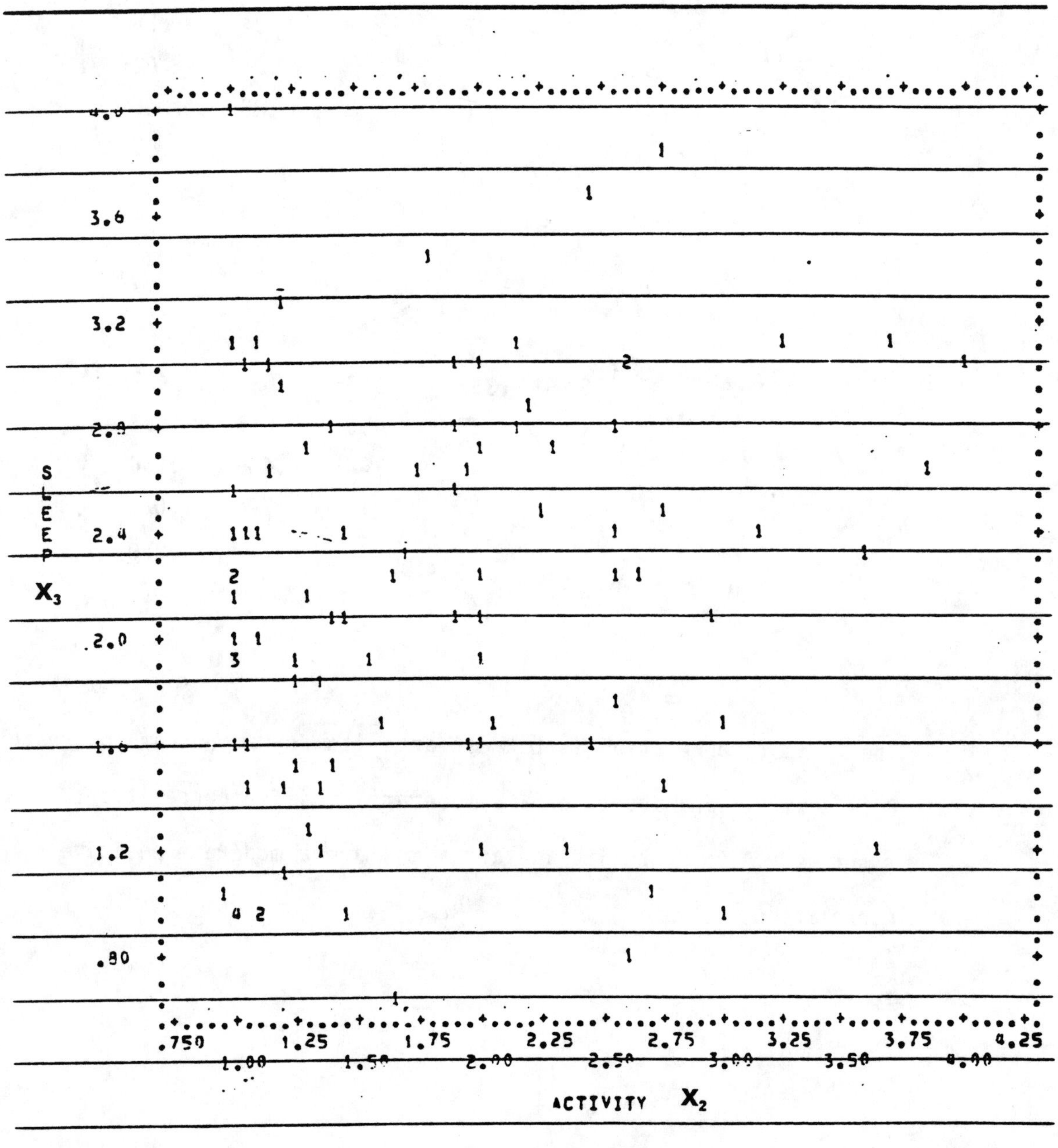

b)

$$\bar{\mathbf{x}} = \begin{pmatrix} 3.54 \\ 1.81 \\ 2.14 \\ 2.21 \\ 2.58 \\ 1.27 \end{pmatrix}$$

1.15

$$S_n = \begin{pmatrix} 4.61 & .92 & .58 & .27 & 1.06 & .15 \\ & .61 & .11 & .12 & .39 & -.02 \\ & & .57 & .09 & .34 & .11 \\ & & & .11 & .21 & .02 \\ & & & & .85 & -.01 \\ & \text{(symmetric)} & & & & .85 \end{pmatrix}$$

$$R = \begin{pmatrix} 1 & .551 & .362 & .386 & .537 & .077 \\ & 1 & .187 & .455 & .535 & -.035 \\ & & 1 & .346 & .496 & .156 \\ & & & 1 & .704 & .071 \\ & & & & 1 & -.010 \\ & \text{(symmetric)} & & & & 1 \end{pmatrix}$$

The largest correlation is between appetite and amount of food eaten. Both activity and appetite have moderate positive correlations with symptoms. Also, appetite and activity have a moderate positive correlation.

1.16

There are significant positive correlations among all variables. The lowest correlation is 0.4420 between Dominant humerus and Ulna, and the highest correlation is 0.89365 bewteen Dominant hemerus and Hemerus.

$$\bar{\mathbf{x}} = \begin{pmatrix} 0.8438 \\ 0.8183 \\ 1.7927 \\ 1.7348 \\ 0.7044 \\ 0.6938 \end{pmatrix}, \quad \mathbf{R} = \begin{pmatrix} 1.00000 & 0.85181 & 0.69146 & 0.66826 & 0.74369 & 0.67789 \\ 0.85181 & 1.00000 & 0.61192 & 0.74909 & 0.74218 & 0.80980 \\ 0.69146 & 0.61192 & 1.00000 & 0.89365 & 0.55222 & 0.44020 \\ 0.66826 & 0.74909 & 0.89365 & 1.00000 & 0.62555 & 0.61882 \\ 0.74369 & 0.74218 & 0.55222 & 0.62555 & 1.00000 & 0.72889 \\ 0.67789 & 0.80980 & 0.44020 & 0.61882 & 0.72889 & 1.00000 \end{pmatrix},$$

$$\mathbf{S}_n = \begin{pmatrix} 0.0124815 & 0.0099633 & 0.0214560 & 0.0192822 & 0.0087559 & 0.0076395 \\ 0.0099633 & 0.0109612 & 0.0177938 & 0.0202555 & 0.0081886 & 0.0085522 \\ 0.0214560 & 0.0177938 & 0.0771429 & 0.0641052 & 0.0161635 & 0.0123332 \\ 0.0192822 & 0.0202555 & 0.0641052 & 0.0667051 & 0.0170261 & 0.0161219 \\ 0.0087559 & 0.0081886 & 0.0161635 & 0.0170261 & 0.0111057 & 0.0077483 \\ 0.0076395 & 0.0085522 & 0.0123332 & 0.0161219 & 0.0077483 & 0.0101752 \end{pmatrix}.$$

1.17

There are high positive correlations among all variables. The lowest correlation is 0.68557 between 200m and Marathon, and the highest correlation is 0.96917 bewteen 1500m and 3000m.

$$\bar{\mathbf{x}} = \begin{pmatrix} 11.6185 \\ 23.6416 \\ 53.4058 \\ 2.0764 \\ 4.3255 \\ 9.4476 \\ 173.2533 \end{pmatrix}, \quad \mathbf{S}_n = \begin{pmatrix} 0.2008 & 0.4700 & 0.9926 & 0.0350 & 0.1075 & 0.2715 & 9.2726 \\ 0.4700 & 1.2120 & 2.5038 & 0.0855 & 0.2532 & 0.6383 & 22.7572 \\ 0.9926 & 2.5038 & 7.0431 & 0.2557 & 0.6887 & 1.6857 & 56.4471 \\ 0.0350 & 0.0855 & 0.2557 & 0.0115 & 0.0318 & 0.0756 & 2.5197 \\ 0.1075 & 0.2532 & 0.6887 & 0.0318 & 0.1085 & 0.2608 & 8.7193 \\ 0.2715 & 0.6383 & 1.6857 & 0.0756 & 0.2608 & 0.6672 & 22.1613 \\ 9.2726 & 22.7572 & 56.4471 & 2.5197 & 8.7193 & 22.1613 & 909.1216 \end{pmatrix},$$

$$\mathbf{R} = \begin{pmatrix} 1.00000 & 0.95279 & 0.83469 & 0.72769 & 0.72837 & 0.74170 & 0.68634 \\ 0.95279 & 1.00000 & 0.85696 & 0.72406 & 0.69836 & 0.70987 & 0.68557 \\ 0.83469 & 0.85696 & 1.00000 & 0.89841 & 0.78784 & 0.77764 & 0.70542 \\ 0.72769 & 0.72406 & 0.89841 & 1.00000 & 0.90161 & 0.86357 & 0.77929 \\ 0.72837 & 0.69836 & 0.78784 & 0.90161 & 1.00000 & 0.96917 & 0.87793 \\ 0.74170 & 0.70987 & 0.77764 & 0.86357 & 0.96917 & 1.00000 & 0.89984 \\ 0.68634 & 0.68557 & 0.70542 & 0.77929 & 0.87793 & 0.89984 & 1.00000 \end{pmatrix}.$$

1.19 (a)

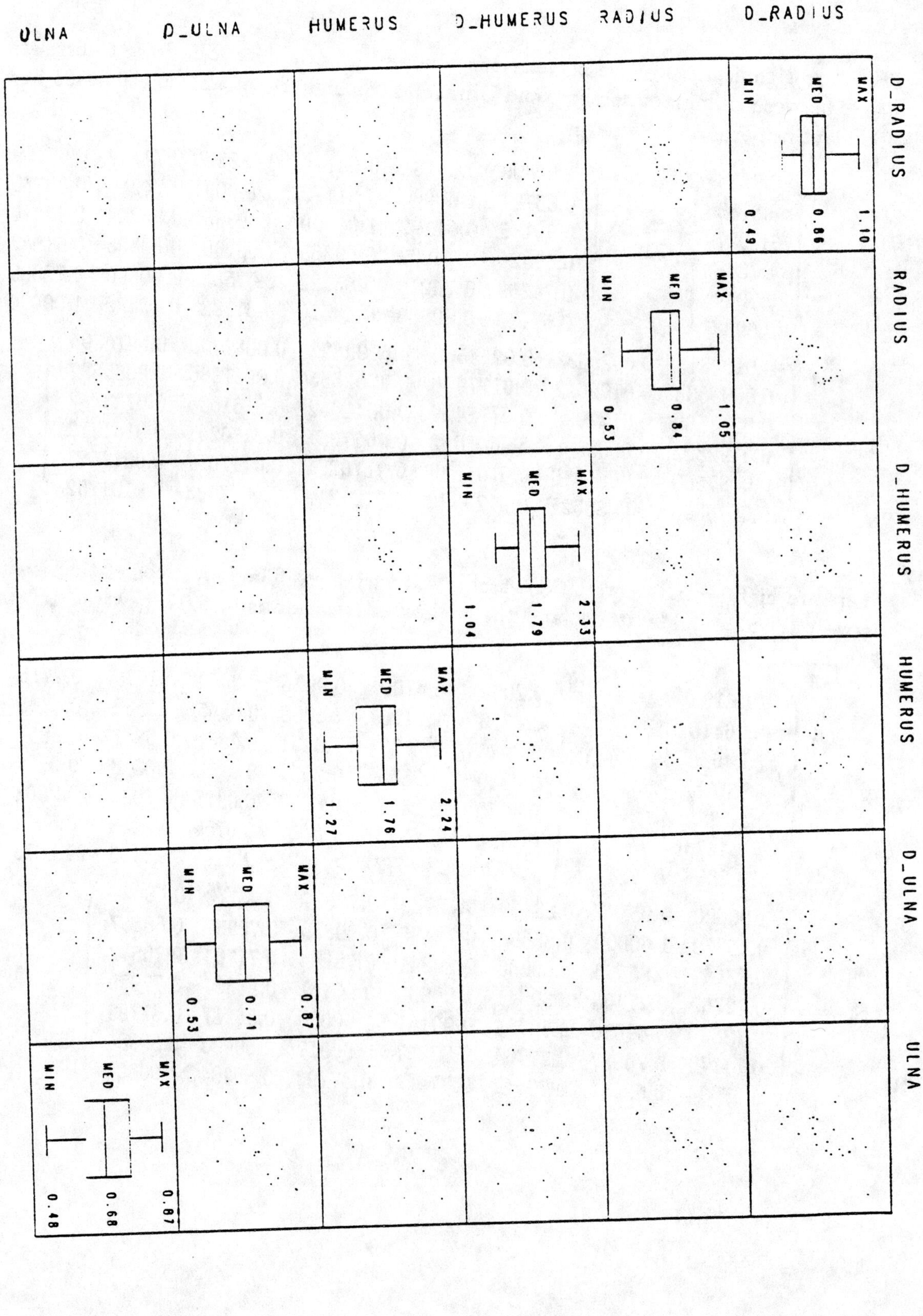

ULNA
D_ULNA
HUMERUS
D_HUMERUS
RADIUS
D_RADIUS
MAX 1.10 MED 0.86 MIN 0.49
MAX 1.05 MED 0.84 MIN 0.53
MAX 2.33 MED 1.79 MIN 1.04
MAX 2.24 MED 1.76 MIN 1.27
MAX 0.87 MED 0.71 MIN 0.53
MAX 0.87 MED 0.68 MIN 0.48

1.19

(b)

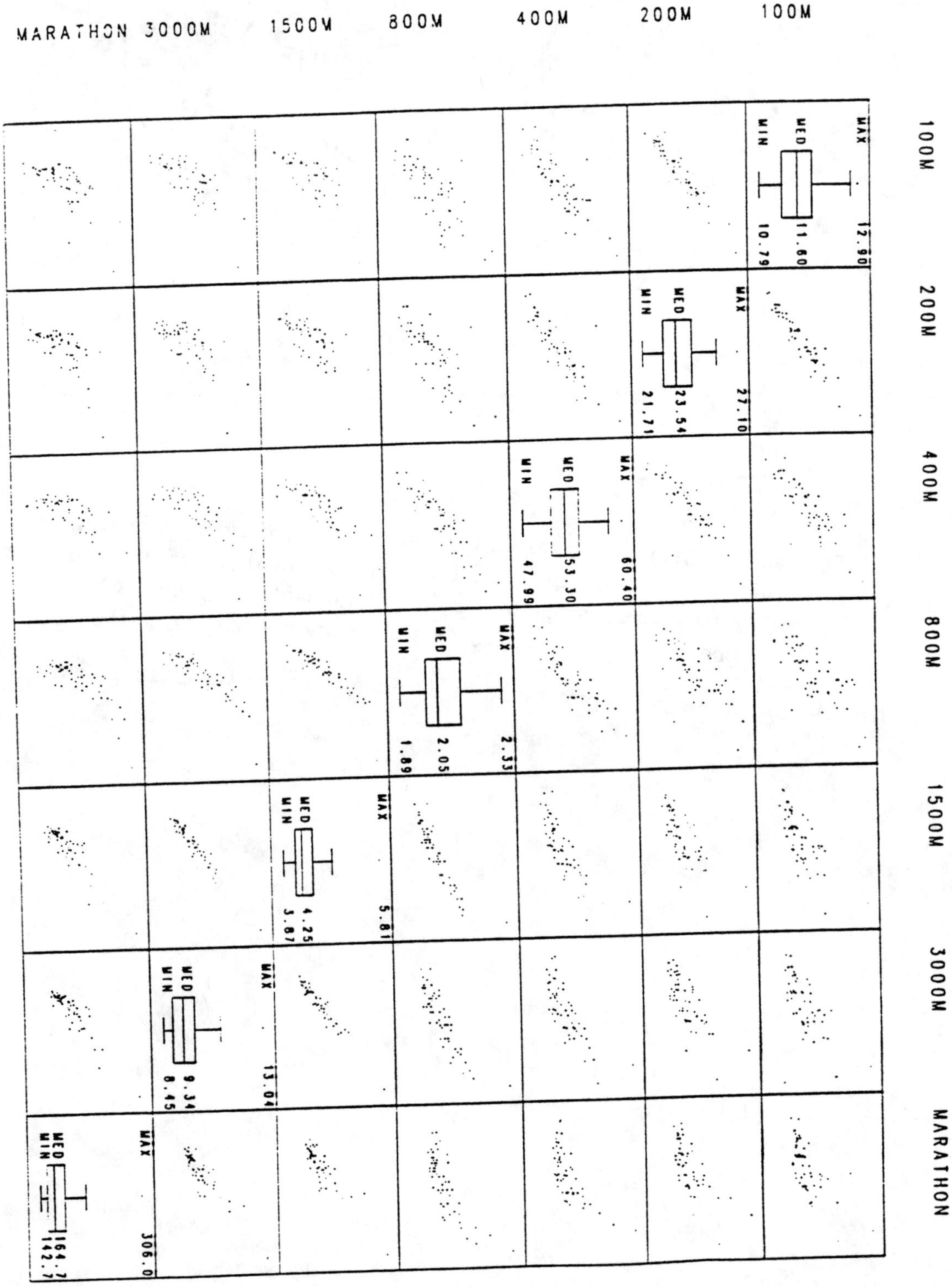
MARATHON
3000M
1500M
800M
400M
200M
100M
MAX
MED
MIN
12.90
11.60
10.79
27.10
23.54
21.71
60.40
53.30
47.99
2.33
2.05
1.89
5.81
4.25
3.87
13.04
9.34
8.45
306.0
164.7
142.7

1.20

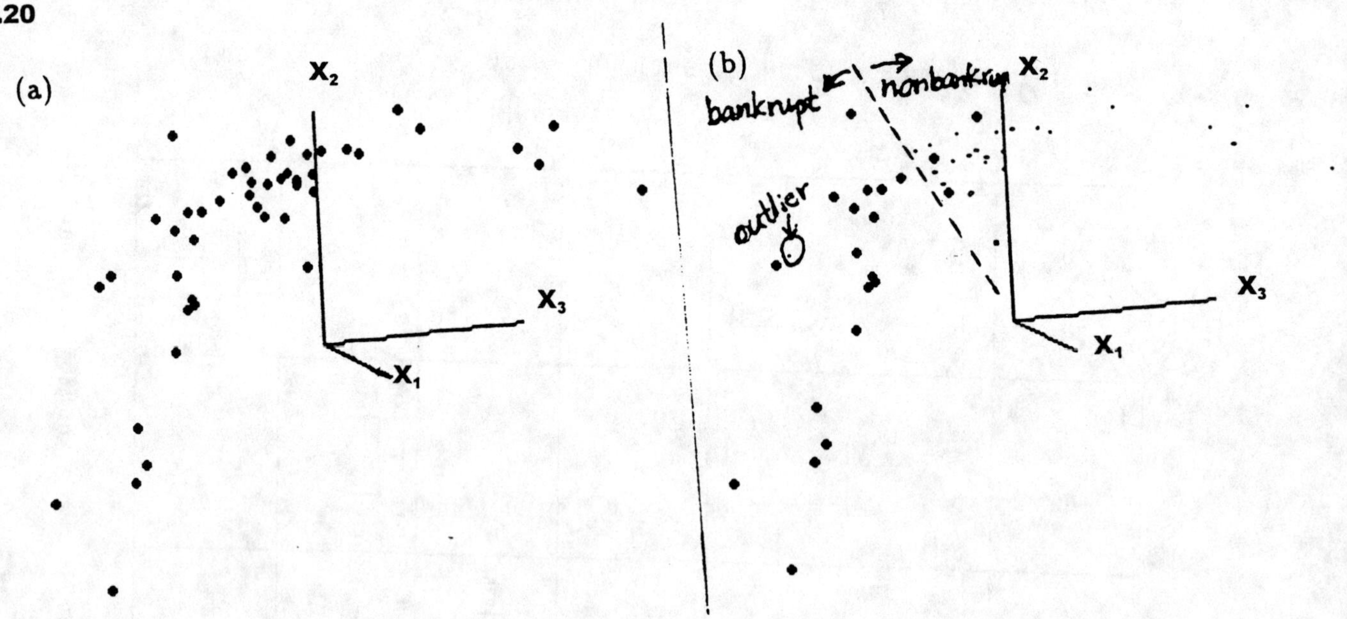

(a) The plot looks like a cigar shape, but bent. Some observations in the lower left hand part could be outliers. From the highlighted plot in (b) (actually non-bankrupt group not highlighted), there is one outlier in the nonbankrupt group, which is apparently located in the bankrupt group, besides the strung out pattern to the right.

(b) The dotted line in the plot would be an orientation for the classification.

1.21

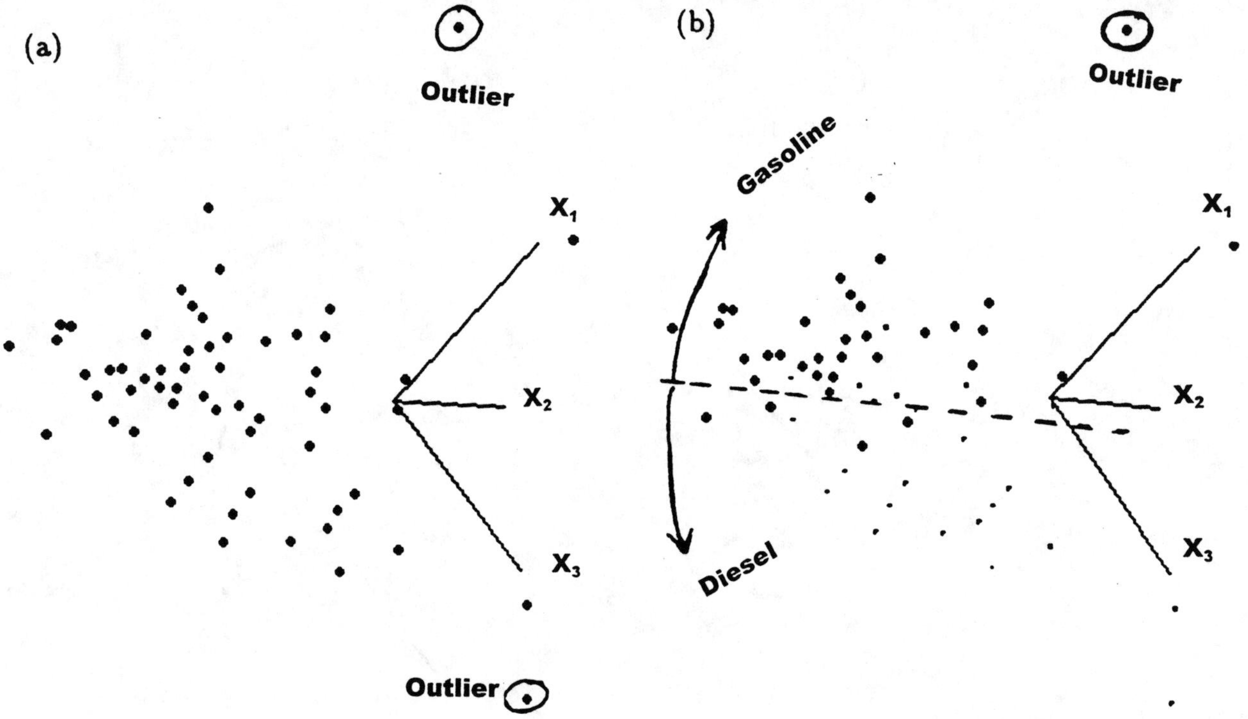

(a) There are two outliers in the upper right and lower right corners of the plot.

(b) Only the points in the gasoline group are highlighted. The observation in the upper right is the outlier. As indicated in the plot, there is an orientation to classify into two groups.

1.22 Possible outliers are indicated.

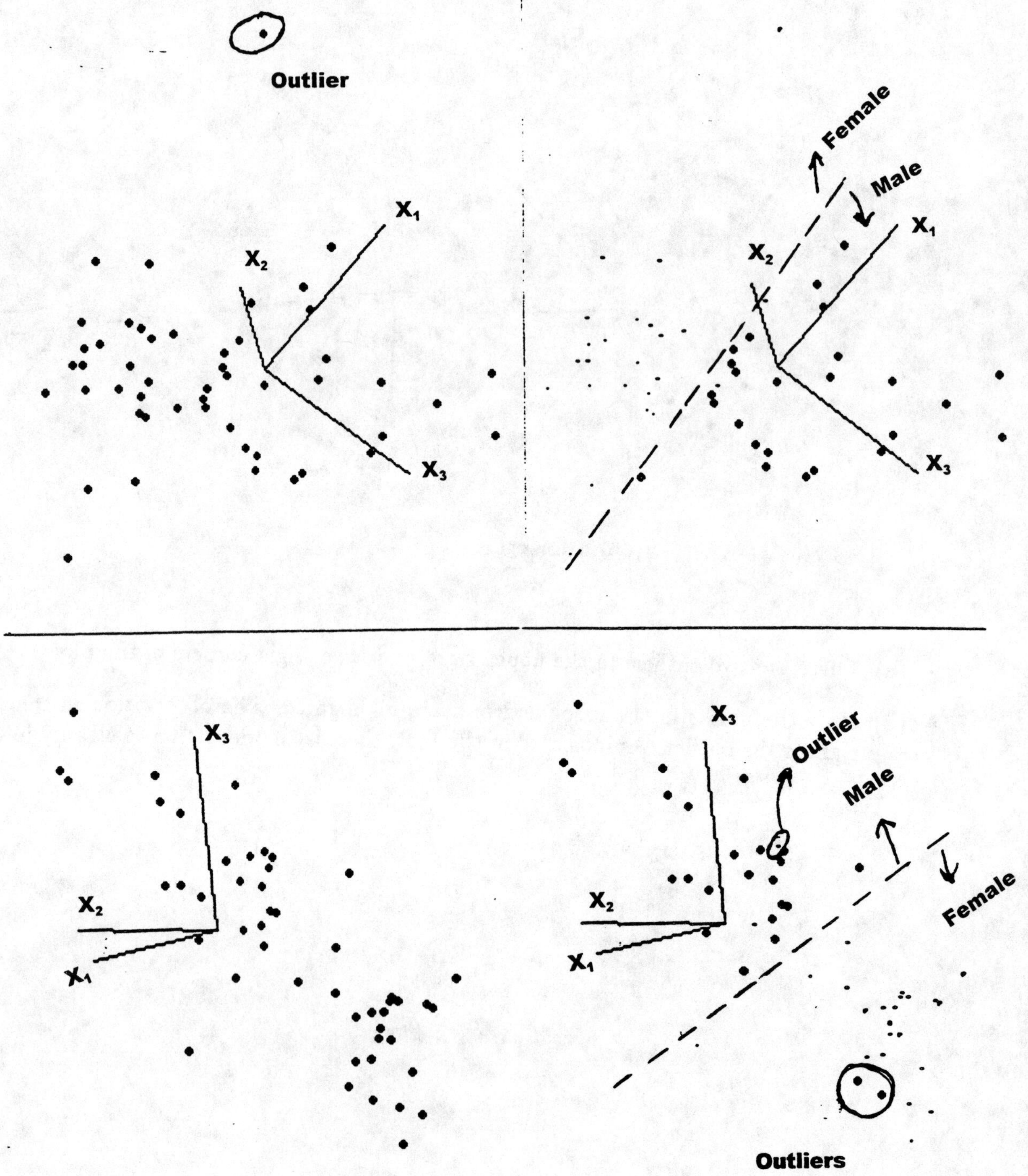

1.23 b) A visual clustering using Chernoff faces is given below.

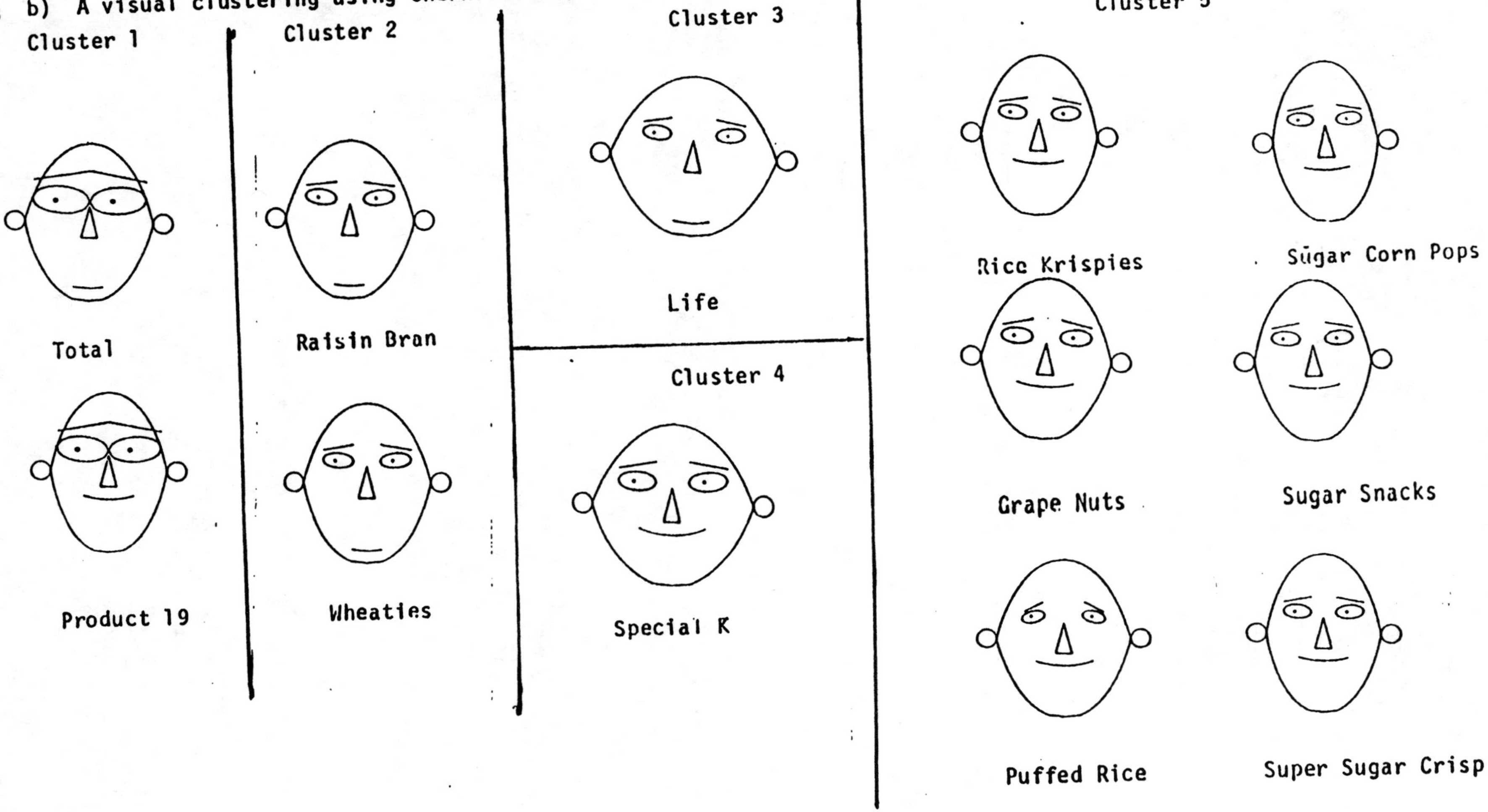

1.24

Cluster 1

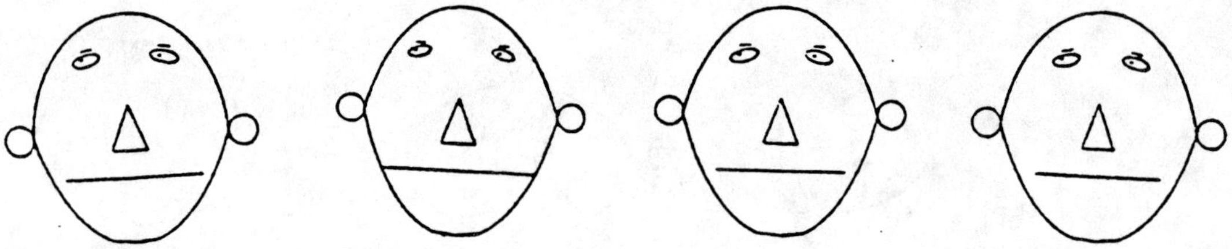

20 13 10 4

Cluster 2

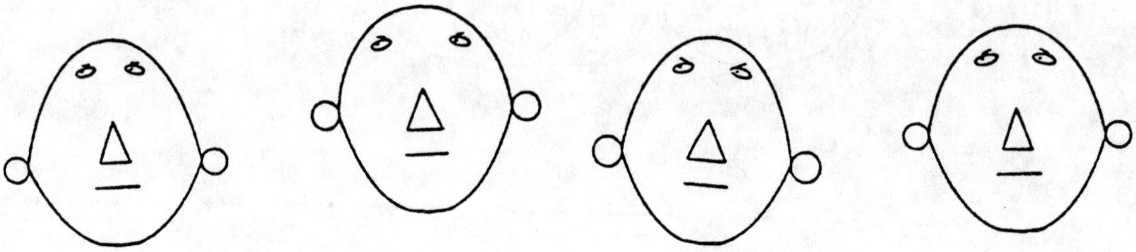

14 9 3 1

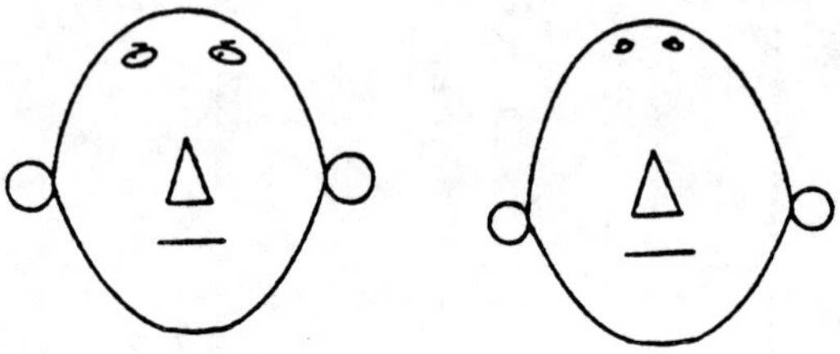

18 19

Cluster 3

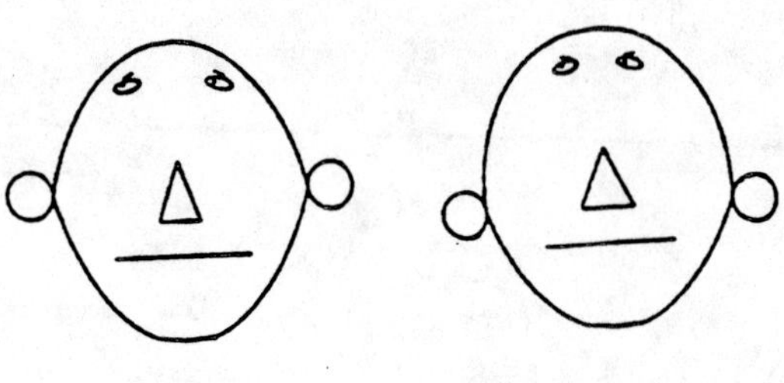

22 6

Cluster 4

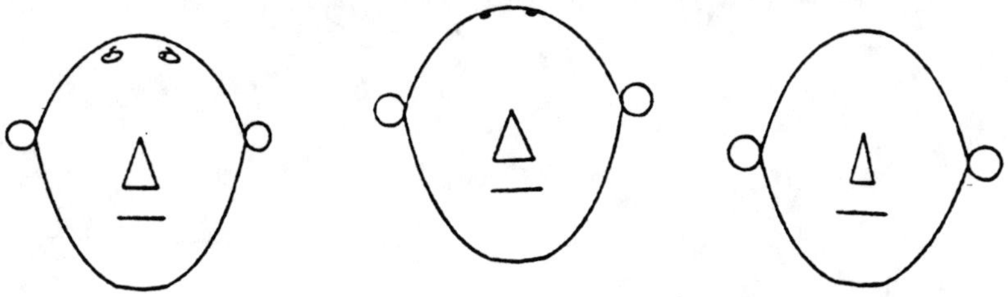

8 16 11

Cluster 5

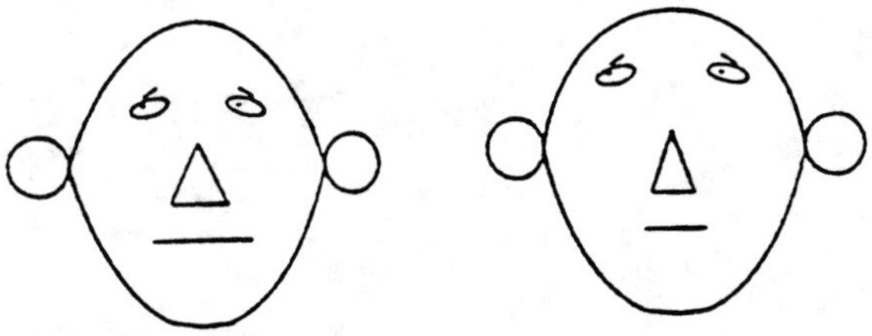

5 21

Cluster 6

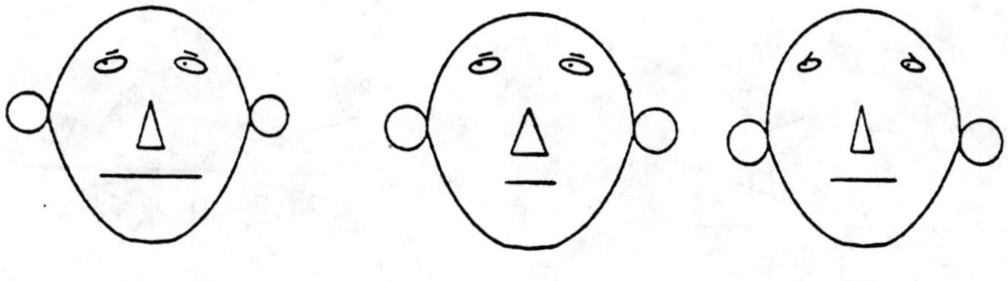

2 12 17

Cluster 7

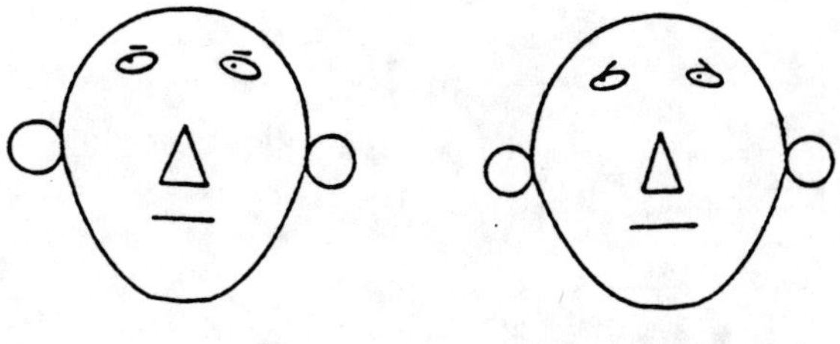

7 15

We have clustered these faces in the same manner as those in Example 1.9 . Note, however, other groupings are equally plausible. For instance, utilities 9 and 18 might be switched from Cluster 2 to Cluster 3 and so forth.

1.25 We illustrate one cluster of "stars". The remaining stars (not shown) can be grouped in 3 or 4 additional clusters.

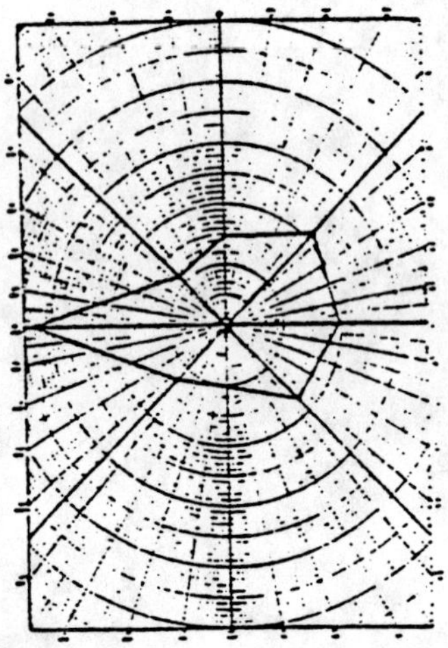

4

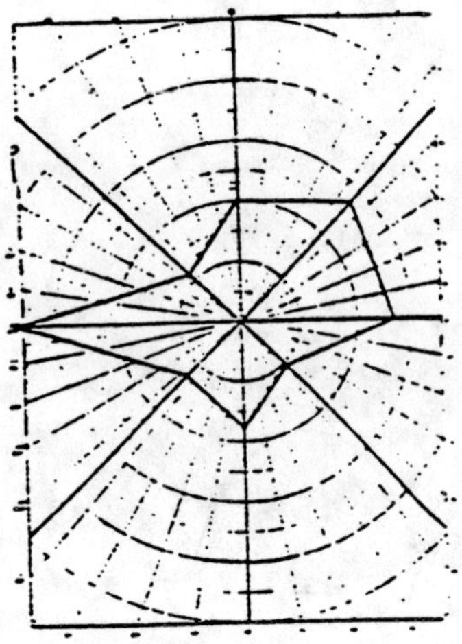

10

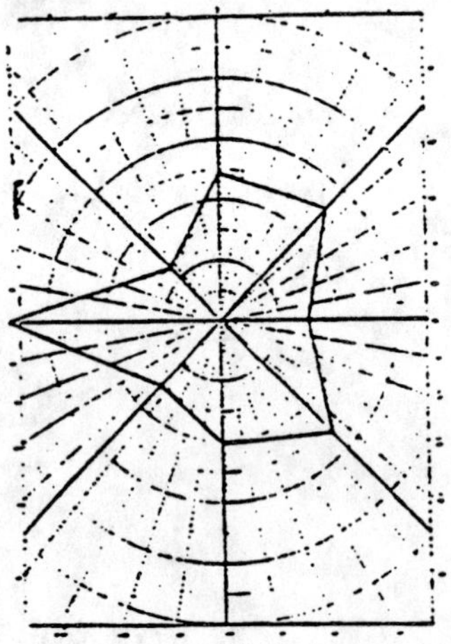

20

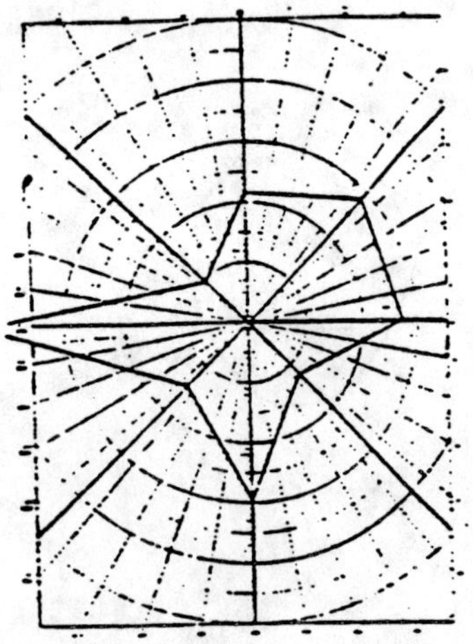

13

1.26 Bull data

(a) XBAR R

XBAR	Breed	SalePr	YrHgt	FtFrBody	PrctFFB	Frame	BkFat	SaleHt	SaleWt
4.3816	1.000	-0.224	0.525	0.409	0.472	0.434	-0.615	0.487	0.116
1742.4342	-0.224	1.000	0.423	0.102	-0.113	0.479	0.277	0.390	0.317
50.5224	0.525	0.423	1.000	0.624	0.523	0.940	-0.344	0.860	0.368
995.9474	0.409	0.102	0.624	1.000	0.691	0.605	-0.168	0.699	0.555
70.8816	0.472	-0.113	0.523	0.691	1.000	0.482	-0.488	0.521	0.198
6.3158	0.434	0.479	0.940	0.605	0.482	1.000	-0.260	0.801	0.368
0.1967	-0.615	0.277	-0.344	-0.168	-0.488	-0.260	1.000	-0.282	0.208
54.1263	0.487	0.390	0.860	0.699	0.521	0.801	-0.282	1.000	0.566
1555.2895	0.116	0.317	0.368	0.555	0.198	0.368	0.208	0.566	1.000

Sn

Breed	SalePr	YrHgt	FtFrBody	PrctFFB	Frame	BkFat	SaleHt	SaleWt
9.55	-429.02	2.79	116.28	4.73	1.23	-0.17	3.00	46.32
-429.02	383026.64	450.47	5813.09	-226.46	272.78	15.24	480.56	25308.44
2.79	450.47	2.96	98.81	2.92	1.49	-0.05	2.94	81.72
116.28	5813.09	98.81	8481.26	206.75	51.27	-1.38	128.23	6592.41
4.73	-226.46	2.92	206.75	10.55	1.44	-0.14	3.37	82.82
1.23	272.78	1.49	51.27	1.44	0.85	-0.02	1.47	43.74
-0.17	15.24	-0.05	-1.38	-0.14	-0.02	0.01	-0.05	2.38
3.00	480.56	2.94	128.23	3.37	1.47	-0.05	3.97	145.35
46.32	25308.44	81.72	6592.41	82.82	43.74	2.38	145.35	16628.94

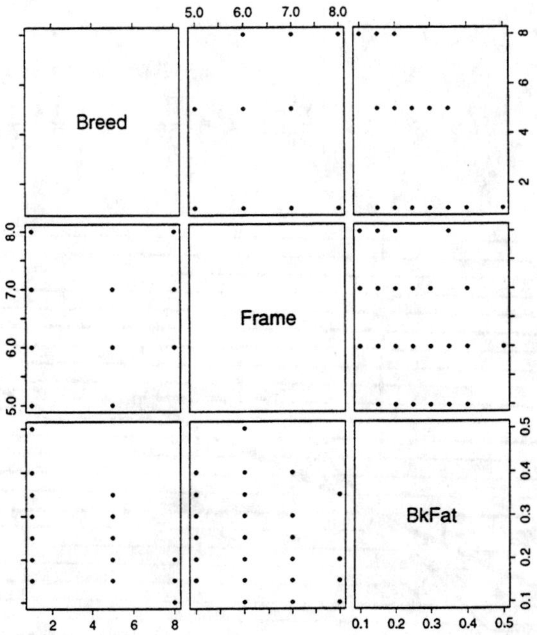

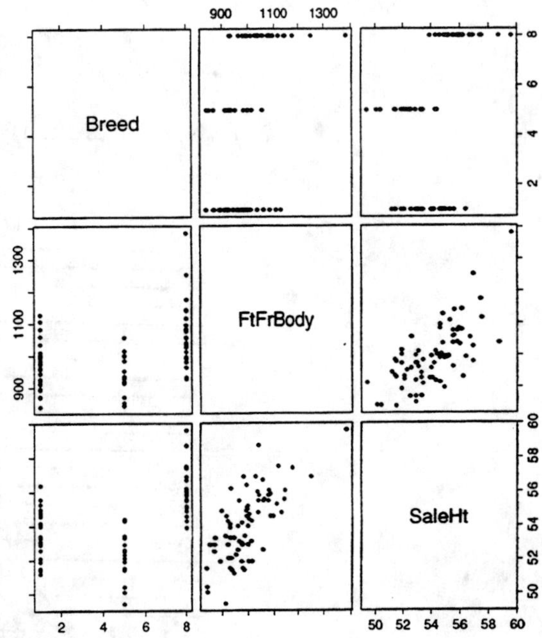

Chapter 2

2.1 a)

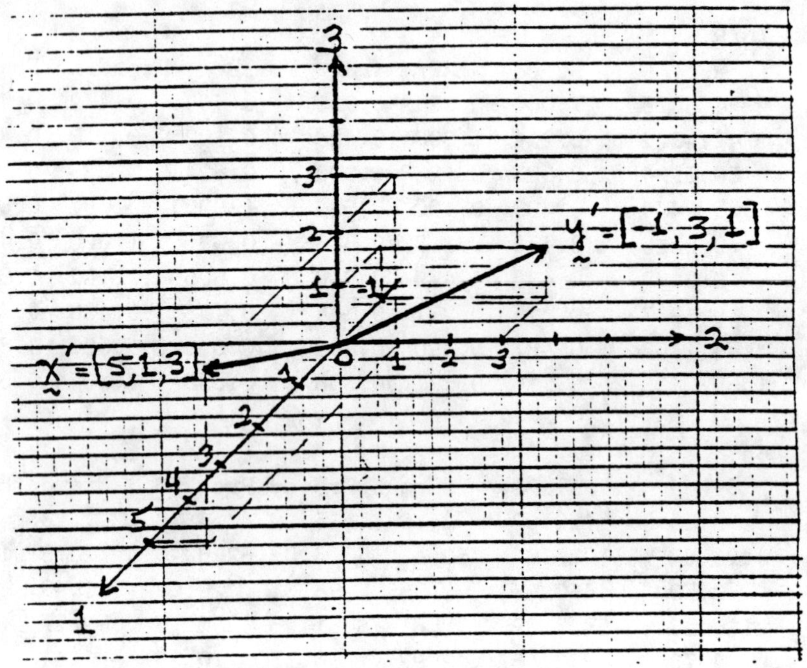

b) i) $L_{\underset{\sim}{x}} = \sqrt{\underset{\sim}{x}'\underset{\sim}{x}} = \sqrt{35} = 5.916$

ii) $\cos(\theta) = \dfrac{\underset{\sim}{x}'\underset{\sim}{y}}{L_{\underset{\sim}{x}} L_{\underset{\sim}{y}}} = \dfrac{1}{19.621} = .051$

$\theta = \text{arc cos } (.051) \doteq 87°$

iii) projection of $\underset{\sim}{y}$ on $\underset{\sim}{x}$ is $\left|\dfrac{\underset{\sim}{y}'\underset{\sim}{x}}{\underset{\sim}{x}'\underset{\sim}{x}}\right| \underset{\sim}{x} = \dfrac{1}{35}\underset{\sim}{x} = \left[\dfrac{1}{7}, \dfrac{1}{35}, \dfrac{3}{35}\right]'$

c)

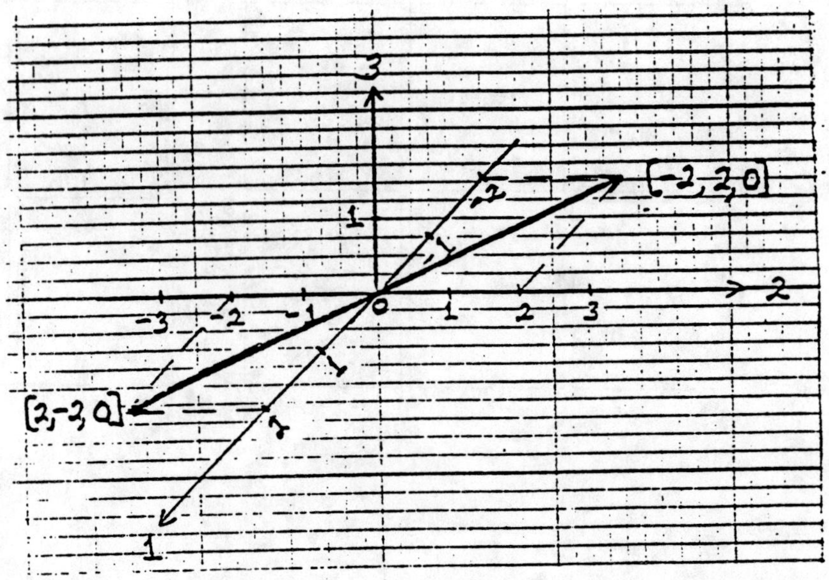

2.2 a) $5A = \begin{bmatrix} -5 & 15 \\ 20 & 10 \end{bmatrix}$ b) $BA = \begin{bmatrix} -16 & 6 \\ -9 & -1 \\ 2 & -6 \end{bmatrix}$

c) $A'B' = \begin{bmatrix} -16 & -9 & 2 \\ 6 & -1 & -6 \end{bmatrix}$ d) $C'B = [12, -7]$

e) No.

2.3 a) $A' = \begin{bmatrix} 2 & 1 \\ 1 & 3 \end{bmatrix} = A$ so $(A')' = A' = A$

b) $C' = \begin{bmatrix} 1 & 3 \\ 4 & 2 \end{bmatrix}$; $(C')^{-1} = \begin{bmatrix} -\frac{2}{10} & \frac{3}{10} \\ \frac{4}{10} & -\frac{1}{10} \end{bmatrix}$

$C^{-1} = \begin{bmatrix} -\frac{2}{10} & \frac{4}{10} \\ \frac{3}{10} & -\frac{1}{10} \end{bmatrix}$; $(C^{-1})' = \begin{bmatrix} -\frac{2}{10} & \frac{3}{10} \\ \frac{4}{10} & -\frac{1}{10} \end{bmatrix} = (C')^{-1}$

c)

$$(AB)' = \begin{bmatrix} 7 & 8 & 7 \\ 16 & 4 & 11 \end{bmatrix}' = \begin{bmatrix} 7 & 16 \\ 8 & 4 \\ 7 & 11 \end{bmatrix}$$

$$B'A' = \begin{bmatrix} 1 & 5 \\ 4 & 0 \\ 2 & 3 \end{bmatrix} \begin{bmatrix} 2 & 1 \\ 1 & 3 \end{bmatrix} = \begin{bmatrix} 7 & 16 \\ 8 & 4 \\ 7 & 11 \end{bmatrix} = (AB)'$$

d) AB has (i,j)th entry

$$a_{ij} = a_{i1}b_{1j} + a_{i2}b_{2j} + \cdots + a_{ik}b_{kj} = \sum_{\ell=1}^{k} a_{i\ell}b_{\ell j}$$

Consequently, $(AB)'$ has $(i,j)^{th}$ entry

$$c_{ji} = \sum_{\ell=1}^{k} a_{j\ell}b_{\ell i} .$$

Next B' has i^{th} row $[b_{1i}, b_{2i}, \cdots, b_{ki}]$ and A' has j^{th}

column $[a_{j1}, a_{j2}, \cdots, a_{jk}]'$ so $B'A'$ has $(i,j)^{th}$ entry

$$b_{1i}a_{j1} + b_{2i}b_{j2} + \cdots + b_{ki}a_{jk} = \sum_{\ell=1}^{k} a_{j\ell}b_{\ell i} = c_{ji}$$

Since i and j were arbitrary choices, $(AB)' = B'A'$.

2.4 a) $I = I'$ and $AA^{-1} = I = A^{-1}A$. Thus $I' = I = (AA^{-1})' = (A^{-1})'A'$ and $I = (A^{-1}A)' = A'(A^{-1})'$. Consequently, $(A^{-1})'$ is the inverse of A' or $(A')^{-1} = (A^{-1})'$.

b) $(B^{-1}A^{-1})AB = B^{-1}\underbrace{(A^{-1}A)}_{I}B = B^{-1}B = I$ so AB has inverse $(AB)^{-1} = B^{-1}A^{-1}$. It was sufficient to check for a left inverse but we may also verify $AB(B^{-1}A^{-1}) = A\underbrace{(BB^{-1})}_{I}A^{-1} = AA^{-1} = I$.

2.5

$$QQ' = \begin{bmatrix} \frac{5}{13} & \frac{12}{13} \\ \frac{-12}{13} & \frac{5}{13} \end{bmatrix} \begin{bmatrix} \frac{5}{13} & \frac{-12}{13} \\ \frac{12}{13} & \frac{5}{13} \end{bmatrix} = \begin{bmatrix} \frac{169}{169} & 0 \\ 0 & \frac{169}{169} \end{bmatrix} = \begin{bmatrix} 1 & 0 \\ 0 & 1 \end{bmatrix} = Q'Q .$$

2.6 a) Since $A = A'$, A is symmetric.

b) Since the quadratic form

$$\underset{\sim}{x}'A\underset{\sim}{x} = [x_1, x_2]\begin{bmatrix} 9 & -2 \\ -2 & 6 \end{bmatrix}\begin{bmatrix} x_1 \\ x_2 \end{bmatrix} = 9x_1^2 - 4x_1x_2 + 6x_2^2$$

$$= (2x_1 - x_2)^2 + 5(x_1^2 + x_2^2) > 0 \text{ for } [x_1, x_2] \neq [0,0]$$

we conclude that A is positive definite.

2.7 a) Eigenvalues: $\lambda_1 = 10$, $\lambda_2 = 5$.

Normalized eigenvectors: $\underset{\sim}{e}_1' = [2/\sqrt{5}, -1/\sqrt{5}] = [.894, -.447]$

$\underset{\sim}{e}_2' = [1/\sqrt{5}, 2/\sqrt{5}] = [.447, .894]$

b) $A = \begin{bmatrix} 9 & -2 \\ -2 & 9 \end{bmatrix} = 10 \begin{bmatrix} 2/\sqrt{5} \\ -1/\sqrt{5} \end{bmatrix} [2/\sqrt{5}, \ -1/\sqrt{5}] + 5 \begin{bmatrix} 1/\sqrt{5} \\ 2/\sqrt{5} \end{bmatrix} [1/\sqrt{5}, \ 2/\sqrt{5}]$

c) $A^{-1} = \frac{1}{9(6)-(-2)(-2)} \begin{bmatrix} 6 & 2 \\ 2 & 9 \end{bmatrix} = \begin{bmatrix} .12 & .04 \\ .04 & .18 \end{bmatrix}$

d) Eigenvalues: $\lambda_1 = .2, \ \lambda_2 = .1$

Normalized eigenvectors: $\underset{\sim}{e}_1' = [1/\sqrt{5}, \ 2/\sqrt{5}]$

$\underset{\sim}{e}_2' = [2/\sqrt{5}, \ -1/\sqrt{5}]$

2.8 Eigenvalues: $\lambda_1 = 2, \ \lambda_2 = -3$

Normalized eigenvectors: $\underset{\sim}{e}_1' = [2/\sqrt{5}, \ 1/\sqrt{5}]$

$\underset{\sim}{e}_2' = [1/\sqrt{5}, \ -2/\sqrt{5}]$

$A = \begin{bmatrix} 1 & 2 \\ 2 & -2 \end{bmatrix} = 2 \begin{bmatrix} 2/\sqrt{5} \\ 1/\sqrt{5} \end{bmatrix} [2/\sqrt{5}, \ 1/\sqrt{5}] - 3 \begin{bmatrix} 1/\sqrt{5} \\ -2/\sqrt{5} \end{bmatrix} [1/\sqrt{5}, \ -2/\sqrt{5}]$

2.9 a) $A^{-1} = \frac{1}{1(-2)-2(2)} \begin{bmatrix} -2 & -2 \\ -2 & 1 \end{bmatrix} = \begin{bmatrix} \frac{1}{3} & \frac{1}{3} \\ \frac{1}{3} & -\frac{1}{6} \end{bmatrix}$

b) Eigenvalues: $\lambda_1 = 1/2, \ \lambda_2 = -1/3$

Normalized eigenvectors: $\underset{\sim}{e}_1' = [2/\sqrt{5}, \ 1/\sqrt{5}]$

$\underset{\sim}{e}_2' = [1/\sqrt{5}, \ -2/\sqrt{5}]$

c) $A^{-1} = \begin{bmatrix} \frac{1}{3} & \frac{1}{3} \\ \frac{1}{3} & -\frac{1}{6} \end{bmatrix} = \frac{1}{2} \begin{bmatrix} 2/\sqrt{5} \\ 1/\sqrt{5} \end{bmatrix} [2/\sqrt{5}, \ 1/\sqrt{5}] - \frac{1}{3} \begin{bmatrix} 1/\sqrt{5} \\ -2/\ 5 \end{bmatrix} [1/\sqrt{5}, \ -2/\sqrt{5}]$

2.10

$$B^{-1} = \frac{1}{4(4.002001)-(4.001)^2}\begin{bmatrix} 4.002001 & -4.001 \\ -4.001 & 4 \end{bmatrix}$$

$$= 333,333 \begin{bmatrix} 4.002001 & -4.001 \\ -4.001 & 4 \end{bmatrix}$$

$$A^{-1} = \frac{1}{4(4.002)-(4.001)^2}\begin{bmatrix} 4.002 & -4.001 \\ -4.001 & 4 \end{bmatrix}$$

$$= -1,000,000 \begin{bmatrix} 4.002 & -4.001 \\ -4.001 & 4 \end{bmatrix}$$

Thus $A^{-1} \doteq (-3)B^{-1}$

2.11 With $p = 1$, $|a_{11}| = a_{11}$ and with $p = 2$

$$\begin{vmatrix} a_{11} & 0 \\ 0 & a_{22} \end{vmatrix} = a_{11}a_{22} - 0(0) = a_{11}a_{22}$$

Proceeding by induction, we assume the result holds for any $(p-1)\times(p-1)$ diagonal matrix A_{11}. Then writing

$$\underset{(p\times p)}{A} = \begin{bmatrix} a_{11} & 0 & \cdots & 0 \\ 0 & & & \\ \vdots & & A_{11} & \\ 0 & & & \end{bmatrix}$$

we expand $|A|$ according to Definition 2A.24 to find $|A| = a_{11}\ |A_{11}| + 0 + \cdots + 0$. Since $|A_{11}| = a_{22}a_{33} \cdots a_{pp}$ by the induction hypothesis, $|A| = a_{11}(a_{22}a_{33} \cdots a_{pp}) = a_{11}a_{22}a_{33} \cdots a_{pp}$.

2.12 By (2-20), $\tilde{A} = P\Lambda P'$ with $PP' = P'P = I$. From Result 2A.11(e) $|A| = |P|\ |\Lambda|\ |P'| = |\Lambda|$. Since Λ is a diagonal matrix with diagonal elements $\lambda_1, \lambda_2, \ldots, \lambda_p$, we can apply Exercise 2.11 to get $|A| = |\Lambda| = \prod_{i=1}^{p} \lambda_i$.

2.14 Let λ be an eigenvalue of A. Thus $0 = |A-\lambda I|$. If Q is orthogonal, $QQ' = I$ and $|Q||Q'| = 1$ by Exercise 2.13. Using Result 2A.11(e) we can then write

$$0 = |Q|\ |A-\lambda I|\ |Q'| = |QAQ'-\lambda I|$$

and it follows that λ is also an eigenvalue of QAQ' if Q is orthogonal.

2.16 $(A'A)' = A'(A')' = A'A$ showing $A'A$ is symmetric.

$$\underset{\sim}{y} = \begin{bmatrix} y_1 \\ y_2 \\ \vdots \\ y_p \end{bmatrix} = A\underset{\sim}{x}. \text{ Then } 0 \le y_1^2+y_2^2+ \cdots + y_p^2 = \underset{\sim}{y}'\underset{\sim}{y} = \underset{\sim}{x}'A'A\underset{\sim}{x}$$

and $A'A$ is non-negative definite by definition.

2.18 Write $c^2 = \underset{\sim}{x}'A\underset{\sim}{x}$ with $A = \begin{bmatrix} 4 & \sqrt{2} \\ \sqrt{2} & 3 \end{bmatrix}$. The eigenvalue-normalized eigenvector pairs for A are:

$$\lambda_1 = 2, \quad \underset{\sim}{e}_1' = [.577, \ .816]$$

$$\lambda_2 = 5, \quad \underset{\sim}{e}_2' = [.816, \ -.577]$$

For $c^2 = 1$, the half lengths of the major and minor axes of the ellipse of constant distance are

$$\frac{c}{\sqrt{\lambda_1}} = \frac{1}{\sqrt{2}} = .707 \text{ and } \frac{c}{\sqrt{\lambda_2}} = \frac{1}{\sqrt{5}} = .447$$

respectively. These axes lie in the directions of the vectors $\underset{\sim}{e}_1$ and $\underset{\sim}{e}_2$ respectively.

For $c^2 = 4$, the half lengths of the major and minor axes are

$$\frac{c}{\sqrt{\lambda_1}} = \frac{2}{\sqrt{2}} = 1.414 \quad \text{and} \quad \frac{c}{\sqrt{\lambda_2}} = \frac{2}{\sqrt{5}} = .894 \ .$$

As c^2 increases the lengths of the major and minor axes increase.

2.20 Using matrix A in Exercise 2.3, we determine

$$\lambda_1 = 1.382, \quad \underset{\sim}{e}_1 = [.8507, \quad -.5257]'$$

$$\lambda_2 = 3.618, \quad \underset{\sim}{e}_2 = [.5257, \quad .8507]'$$

We know

$$A^{1/2} = \sqrt{\lambda_1}\, \underset{\sim}{e}_1 \underset{\sim}{e}_1' + \sqrt{\lambda_2}\, \underset{\sim}{e}_2 \underset{\sim}{e}_2' = \begin{bmatrix} 1.376 & .325 \\ .325 & 1.701 \end{bmatrix}$$

$$A^{-1/2} = \frac{1}{\sqrt{\lambda_1}} \underset{\sim}{e}_1 \underset{\sim}{e}_1' + \frac{1}{\sqrt{\lambda_2}} \underset{\sim}{e}_2 \underset{\sim}{e}_2' = \begin{bmatrix} .7608 & -.1453 \\ -.1453 & .6155 \end{bmatrix}$$

We check

$$A^{1/2} A^{-1/2} = \begin{bmatrix} 1 & 0 \\ 0 & 1 \end{bmatrix} = A^{-1/2} A^{1/2}$$

2.21 (a)

$$\mathbf{A}'\mathbf{A} = \begin{bmatrix} 1 & 2 & 2 \\ 1 & -2 & 2 \end{bmatrix} \begin{bmatrix} 1 & 1 \\ 2 & -2 \\ 2 & 2 \end{bmatrix} = \begin{bmatrix} 9 & 1 \\ 1 & 9 \end{bmatrix}$$

$0 = |\mathbf{A}'\mathbf{A} - \lambda\,\mathbf{I}\,| = (9-\lambda)^2 - 1 = (10-\lambda)(8-\lambda)$, so $\lambda_1 = 10$ and $\lambda_2 = 8$.
Next,

$$\begin{bmatrix} 1 & 1 \\ 1 & 9 \end{bmatrix} \begin{bmatrix} e_1 \\ e_2 \end{bmatrix} = 10 \begin{bmatrix} e_1 \\ e_2 \end{bmatrix} \quad \text{gives} \quad \boldsymbol{e}_1 = \begin{bmatrix} 1/\sqrt{2} \\ 1/\sqrt{2} \end{bmatrix}$$

$$\begin{bmatrix} 1 & 1 \\ 1 & 9 \end{bmatrix} \begin{bmatrix} e_1 \\ e_2 \end{bmatrix} = 8 \begin{bmatrix} e_1 \\ e_2 \end{bmatrix} \quad \text{gives} \quad \boldsymbol{e}_2 = \begin{bmatrix} 1/\sqrt{2} \\ -1/\sqrt{2} \end{bmatrix}$$

(b)

$$\mathbf{A}\mathbf{A}' = \begin{bmatrix} 1 & 1 \\ 2 & -2 \\ 2 & 2 \end{bmatrix} \begin{bmatrix} 1 & 2 & 2 \\ 1 & -2 & 2 \end{bmatrix} = \begin{bmatrix} 2 & 0 & 4 \\ 0 & 8 & 0 \\ 4 & 0 & 8 \end{bmatrix}$$

$$0 = |\mathbf{A}\mathbf{A}' - \lambda\mathbf{I}\,| = \begin{vmatrix} 2-\lambda & 0 & 4 \\ 0 & 8-\lambda & 0 \\ 4 & 0 & 8-\lambda \end{vmatrix}$$

$= (2-\lambda)(8-\lambda)^2 - 4^2(8-\lambda) = (8-\lambda)(\lambda-10)\lambda$ so $\lambda_1 = 10, \lambda_2 = 8$, and $\lambda_3 = 0$.

$$\begin{bmatrix} 2 & 0 & 4 \\ 0 & 8 & 0 \\ 4 & 0 & 8 \end{bmatrix} \begin{bmatrix} e_1 \\ e_2 \\ e_3 \end{bmatrix} = 10 \begin{bmatrix} e_1 \\ e_2 \\ e_2 \end{bmatrix}$$

$$\text{gives} \quad \begin{matrix} 4e_3 & = & 8e_1 \\ 8e_2 & = & 10e_2 \end{matrix} \quad \text{so } \boldsymbol{e}_1 = \frac{1}{\sqrt{5}} \begin{bmatrix} 1 \\ 0 \\ 2 \end{bmatrix}$$

$$\begin{bmatrix} 2 & 0 & 4 \\ 0 & 8 & 0 \\ 4 & 0 & 8 \end{bmatrix} \begin{bmatrix} e_1 \\ e_2 \\ e_3 \end{bmatrix} = 8 \begin{bmatrix} e_1 \\ e_2 \\ e_2 \end{bmatrix}$$

$$\text{gives} \quad \begin{matrix} 4e_3 & = & 6e_1 \\ 4e_1 & = & 0 \end{matrix} \quad \text{so } \boldsymbol{e}_2 = \begin{bmatrix} 0 \\ 1 \\ 0 \end{bmatrix}$$

Also, $\boldsymbol{e}_3 = [-2/\sqrt{5}, 0, 1/\sqrt{5}\,]'$.

(c)

$$\begin{bmatrix} 1 & 1 \\ 2 & -2 \\ 2 & 2 \end{bmatrix} = \sqrt{10}\begin{bmatrix} \frac{1}{\sqrt{5}} \\ 0 \\ \frac{2}{\sqrt{5}} \end{bmatrix} [\ \tfrac{1}{\sqrt{2}}\ ,\ \tfrac{1}{\sqrt{2}}\ ] + \sqrt{8}\begin{bmatrix} 0 \\ 1 \\ 0 \end{bmatrix} [\ \tfrac{1}{\sqrt{2}}\ ,\ -\tfrac{1}{\sqrt{2}}\ ]$$

2.22 (a)

$$\mathbf{AA}' = \begin{bmatrix} 4 & 8 & 8 \\ 3 & 6 & -9 \end{bmatrix} \begin{bmatrix} 4 & 3 \\ 8 & 6 \\ 8 & -9 \end{bmatrix} = \begin{bmatrix} 144 & -12 \\ -12 & 126 \end{bmatrix}$$

$0 = |\mathbf{AA}' - \lambda\,\mathbf{I}\,| = (144 - \lambda)(126 - \lambda) - (12)^2 = (150 - \lambda)(120 - \lambda)$, so $\lambda_1 = 150$ and $\lambda_2 = 120$. Next,

$$\begin{bmatrix} 144 & -12 \\ -12 & 126 \end{bmatrix} \begin{bmatrix} e_1 \\ e_2 \end{bmatrix} = 150 \begin{bmatrix} e_1 \\ e_2 \end{bmatrix} \quad \text{gives} \quad \boldsymbol{e}_1 = \begin{bmatrix} 2/\sqrt{5} \\ -1/\sqrt{5} \end{bmatrix}$$

and $\lambda_2 = 120$ gives $\boldsymbol{e}_2 = [1/\sqrt{5}, 2/\sqrt{5}\,]'$.

(b)

$$\mathbf{A}'\mathbf{A} = \begin{bmatrix} 4 & 3 \\ 8 & 6 \\ 8 & -9 \end{bmatrix} \begin{bmatrix} 4 & 8 & 8 \\ 3 & 6 & -9 \end{bmatrix} = \begin{bmatrix} 25 & 50 & 5 \\ 50 & 100 & 10 \\ 5 & 10 & 145 \end{bmatrix}$$

$$0 = |\mathbf{A}'\mathbf{A} - \lambda\mathbf{I}\,| = \begin{vmatrix} 25-\lambda & 50 & 5 \\ 50 & 100-\lambda & 10 \\ 5 & 10 & 145-\lambda \end{vmatrix} = (150-\lambda)(\lambda-120)\lambda$$

so $\lambda_1 = 150$, $\lambda_2 = 120$, and $\lambda_3 = 0$. Next,

$$\begin{bmatrix} 25 & 50 & 5 \\ 50 & 100 & 10 \\ 5 & 10 & 145 \end{bmatrix} \begin{bmatrix} e_1 \\ e_2 \\ e_3 \end{bmatrix} = 150 \begin{bmatrix} e_1 \\ e_2 \\ e_2 \end{bmatrix}$$

$$\text{gives} \quad \begin{array}{rcrl} -120e_1 & + & 60e_2 & = 0 \\ -25e_1 & + & 5e_3 & = 0 \end{array} \quad \text{or } \boldsymbol{e}_1 = \frac{1}{\sqrt{30}} \begin{bmatrix} 1 \\ 2 \\ 5 \end{bmatrix}$$

$$\begin{bmatrix} 25 & 50 & 5 \\ 50 & 100 & 10 \\ 5 & 10 & 145 \end{bmatrix} \begin{bmatrix} e_1 \\ e_2 \\ e_3 \end{bmatrix} = 120 \begin{bmatrix} e_1 \\ e_2 \\ e_2 \end{bmatrix}$$

$$\text{gives} \quad \begin{array}{rcrl} 60e_1 & + & 60e_3 & = 0 \\ -120e_2 & + & -240e_3 & = 0 \end{array} \quad \text{or } e_2 = \frac{1}{\sqrt{6}}\begin{bmatrix} 1 \\ 2 \\ -1 \end{bmatrix}$$

Also, $e_3 = [2/\sqrt{5}, -1/\sqrt{5}, 0]'$.

(c)

$$\begin{bmatrix} 4 & 8 & 8 \\ 3 & 6 & -9 \end{bmatrix}$$

$$= \sqrt{150}\begin{bmatrix} \frac{2}{\sqrt{5}} \\ -\frac{1}{\sqrt{5}} \end{bmatrix}\left[\begin{array}{ccc} \frac{1}{\sqrt{30}} & \frac{2}{\sqrt{30}} & \frac{5}{\sqrt{30}} \end{array}\right] + \sqrt{120}\begin{bmatrix} \frac{1}{\sqrt{5}} \\ \frac{2}{\sqrt{5}} \end{bmatrix}\left[\begin{array}{ccc} \frac{1}{\sqrt{6}} & \frac{2}{\sqrt{6}} & -\frac{1}{\sqrt{6}} \end{array}\right]$$

2.24

a) $\boldsymbol{\Sigma}^{-1} = \begin{bmatrix} \frac{1}{4} & 0 & 0 \\ 0 & \frac{1}{9} & 0 \\ 0 & 0 & 1 \end{bmatrix}$

b) $\lambda_1 = 4, \quad \mathbf{e}_1 = [1,0,0]'$
$\lambda_2 = 9, \quad \mathbf{e}_2 = [0,1,0]'$
$\lambda_3 = 1, \quad \mathbf{e}_3 = [0,0,1]'$

c) For $\boldsymbol{\Sigma}^{-1}$: $\lambda_1 = 1/4, \quad \mathbf{e}_1' = [1,0,0]'$
$\lambda_2 = 1/9, \quad \mathbf{e}_2' = [0,1,0]'$
$\lambda_3 = 1, \quad \mathbf{e}_3' = [0,0,1]'$

2.25

a) $V^{1/2} = \begin{bmatrix} 5 & 0 & 0 \\ 0 & 2 & 0 \\ 0 & 0 & 3 \end{bmatrix}$; $\boldsymbol{\rho} = \begin{bmatrix} 1 & -1/5 & 4/15 \\ -1/5 & 1 & 1/6 \\ 4/15 & 1/6 & 1 \end{bmatrix} = \begin{bmatrix} 1 & -.2 & .267 \\ -.2 & 1 & .167 \\ .267 & .167 & 1 \end{bmatrix}$

b) $V^{1/2} \boldsymbol{\rho} V^{1/2} =$

$$\begin{bmatrix} 5 & 0 & 0 \\ 0 & 2 & 0 \\ 0 & 0 & 3 \end{bmatrix} \begin{bmatrix} 1 & -1/5 & 4/15 \\ -1/5 & 1 & 1/6 \\ 4/15 & 1/6 & 1 \end{bmatrix} \begin{bmatrix} 5 & 0 & 0 \\ 0 & 2 & 0 \\ 0 & 0 & 3 \end{bmatrix} = \begin{bmatrix} 5 & -1 & 4/3 \\ -2/5 & 2 & 1/3 \\ 4/5 & 1/2 & 3 \end{bmatrix} \begin{bmatrix} 5 & 0 & 0 \\ 0 & 2 & 0 \\ 0 & 0 & 3 \end{bmatrix}$$

$$= \begin{bmatrix} 25 & -2 & 4 \\ -2 & 4 & 1 \\ 4 & 1 & 9 \end{bmatrix} = \boldsymbol{\Sigma}$$

2.26 a) $\rho_{13} = \sigma_{13}/\sigma_{11}^{1/2}\sigma_{22}^{1/2} = 4/\sqrt{25}\,\sqrt{9} = 4/15 = .267$

b) Write $X_1 = 1 \cdot X_1 + 0 \cdot X_2 + 0 \cdot X_3 = \mathbf{c}_1' \mathbf{X}$ with $\mathbf{c}_1' = [1,0,0]$

$$\frac{1}{2}X_2 + \frac{1}{2}X_3 = \mathbf{c}_2' \mathbf{X} \quad \text{with} \quad \mathbf{c}_2' = [0, \tfrac{1}{2}, \tfrac{1}{2}]$$

Then $\text{Var}(X_1) = \sigma_{11} = 25$. By (2-43),

$$\text{Var}(\tfrac{1}{2}X_2 + \tfrac{1}{2}X_3) = \mathbf{c}_2' \boldsymbol{\Sigma} \mathbf{c}_2 = \frac{1}{4}\sigma_{22} + \frac{2}{4}\sigma_{23} + \frac{1}{4}\sigma_{33} = 1 + \frac{1}{2} + \frac{9}{4}$$

$$= \frac{15}{4} = 3.75$$

By (2-45), (see also hint to Exercise 2.28),

$$\text{Cov}(X_1, \tfrac{1}{2}X_1 + \tfrac{1}{2}X_2) = \mathbf{c}_1' \boldsymbol{\Sigma} \mathbf{c}_2 = \frac{1}{2}\sigma_{12} + \frac{1}{2}\sigma_{13} = -1 + 2 = 1$$

so

$$\text{Corr}(X_1, \tfrac{1}{2}X_1 + \tfrac{1}{2}X_2) = \frac{\text{Cov}(X_1, \tfrac{1}{2}X_1 + \tfrac{1}{2}X_2)}{\sqrt{\text{Var}(X_1)}\ \sqrt{\text{Var}(\tfrac{1}{2}X_1 + \tfrac{1}{2}X_2)}} = \frac{1}{5\sqrt{3.75}} = .103$$

2.27 a) $\mu_1 - 2\mu_2$, $\sigma_{11} + 4\sigma_{22} - 4\sigma_{12}$

b) $-\mu_1 + 3\mu_2$, $\sigma_{11} + 9\sigma_{22} - 6\sigma_{12}$

c) $\mu_1 + \mu_2 + \mu_3$, $\sigma_{11} + \sigma_{22} + \sigma_{33} + 2\sigma_{12} + 2\sigma_{13} + 2\sigma_{23}$

d) $\mu_1 + 2\mu_2 - \mu_3$, $\sigma_{11} + 4\sigma_{22} + \sigma_{33} + 4\sigma_{12} - 2\sigma_{13} - 4\sigma_{23}$

e) $3\mu_1 - 4\mu_2$, $9\sigma_{11} + 16\sigma_{22}$ since $\sigma_{12} = 0$.

2.29

$$\boldsymbol{\Sigma} = \left[\begin{array}{cc|ccc} \sigma_{11} & \sigma_{12} & \sigma_{13} & \sigma_{14} & \sigma_{15} \\ \sigma_{21} & \sigma_{22} & \sigma_{23} & \sigma_{24} & \sigma_{25} \\ \hline \sigma_{31} & \sigma_{32} & \sigma_{33} & \sigma_{34} & \sigma_{35} \\ \sigma_{41} & \sigma_{42} & \sigma_{43} & \sigma_{44} & \sigma_{45} \\ \sigma_{51} & \sigma_{52} & \sigma_{53} & \sigma_{54} & \sigma_{55} \end{array}\right] = \left[\begin{array}{c|c} \boldsymbol{\Sigma}_{11} & \boldsymbol{\Sigma}_{12} \\ \hline \boldsymbol{\Sigma}_{21} & \boldsymbol{\Sigma}_{22} \end{array}\right]$$

2.31 (a)

$$E[\boldsymbol{X}^{(1)}] = \boldsymbol{\mu}^{(1)} = \begin{bmatrix} 4 \\ 3 \end{bmatrix} \quad \text{(b)} \quad \mathbf{A}\boldsymbol{\mu}^{(1)} = [\ 1 \ \ -1\ ] \begin{bmatrix} 4 \\ 3 \end{bmatrix} = 1$$

(c)

$$\text{Cov}(\boldsymbol{X}^{(1)}) = \boldsymbol{\Sigma}_{11} = \begin{bmatrix} 3 & 0 \\ 0 & 1 \end{bmatrix}$$

(d)

$$\text{Cov}(\mathbf{A}\boldsymbol{X}^{(1)}) = \mathbf{A}\boldsymbol{\Sigma}_{11}\mathbf{A}' = [\ 1 \ \ -1\ ] \begin{bmatrix} 3 & 0 \\ 0 & 1 \end{bmatrix} \begin{bmatrix} 1 \\ -1 \end{bmatrix} = 4$$

(e)

$$E[\boldsymbol{X}^{(2)}] = \boldsymbol{\mu}^{(2)} = \begin{bmatrix} 2 \\ 1 \end{bmatrix} \quad \text{(f)} \quad \mathbf{B}\boldsymbol{\mu}^{(2)} = \begin{bmatrix} 2 & -1 \\ 0 & 1 \end{bmatrix} \begin{bmatrix} 2 \\ 1 \end{bmatrix} = \begin{bmatrix} 3 \\ 1 \end{bmatrix}$$

(g)

$$\text{Cov}(\boldsymbol{X}^{(2)}) = \boldsymbol{\Sigma}_{22} = \begin{bmatrix} 9 & -2 \\ -2 & 4 \end{bmatrix}$$

(h)

$$\text{Cov}(\mathbf{B}\boldsymbol{X}^{(2)}) = \mathbf{B}\boldsymbol{\Sigma}_{22}\mathbf{B}' = \begin{bmatrix} 2 & -1 \\ 0 & 1 \end{bmatrix} \begin{bmatrix} 9 & -2 \\ -2 & 4 \end{bmatrix} \begin{bmatrix} 2 & 0 \\ -1 & 1 \end{bmatrix} = \begin{bmatrix} 48 & -8 \\ -8 & 4 \end{bmatrix}$$

(i)

$$\text{Cov}(\boldsymbol{X}^{(1)}, \boldsymbol{X}^{(2)}) = \begin{bmatrix} 2 & 2 \\ 1 & 0 \end{bmatrix}$$

(j)

$$\text{Cov}(\mathbf{A}\boldsymbol{X}^{(1)}, \mathbf{B}\boldsymbol{X}^{(2)}) = \mathbf{A}\boldsymbol{\Sigma}_{12}\mathbf{B}' = [\ 1 \ \ -1\ ] \begin{bmatrix} 2 & 2 \\ 1 & 0 \end{bmatrix} \begin{bmatrix} 2 & 0 \\ -1 & 1 \end{bmatrix} = [\ 0 \ \ 2\ ]$$

2.32 (a)

$$E[\boldsymbol{X}^{(1)}] = \boldsymbol{\mu}^{(1)} = \begin{bmatrix} 2 \\ 4 \end{bmatrix} \quad \text{(b)} \quad \mathbf{A}\boldsymbol{\mu}^{(1)} = \begin{bmatrix} 1 & -1 \\ 1 & 1 \end{bmatrix} \begin{bmatrix} 2 \\ 4 \end{bmatrix} = \begin{bmatrix} -2 \\ 6 \end{bmatrix}$$

(c)

$$\text{Cov}(\boldsymbol{X}^{(1)}\ ) = \boldsymbol{\Sigma}_{11} = \begin{bmatrix} 4 & -1 \\ -1 & 3 \end{bmatrix}$$

(d)

$$\text{Cov}(\mathbf{A}\boldsymbol{X}^{(1)}\ ) = \mathbf{A}\boldsymbol{\Sigma}_{11}\mathbf{A}' = \begin{bmatrix} 1 & -1 \\ 1 & 1 \end{bmatrix} \begin{bmatrix} 4 & -1 \\ -1 & 3 \end{bmatrix} \begin{bmatrix} 1 & 1 \\ -1 & 1 \end{bmatrix} = \begin{bmatrix} 9 & 1 \\ 1 & 5 \end{bmatrix}$$

(e)

$$E[\boldsymbol{X}^{(2)}] = \boldsymbol{\mu}^{(2)} = \begin{bmatrix} -1 \\ 0 \\ 3 \end{bmatrix} \quad \text{(f)} \quad \mathbf{B}\boldsymbol{\mu}^{(2)} = \begin{bmatrix} 1 & 1 & 1 \\ 1 & 1 & -2 \end{bmatrix} \begin{bmatrix} -1 \\ 0 \\ 3 \end{bmatrix} = \begin{bmatrix} 2 \\ -7 \end{bmatrix}$$

(g)

$$\text{Cov}(\boldsymbol{X}^{(2)}\ ) = \boldsymbol{\Sigma}_{22} = \begin{bmatrix} 6 & 1 & -1 \\ 1 & 4 & 0 \\ -1 & 0 & 2 \end{bmatrix}$$

(h)

$$\text{Cov}(\mathbf{B}\boldsymbol{X}^{(2)}\ ) = \mathbf{B}\boldsymbol{\Sigma}_{22}\mathbf{B}'$$

$$= \begin{bmatrix} 1 & 1 & 1 \\ 1 & 1 & -2 \end{bmatrix} \begin{bmatrix} 6 & 1 & -1 \\ 1 & 4 & 0 \\ -1 & 0 & 2 \end{bmatrix} \begin{bmatrix} 1 & 1 \\ 1 & 1 \\ 1 & -2 \end{bmatrix} = \begin{bmatrix} 12 & 9 \\ 9 & 24 \end{bmatrix}$$

(i)

$$\text{Cov}(\boldsymbol{X}^{(1)}, \boldsymbol{X}^{(2)}) = \begin{bmatrix} \frac{1}{2} & -\frac{1}{2} & 0 \\ 1 & -1 & 0 \end{bmatrix}$$

(j)

$$\text{Cov}(\mathbf{A}\boldsymbol{X}^{(1)}, \mathbf{B}\boldsymbol{X}^{(2)}) = \mathbf{A}\boldsymbol{\Sigma}_{12}\mathbf{B}'$$

$$= \begin{bmatrix} 1 & 1 \\ 1 & 1 \\ 1 & -2 \end{bmatrix} \begin{bmatrix} \frac{1}{2} & -\frac{1}{2} & 0 \\ 1 & -1 & 0 \end{bmatrix} \begin{bmatrix} 1 & 1 \\ 1 & 1 \\ 1 & -2 \end{bmatrix} = \begin{bmatrix} 0 & 0 \\ 0 & 0 \end{bmatrix}$$

2.33 (a)

$$E[\boldsymbol{X}^{(1)}] = \boldsymbol{\mu}^{(1)} = \begin{bmatrix} 2 \\ 4 \\ -1 \end{bmatrix} \quad \text{(b)} \quad \mathbf{A}\boldsymbol{\mu}^{(1)} = \begin{bmatrix} 2 & -1 & 0 \\ 1 & 1 & 3 \end{bmatrix} \begin{bmatrix} 2 \\ 4 \\ -1 \end{bmatrix} = \begin{bmatrix} 0 \\ 3 \end{bmatrix}$$

(c)

$$\text{Cov}(\boldsymbol{X}^{(1)}) = \boldsymbol{\Sigma}_{11} = \begin{bmatrix} 4 & -1 & \frac{1}{2} \\ -1 & 3 & 1 \\ \frac{1}{2} & 1 & 6 \end{bmatrix}$$

(d)

$$\text{Cov}(\mathbf{A}\boldsymbol{X}^{(1)}) = \mathbf{A}\boldsymbol{\Sigma}_{11}\mathbf{A}'$$

$$= \begin{bmatrix} 2 & -1 & 0 \\ 1 & 1 & 3 \end{bmatrix} \begin{bmatrix} 4 & -1 & \frac{1}{2} \\ -1 & 3 & 1 \\ \frac{1}{2} & 1 & 6 \end{bmatrix} \begin{bmatrix} 2 & 1 \\ -1 & 1 \\ 0 & 3 \end{bmatrix} = \begin{bmatrix} 23 & 4 \\ 4 & 63 \end{bmatrix}$$

(e)

$$E[\boldsymbol{X}^{(2)}] = \boldsymbol{\mu}^{(2)} = \begin{bmatrix} 3 \\ 0 \end{bmatrix} \quad \text{(f)} \quad \mathbf{B}\boldsymbol{\mu}^{(2)} = \begin{bmatrix} 1 & 2 \\ 1 & -1 \end{bmatrix} \begin{bmatrix} 3 \\ 0 \end{bmatrix} = \begin{bmatrix} 3 \\ 3 \end{bmatrix}$$

(g)

$$\text{Cov}(\boldsymbol{X}^{(2)}) = \boldsymbol{\Sigma}_{22} = \begin{bmatrix} 4 & 0 \\ 0 & 2 \end{bmatrix}$$

(h)

$$\text{Cov}(\mathbf{B}\boldsymbol{X}^{(2)}) = \mathbf{B}\boldsymbol{\Sigma}_{22}\mathbf{B}' = \begin{bmatrix} 1 & 2 \\ 1 & -1 \end{bmatrix} \begin{bmatrix} 4 & 0 \\ 0 & 2 \end{bmatrix} \begin{bmatrix} 1 & 1 \\ 2 & -1 \end{bmatrix} = \begin{bmatrix} 12 & 0 \\ 0 & 6 \end{bmatrix}$$

(i)

$$\text{Cov}(\boldsymbol{X}^{(1)}, \boldsymbol{X}^{(2)}) = \begin{bmatrix} -\frac{1}{2} & 0 \\ -1 & 0 \\ 1 & -1 \end{bmatrix}$$

(j)

$$\text{Cov}(\mathbf{A}\boldsymbol{X}^{(1)}, \mathbf{B}\boldsymbol{X}^{(2)}) = \mathbf{A}\boldsymbol{\Sigma}_{12}\mathbf{B}'$$

$$= \begin{bmatrix} 2 & -1 & 0 \\ 1 & 1 & 3 \end{bmatrix} \begin{bmatrix} -\frac{1}{2} & 0 \\ -1 & 0 \\ 1 & -1 \end{bmatrix} \begin{bmatrix} 1 & 1 \\ 2 & -1 \end{bmatrix} = \begin{bmatrix} 0 & 0 \\ -4.5 & 4.5 \end{bmatrix}$$

2.34 $\underset{\sim}{b}'\underset{\sim}{b} = 4 + 1 + 16 + 0 = 21$, $\underset{\sim}{d}'\underset{\sim}{d} = 15$ and $\underset{\sim}{b}'\underset{\sim}{d} = -2 - 3 - 8 + 0 = -13$

$$(\underset{\sim}{b}'\underset{\sim}{d})^2 = 169 \le 21(15) = 315$$

2.35 $\underset{\sim}{b}'\underset{\sim}{d} = -4 + 3 = -1$

$$\underset{\sim}{b}'B\underset{\sim}{b} = [-4,\ 3]\begin{bmatrix} 2 & -2 \\ -2 & 5 \end{bmatrix}\begin{bmatrix} -4 \\ 3 \end{bmatrix} = [-14 \quad 23]\begin{bmatrix} -4 \\ 3 \end{bmatrix} = 125$$

$$\underset{\sim}{d}'B^{-1}\underset{\sim}{d} = [1,1]\begin{bmatrix} 5/6 & 2/6 \\ 2/6 & 2/6 \end{bmatrix}\begin{bmatrix} 1 \\ 1 \end{bmatrix} = 11/6$$

so $1 = (\underset{\sim}{b}'\underset{\sim}{d})^2 \le 125\ (11/6) = 229.17$

2.36

$$4x_1^2 + 4x_2^2 + 6x_1x_2 = \mathrm{x}'A\mathrm{x} \text{ where } A = \begin{pmatrix} 4 & 3 \\ 3 & 4 \end{pmatrix}.$$

$(4-\lambda)^2 - 3^2 = 0$ gives $\lambda_1 = 7, \lambda_2 = 1$. Hence the maximum is 7 and the minimum is 1.

2.37 From (2-51), $\max\limits_{\underset{\sim}{x}'\underset{\sim}{x}=1} \underset{\sim}{x}'A\underset{\sim}{x} = \max\limits_{\underset{\sim}{x}\neq\underset{\sim}{0}} \dfrac{\underset{\sim}{x}'A\underset{\sim}{x}}{\underset{\sim}{x}'\underset{\sim}{x}} = \lambda_1$

where λ_1 is the largest eigenvalue of A. For A given in Exercise 2.6, we have from Exercise 2.7, $\lambda_1 = 10$ and $\underset{\sim}{e}_1' = [.894, -.447]$. Therefore $\max\limits_{\underset{\sim}{x}'\underset{\sim}{x}=1} \underset{\sim}{x}'A\underset{\sim}{x} = 10$ and this maximum is attained for $\underset{\sim}{x} = \underset{\sim}{e}_1$.

2.38
Using computer, $\lambda_1 = 18$, $\lambda_2 = 9$, $\lambda_3 = 9$. Hence the maximum is 18 and the minimum is 9.

Chapter 3

3.1

a) $\bar{\mathbf{x}} = \begin{bmatrix} 5 \\ 2 \end{bmatrix}$

b) $\mathbf{e}_1 = \mathbf{y}_1 - \bar{x}_1\mathbf{1} = [4, 0, -4]'$

$\mathbf{e}_2 = \mathbf{y}_2 - \bar{x}_2\mathbf{1} = [-1, 1, 0]'$

c) $L_{\mathbf{e}_1} = \sqrt{32}$; $L_{\mathbf{e}_2} = \sqrt{2}$

Let θ be the angle between $\mathbf{e}_1$ and $\mathbf{e}_2$, then $\cos(\theta) = -4/\sqrt{32 \times 2} = -.5$

Therefore $n\, s_{11} = L^2_{\mathbf{e}_1}$ or $s_{11} = 32/3$; $n\, s_{22} = L^2_{\mathbf{e}_2}$ or $s_{22} = 2/3$; $n\, s_{12} = \mathbf{e}_1'\mathbf{e}_2$ or $s_{12} = -4/3$. Also, $r_{12} = \cos(\theta) = -.5$. Consequently $S_n = \begin{bmatrix} 32/3 & -4/3 \\ -4/3 & 2/3 \end{bmatrix}$ and $R = \begin{bmatrix} 1 & -.5 \\ -.5 & 1 \end{bmatrix}$.

3.2

a) $\bar{\mathbf{x}} = \begin{bmatrix} 4 \\ 1 \end{bmatrix}$

b) $\mathbf{e}_1 = \mathbf{y}_1 - \bar{x}_1\mathbf{1} = [-1, 2, -1]'$

$\mathbf{e}_2 = \mathbf{y}_2 - \bar{x}_2\mathbf{1} = [3, -3, 0]'$

c) $L_{\mathbf{e}_1} = \sqrt{6}$; $L_{\mathbf{e}_2} = \sqrt{18}$

Let θ be the angle between $\mathbf{e}_1$ and $\mathbf{e}_2$, then $\cos(\theta) = -9/\sqrt{6 \times 18} = -.866$.

Therefore $n\, s_{11} = L^2_{\mathbf{e}_1}$ or $s_{11} = 6/3 = 2$; $n\, s_{22} = L^2_{\mathbf{e}_2}$ or $s_{22} = 18/3 = 6$; $n\, s_{12} = \mathbf{e}_1'\mathbf{e}_2$ or $s_{12} = -9/3 = -3$. Also, $r_{12} = \cos(\theta) = -.866$. Consequently $S_n = \begin{bmatrix} 2 & -3 \\ -3 & 6 \end{bmatrix}$ and $R = \begin{bmatrix} 1 & -.866 \\ -.866 & 1 \end{bmatrix}$

3.3 $\underset{\sim}{y}_1 = [1, 4, 4]'$; $\bar{x}_1 \underset{\sim}{1} = [3, 3, 3]$; $\underset{\sim}{y}_1 - \bar{x}_1 \underset{\sim}{1} = [-2, 1, 1]'$

Thus

$$\underset{\sim}{y}_1 = \begin{bmatrix} 1 \\ 4 \\ 4 \end{bmatrix} = \begin{bmatrix} 3 \\ 3 \\ 3 \end{bmatrix} + \begin{bmatrix} -2 \\ 1 \\ 1 \end{bmatrix} = \bar{x}_1 \underset{\sim}{1} + (\underset{\sim}{y}_1 - \bar{x}_1 \underset{\sim}{1})$$

3.5 a) $\mathbf{X}' = \begin{bmatrix} 9 & 5 & 1 \\ 1 & 3 & 2 \end{bmatrix}$; $\underset{\sim}{\bar{x}} \underset{\sim}{1}' = \begin{bmatrix} 5 & 5 & 5 \\ 2 & 2 & 2 \end{bmatrix}$

$$2\,S = (\mathbf{X} - \underset{\sim}{\bar{x}} \underset{\sim}{1}')(\mathbf{X} - \underset{\sim}{\bar{x}} \underset{\sim}{1}')' = \begin{bmatrix} 4 & 0 & -4 \\ -1 & 1 & 0 \end{bmatrix} \begin{bmatrix} 4 & -1 \\ 0 & 1 \\ -4 & 0 \end{bmatrix} = \begin{bmatrix} 32 & -4 \\ -4 & 2 \end{bmatrix}$$

so $S = \begin{bmatrix} 16 & -2 \\ -2 & 1 \end{bmatrix}$ and $|S| = 12$

b) $\mathbf{X}' = \begin{bmatrix} 3 & 6 & 3 \\ 4 & -2 & 1 \end{bmatrix}$; $\underset{\sim}{\bar{x}} \underset{\sim}{1}' = \begin{bmatrix} 4 & 4 & 4 \\ 1 & 1 & 1 \end{bmatrix}$

$$2\,S = (\mathbf{X} - \underset{\sim}{1} \underset{\sim}{\bar{x}}')'(\mathbf{X} - \underset{\sim}{1} \underset{\sim}{\bar{x}}') = \begin{bmatrix} -1 & 2 & -1 \\ 3 & -3 & 0 \end{bmatrix} \begin{bmatrix} -1 & 3 \\ 2 & -3 \\ -1 & 0 \end{bmatrix} = \begin{bmatrix} 6 & -9 \\ -9 & 18 \end{bmatrix}$$

so $S = \begin{bmatrix} 3 & -9/2 \\ -9/2 & 9 \end{bmatrix}$ and $|S| = 27/4$

3.6 a) $\mathbf{X}' - \underset{\sim}{1} \underset{\sim}{\bar{x}}' = \begin{bmatrix} -3 & 0 & -3 \\ 0 & 1 & 1 \\ 3 & -1 & 2 \end{bmatrix}$. Thus $\underset{\sim}{d}'_1 = [-3, 0, -3]$,

$\underset{\sim}{d}'_2 = [0, 1, -1]$ and $\underset{\sim}{d}'_3 = [-3, 1, 2]$.

Since $\underset{\sim}{d}_1 = \underset{\sim}{d}_2 = \underset{\sim}{d}_3$, the matrix of deviations is not of full rank.

b)

$$2S = (\mathbf{X} - \underset{\sim}{1}\,\bar{\underset{\sim}{x}}')'(\mathbf{X} - \underset{\sim}{1}\,\bar{\underset{\sim}{x}}') = \begin{bmatrix} 18 & -3 & 15 \\ -3 & 2 & -1 \\ 15 & -1 & 14 \end{bmatrix}$$

So

$$S = \begin{bmatrix} 9 & -3/2 & 15/2 \\ -3/2 & 1 & -1/2 \\ 15/2 & -1/2 & 7 \end{bmatrix}$$

$|S| = 0$ (Verify). The 3 deviation vectors lie in a 2-dimensional subspace. The 3-dimensional volume enclosed by the deviation vectors is zero.

c) Total sample variance = 9 + 1 + 7 = 17 .

3.7 All ellipses are centered at $\bar{\underset{\sim}{x}}$.

i) For $S = \begin{bmatrix} 5 & 4 \\ 4 & 5 \end{bmatrix}$, $S^{-1} = \begin{bmatrix} 5/9 & -4/9 \\ -4/9 & 5/9 \end{bmatrix}$

Eigenvalue-normalized eigenvector pairs for S^{-1} are:

$$\lambda_1 = 1, \quad \underset{\sim}{e}_1' = [.707, \quad -.707]$$

$$\lambda_2 = 1/9, \quad \underset{\sim}{e}_2' = [.707, \quad .707]$$

Half lengths of axes of ellipse $(\underset{\sim}{x} - \bar{\underset{\sim}{x}})'S^{-1}(\underset{\sim}{x} - \bar{\underset{\sim}{x}}) \le 1$ are $1/\sqrt{\lambda_1} = 1$ and $1/\sqrt{\lambda_2} = 3$ respectively. The major axis of ellipse lies in the direction of $\underset{\sim}{e}_2$; the minor axis lies in the direction of $\underset{\sim}{e}_1$.

ii) For $S = \begin{bmatrix} 5 & -4 \\ -4 & 5 \end{bmatrix}$, $S^{-1} = \begin{bmatrix} 5/9 & 4/9 \\ 4/9 & 5/9 \end{bmatrix}$

Eigenvalue-normalized eigenvectors for S^{-1} are:

$$\lambda_1 = 1, \quad \underset{\sim}{e}_1' = [.707, \quad .707]$$

$$\lambda_2 = 1/9, \quad \underset{\sim}{e}_2' = [.707, \quad -.707]$$

Half lengths of axes of ellipse $(\underset{\sim}{x}-\bar{\underset{\sim}{x}})'S^{-1}(\underset{\sim}{x}-\bar{\underset{\sim}{x}}) \le 1$ are, again, $1/\sqrt{\lambda_1} = 1$ and $1/\sqrt{\lambda_2} = 3$. The major axes of the ellipse lies in the direction of $\underset{\sim}{e}_2$; the minor axis lies in the direction of $\underset{\sim}{e}_1$. Note that $\underset{\sim}{e}_2$ here is $\underset{\sim}{e}_1$ in part (i) above and $\underset{\sim}{e}_1$ here is $\underset{\sim}{e}_2$ in part (i) above.

iii) For $S = \begin{bmatrix} 3 & 0 \\ 0 & 3 \end{bmatrix}$, $S^{-1} = \begin{bmatrix} 1/3 & 0 \\ 0 & 1/3 \end{bmatrix}$

Eigenvalue-normalized eigenvector pairs for S^{-1} are:

$$\lambda_1 = 1/3; \quad \underset{\sim}{e}_1' = [1, 0]$$

$$\lambda_2 = 1/3, \quad \underset{\sim}{e}_2' = [0, 1]$$

Half lengths of axes of ellipse $(\underset{\sim}{x}-\bar{\underset{\sim}{x}})'S^{-1}(\underset{\sim}{x}-\bar{\underset{\sim}{x}}) \le 1$ are equal and given by $1/\sqrt{\lambda_1} = 1/\sqrt{\lambda_2} = \sqrt{3}$. Major and minor axes of ellipse can be taken to lie in the directions of the coordinate axes. Here, the solid ellipse is, in fact, a solid sphere.

Notice for all three cases $|S| = 9$.

3.8 a) Total sample variance in both cases is 3.

b) For $S = \begin{bmatrix} 1 & 0 & 0 \\ 0 & 1 & 0 \\ 0 & 0 & 1 \end{bmatrix}$, $|S| = 1$

For $S = \begin{bmatrix} 1 & -1/2 & -1/2 \\ -1/2 & 1 & -1/2 \\ -1/2 & -1/2 & 1 \end{bmatrix}$, $|S| = 0$

3.9 (a) We calculate $\overline{\boldsymbol{x}} = [16, 18, 34\ ]'$ and

$$\boldsymbol{X}_c = \begin{bmatrix} -4 & -1 & -5 \\ 2 & 2 & 4 \\ -2 & -2 & -4 \\ 4 & 0 & 4 \\ 0 & 1 & 1 \end{bmatrix} \quad \text{and we notice } \text{col}_1(\ \boldsymbol{X}_c\ ) + \text{col}_2(\ \boldsymbol{X}_c\ ) = \text{col}_1(\ \boldsymbol{X}_c\ )$$

so $\boldsymbol{a} = [1, 1, -1\ ]'$ gives $\boldsymbol{X}_c\boldsymbol{a} = \boldsymbol{0}$.

(b)

$$\text{S} = \begin{bmatrix} 10 & 3 & 13 \\ 3 & 2.5 & 5.5 \\ 13 & 5.5 & 18.5 \end{bmatrix} \quad \text{so } |\,\text{S}\,| = \begin{matrix} 10(2.5)(18.5) & + & 39(15.5) & + & 39(15.5) \\ -(13)^2(2.5) & - & 9(18.5) & - & 55(5.5) \end{matrix} = 0$$

As above in a)

$$\textbf{S}\boldsymbol{a} = \begin{bmatrix} 10 + 3 - 13 \\ 3 + 2.5 - 5.5 \\ 13 + 5.5 - 18.5 \end{bmatrix} = \begin{bmatrix} 0 \\ 0 \\ 0 \end{bmatrix}$$

(c) Check.

3.10 (a) We calculate $\overline{\boldsymbol{x}} = [5, 2, 3\ ]'$ and

$$\boldsymbol{X}_c = \begin{bmatrix} -2 & -1 & -3 \\ 1 & 2 & 3 \\ -1 & 0 & -1 \\ 2 & -2 & 0 \\ 0 & 1 & 1 \end{bmatrix} \quad \text{and we notice } \text{col}_1(\ \boldsymbol{X}_c\ ) + \text{col}_2(\ \boldsymbol{X}_c\ ) = \text{col}_1(\ \boldsymbol{X}_c\ )$$

so $\boldsymbol{a} = [1, 1, -1\ ]'$ gives $\boldsymbol{X}_c\boldsymbol{a} = \boldsymbol{0}$.

(b)

$$\boldsymbol{S} = \begin{bmatrix} 2.5 & 0 & 2.5 \\ 0 & 2.5 & 2.5 \\ 2.5 & 2.5 & 5 \end{bmatrix} \quad \text{so } |\,\text{S}\,| = \begin{matrix} 5(2.5)^2 & + & 0 & + & 0 \\ -(2.5)^3 & - & 0 & - & (2.5)^3 \end{matrix} = 0$$

Using the save coefficient vector $\boldsymbol{a}$ as in Part a) $\textbf{S}\boldsymbol{a} = \boldsymbol{0}$.

(c) Setting $\boldsymbol{Xa} = 0$,

$$\begin{array}{lllll} 3a_1 & + & a_2 & & = 0 \\ 7a_1 & & & + \; 3a_3 & = 0 \\ 5a_1 & + & 3a_2 & + \; 4a_3 & = 0 \end{array} \quad \text{so} \quad \begin{array}{lll} a_1 & = & -\frac{3}{7}a_3 \\ 5a_1 & - & 3(3a_1) + 4a_3 = 0 \end{array}$$

so we must have $a_1 = a_3 = 0$ but then, by the first equation in the first set, $a_2 = 0$. The columns of the data matrix are linearly independent.

3.11

$$S = \begin{bmatrix} 14808 & 14213 \\ 14213 & 15538 \end{bmatrix} .$$

Consequently

$$R = \begin{bmatrix} 1 & .9370 \\ .9370 & 1 \end{bmatrix}; \quad D^{1/2} = \begin{bmatrix} 121.6881 & 0 \\ 0 & 124.6515 \end{bmatrix}$$

and $$D^{-1/2} = \begin{bmatrix} .0082 & 0 \\ 0 & .0080 \end{bmatrix}$$

The relationships $R = D^{-1/2}\, S\, D^{-1/2}$ and $S = D^{1/2}\, R\, D^{1/2}$ can now be verified by direct matrix multiplication.

3.14 a) From first principles we have

$$\underset{\sim}{b}'\underset{\sim}{x}_1 = [2 \quad 3]\begin{bmatrix} 9 \\ 1 \end{bmatrix} = 21$$

Similarly $\underset{\sim}{b}'\underset{\sim}{x}_2 = 19$ and $\underset{\sim}{b}'\underset{\sim}{x}_3 = 8$ so

$$\text{sample mean} = \frac{21+19+8}{3} = 16$$

$$\text{sample variance} = \frac{(21-16)^2+(19-16)^2+(8-16)^2}{2} = 49$$

Also $\underset{\sim}{c}'\underset{\sim}{x}_1 = [-1 \quad 2]\begin{bmatrix} 9 \\ 1 \end{bmatrix} = -7$; $\underset{\sim}{c}'\underset{\sim}{x}_2 = 1$ and $\underset{\sim}{c}'\underset{\sim}{x}_3 = 3$

so

sample mean = -1

sample variance = 28

Finally sample covariance = $\frac{(21-16)(-7+1)+(19-16)(1+1)+(8-16)(3+1)}{2} =$ -28.

b) $\bar{\underset{\sim}{x}}' = [5 \quad 2]$ and $S = \begin{bmatrix} 16 & -2 \\ -2 & 1 \end{bmatrix}$

Using (3-36)

sample mean of $\underset{\sim}{b}'\ \underset{\sim}{X} = \underset{\sim}{b}'\ \bar{\underset{\sim}{x}} = [2 \quad 3] \begin{bmatrix} 5 \\ 2 \end{bmatrix} = 16$

sample mean of $\underset{\sim}{c}'\ \underset{\sim}{X} = [-1 \quad 2] \begin{bmatrix} 5 \\ 2 \end{bmatrix} = -1$

sample variance of $\underset{\sim}{b}'\ \underset{\sim}{X} = \underset{\sim}{b}'\ S\underset{\sim}{b} = [2 \quad 3] \begin{bmatrix} 16 & -2 \\ -2 & 1 \end{bmatrix} \begin{bmatrix} 2 \\ 3 \end{bmatrix} = 49$

sample variance of $\underset{\sim}{c}'\ \underset{\sim}{X} = \underset{\sim}{c}'\ S\underset{\sim}{c} = [-1 \quad 2] \begin{bmatrix} 16 & -2 \\ -2 & 1 \end{bmatrix} \begin{bmatrix} -1 \\ 2 \end{bmatrix} = 28$

sample covariance of $\underset{\sim}{b}'\ \underset{\sim}{X}$ and $\underset{\sim}{c}'\ \underset{\sim}{X}$

$$= \underset{\sim}{b}' S\underset{\sim}{c} = [2 \quad 3] \begin{bmatrix} 16 & -2 \\ -2 & 1 \end{bmatrix} \begin{bmatrix} -1 \\ 2 \end{bmatrix} = -28$$

Results same as those in part (a).

3.15

$$\bar{\underset{\sim}{x}} = \begin{bmatrix} 5 \\ 3 \\ 4 \end{bmatrix}, \quad S = \begin{bmatrix} 13 & -2.5 & 1.5 \\ -2.5 & 1 & -1.5 \\ 1.5 & -1.5 & 3 \end{bmatrix}$$

sample mean of $\underset{\sim}{b}'\ \underset{\sim}{X} = 12$

sample mean of $\underset{\sim}{c}'\ \underset{\sim}{X} = -1$

sample variance of $\underset{\sim}{b}'\ \underset{\sim}{X} = 12$

sample variance of $\underset{\sim}{c}'\ \underset{\sim}{X} = 43$

sample covariance of $\underset{\sim}{b}'\ \underset{\sim}{X}$ and $\underset{\sim}{c}'\ \underset{\sim}{X} = -3$

3.16 Since $\Sigma_V = E(\underset{\sim}{V} - \underset{\sim}{\mu}_V)(\underset{\sim}{V} - \underset{\sim}{\mu}_V)'$

$$= E(\underset{\sim}{V}\underset{\sim}{V}' - \underset{\sim}{V}\underset{\sim}{\mu}_V' - \underset{\sim}{\mu}_V\underset{\sim}{V}' + \underset{\sim}{\mu}_V\underset{\sim}{\mu}_V')$$

$$= E(\underset{\sim}{V}\underset{\sim}{V}') - E(\underset{\sim}{V})\underset{\sim}{\mu}_V' - \underset{\sim}{\mu}_V E(\underset{\sim}{V}') + \underset{\sim}{\mu}_V\underset{\sim}{\mu}_V'$$

$$= E(\underset{\sim}{V}\underset{\sim}{V}') - \underset{\sim}{\mu}_V\underset{\sim}{\mu}_V' - \underset{\sim}{\mu}_V\underset{\sim}{\mu}_V' + \underset{\sim}{\mu}_V\underset{\sim}{\mu}_V'$$

$$= E(\underset{\sim}{V}\underset{\sim}{V}') - \underset{\sim}{\mu}_V\underset{\sim}{\mu}_V' ,$$

we have $E(\underset{\sim}{V}\underset{\sim}{V}') = \Sigma + \underset{\sim}{\mu}_V\underset{\sim}{\mu}_V'$.

Chapter 4

4.1 (a) We are given $p = 2$, $\boldsymbol{\mu} = \begin{bmatrix} 1 \\ 3 \end{bmatrix}$, $\boldsymbol{\Sigma} = \begin{bmatrix} 2 & -.8 \times \sqrt{2} \\ -.8 \times \sqrt{2} & 1 \end{bmatrix}$ so $|\,\boldsymbol{\Sigma}\,| = .72$ and

$$\boldsymbol{\Sigma}^{-1} = \begin{bmatrix} \frac{1}{.72} & \frac{\sqrt{2}}{.9} \\ \frac{\sqrt{2}}{.9} & \frac{2}{.72} \end{bmatrix}$$

$$f(\boldsymbol{x}) = \frac{1}{(2\pi)\sqrt{.72}} \exp\left(-\frac{1}{2} \left[\frac{1}{.72}(x_1 - 1)^2 + \frac{2\sqrt{2}}{.9}(x_1 - 1)(x_2 - 3) + \frac{2}{.72}(x_2 - 3)^2 \right] \right)$$

(b)

$$\frac{1}{.72}(x_1 - 1)^2 + \frac{2\sqrt{2}}{.9}(x_1 - 1)(x_2 - 3) + \frac{2}{.72}(x_2 - 3)^2$$

4.2 (a) We are given $p = 2$, $\boldsymbol{\mu} = \begin{bmatrix} 0 \\ 2 \end{bmatrix}$, $\boldsymbol{\Sigma} = \begin{bmatrix} 2 & \frac{1}{\sqrt{2}} \\ \frac{1}{\sqrt{2}} & 1 \end{bmatrix}$ so $|\,\boldsymbol{\Sigma}\,| = 3/2$

and

$$\boldsymbol{\Sigma}^{-1} = \begin{bmatrix} \frac{2}{\sqrt{3}} & -\frac{\sqrt{2}}{3} \\ -\frac{\sqrt{2}}{3} & \frac{2}{\sqrt{3}} \end{bmatrix}$$

$$f(\boldsymbol{x}) = \frac{1}{(2\pi)\sqrt{3/2}} \exp\left(-\frac{1}{2} \left[\frac{2}{3}x_1^2 - \frac{2\sqrt{2}}{3}x_1(x_2 - 2) + \frac{4}{3}(x_2 - 2)^2 \right] \right)$$

(b)

$$\frac{2}{3}x_1^2 - \frac{2\sqrt{2}}{3}x_1(x_2 - 2) + \frac{4}{3}(x_2 - 2)^2$$

(c) $c^2 = \chi_2^2(.5) = 1.39$. Ellipse centered at $[0, 2]'$ with the major axis having half-length $\sqrt{\lambda_1}\ c = \sqrt{2.366}\sqrt{1.39} = 1.81$. The major axis lies in the direction $\boldsymbol{e} = [.888, .460]'$. The minor axis lies in the direction $\boldsymbol{e} = [-.460\ , .888]'$ and has half-length $\sqrt{\lambda_2}\ c = \sqrt{.634}\sqrt{1.39} = .94$.

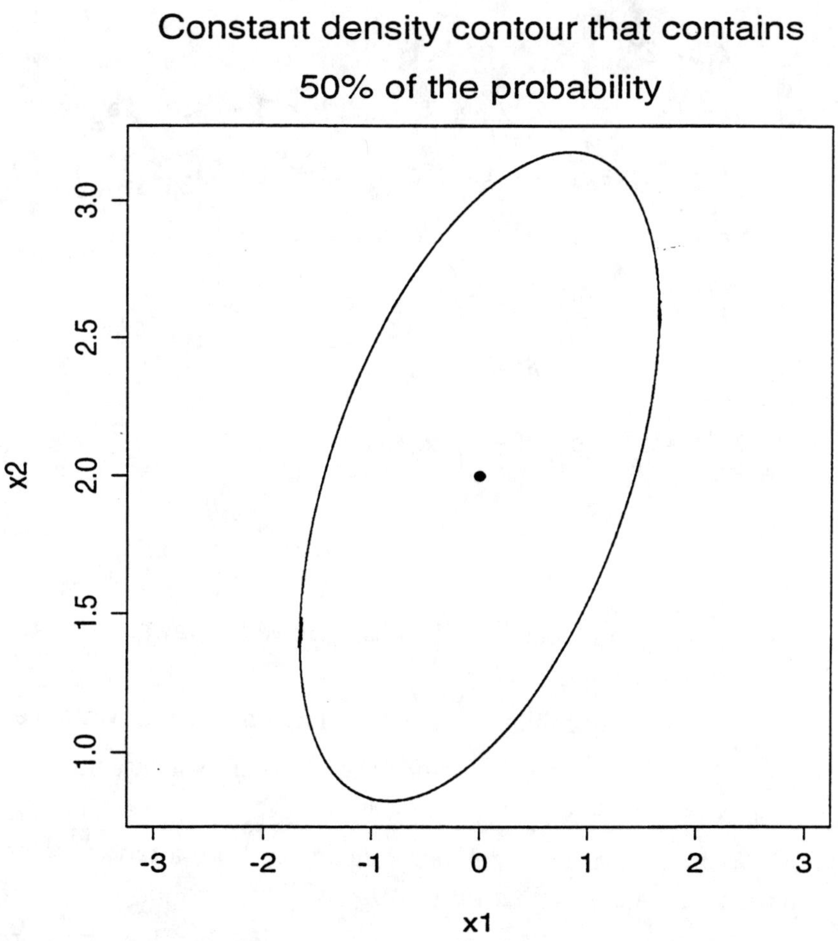

4.3 We apply Result 4.5 that relates zero covariance to statistical independence

a) No, $\sigma_{12} \neq 0$

b) Yes, $\sigma_{23} = 0$

c) Yes, $\sigma_{13} = \sigma_{23} = 0$

d) Yes, by Result 4.3, $(X_1+X_2)/2$ and X_3 are jointly normal and their covariance is $\frac{1}{2}\sigma_{13}+\frac{1}{2}\sigma_{23} = 0$.

e) No, by Result 4.3 with $A = \begin{bmatrix} 0 & 1 & 0 \\ -\frac{5}{2} & 1 & -1 \end{bmatrix}$, form $A \Sigma A'$ to see that the covariance is 10 and not 0.

4.4 a) $3X_1 - 2X_2 + X_3$ is $N(13,9)$

b) Require $\text{Cov}(X_2, X_2 - a_1X_1 - a_3X_3) = 3 - a_1 - 2a_3 = 0$. Thus any $\underset{\sim}{a}' = [a_1, a_3]$ of the form $\underset{\sim}{a}' = [3-2a_3, a_3]$ will meet the requirement. As an example, $\underset{\sim}{a}' = [1,1]$.

4.5 a) $X_1|x_2$ is $N(\frac{1}{\sqrt{2}}(x_2-2), \frac{3}{2})$

b) $X_2|x_1,x_3$ is $N(-2x_1-5, 1)$

c) $X_3|x_1,x_2$ is $N(\frac{1}{2}(x_1+x_2+3), \frac{1}{2})$

4.6 (a) X_1 and X_2 are independent since they have a bivariate normal distribution with covariance $\sigma_{12} = 0$.

(b) X_1 and X_3 are dependent since they have nonzero covariance $\sigma_{13} = -1$.

(c) X_2 and X_3 are independent since they have a bivariate normal distribution with covariance $\sigma_{23} = 0$.

(d) X_1, X_3 and X_2 are independent since they have a trivariate normal distribution where $\sigma_{12} = 0$ and $\sigma_{32} = 0$.

(e) X_1 and $X_1 + 2X_2 - 3X_3$ are dependent since they have nonzero covariance

$$\sigma_{11} + 2\sigma_{12} - 3\sigma_{13} = 4 + 2(0) - 3(-1) = 7$$

4.7 (a) $X_1|x_3$ is $N(1 + .5(x_3 - 2), 3.5)$

(b) $X_1|x_2, x_3$ is $N(1 + .5(x_3 - 2), 3.5)$. Since X_2 is independent of X_1, conditioning further on x_2 does not change the answer from Part a).

4.15 First,

$$\sum_{j=1}^{n} (\bar{\mathbf{x}}-\boldsymbol{\mu})(\mathbf{x}_j-\bar{\mathbf{x}})' = (\bar{\mathbf{x}}-\boldsymbol{\mu})[\sum_{j=1}^{n}(\mathbf{x}_j-\bar{\mathbf{x}})']$$
$$= (\bar{\mathbf{x}}-\boldsymbol{\mu})(\sum_{j=1}^{n} \mathbf{x}_j-n\bar{\mathbf{x}})'$$
$$= (\bar{\mathbf{x}}-\boldsymbol{\mu})(n\bar{\mathbf{x}}-n\bar{\mathbf{x}})'$$
$$= 0$$

Also,

$$\sum_{j=1}^{n} (\mathbf{x}_j-\bar{\mathbf{x}})(\bar{\mathbf{x}}-\boldsymbol{\mu})' = [\sum_{j=1}^{n}(\bar{\mathbf{x}}-\boldsymbol{\mu})(\mathbf{x}_j-\bar{\mathbf{x}})']' = 0' = 0.$$

4.16 (a) By Result 4.8, with $c_1 = c_3 = 1/4$, $c_2 = c_4 = -1/4$ and $\boldsymbol{\mu}_j = \boldsymbol{\mu}$ for $j = 1, ..., 4$ we have $\sum_{j=1}^{4} c_j \boldsymbol{\mu}_j = \mathbf{0}$ and $(\sum_{j=1}^{4} c_j^2)\, \Sigma = \frac{1}{4}\Sigma$. Consequently, $\mathbf{V}_1$ is $N(\mathbf{0}, \frac{1}{4}\Sigma)$. Similarly, setting $b_1 = b_2 = 1/4$ and $b_3 = b_4 = -1/4$, we find that $\mathbf{V}_2$ is $N(\mathbf{0}, \frac{1}{4}\Sigma)$.

(b) Again by Result 4.8, we know that $\mathbf{V}_1$ and $\mathbf{V}_2$ are jointly multivariate normal with covariance

$$(\sum_{j=1}^{4} b_j c_j)\, \Sigma = \left(\frac{1}{4}(\frac{1}{4}) + \frac{-1}{4}(\frac{1}{4}) + \frac{1}{4}(\frac{-1}{4}) + \frac{-1}{4}(\frac{-1}{4}) \right) \Sigma = 0$$

That is,

$$\begin{bmatrix} \mathbf{V}_1 \\ \mathbf{V}_2 \end{bmatrix} \quad \text{is distributed} \quad N_{2p}\left(\mathbf{0}, \begin{bmatrix} \frac{1}{4}\Sigma & 0 \\ 0 & \frac{1}{4}\Sigma \end{bmatrix}\right)$$

so the joint density of the $2p$ variables is

$$f(\mathbf{v}_1, \mathbf{v}_2) = \frac{1}{(2\pi)^p |\, \frac{1}{4}\Sigma\,|} \exp\left(-\frac{1}{2}[\, \mathbf{v}_1', \mathbf{v}_2'\,] \begin{bmatrix} \frac{1}{4}\Sigma & 0 \\ 0 & \frac{1}{4}\Sigma \end{bmatrix}^{-1} \begin{bmatrix} \mathbf{v}_1 \\ \mathbf{v}_1 \end{bmatrix} \right)$$
$$= \frac{1}{(2\pi)^p |\, \frac{1}{4}\Sigma\,|} \exp\left(-\frac{1}{8}(\, \mathbf{v}_1' \Sigma^{-1}\mathbf{v}_1 + \mathbf{v}_2' \Sigma^{-1}\mathbf{v}_2 \,) \right)$$

4.17 By Result 4.8, with $c_1 = c_2 = c_3 = c_4 = c_5 = 1/5$ and $\boldsymbol{\mu}_j = \boldsymbol{\mu}$ for $j = 1, ..., 5$ we find that $\mathbf{V}_1$ has mean $\sum_{j=1}^{5} c_j \boldsymbol{\mu}_j = \boldsymbol{\mu}$ and covariance matrix $(\sum_{j=1}^{5} c_j^2)\, \Sigma = \frac{1}{5}\Sigma$.

Similarly, setting $b_1 = b_3 = b_5 = 1/5$ and $b_2 = b_4 = -1/5$ we find that $\mathbf{V}_2$ has mean $\sum_{j=1}^{5} b_j \boldsymbol{\mu}_j = \frac{1}{5}\boldsymbol{\mu}$ and covariance matrix $(\sum_{j=1}^{5} b_j^2)\, \Sigma = \frac{1}{5}\Sigma$.

Again by Result 4.8, we know that $\mathbf{V}_1$ and $\mathbf{V}_2$ have covariance

$$(\sum_{j=1}^{4} b_j\, c_j)\, \Sigma = \left(\frac{1}{5}(\frac{1}{5}) + \frac{-1}{5}(\frac{1}{5}) + \frac{1}{5}(\frac{1}{5}) + \frac{-1}{5}(\frac{1}{5}) + \frac{1}{5}(\frac{1}{5}) \right) \Sigma = \frac{1}{25}\Sigma$$

4.18 By Result 4.11 we know that the maximum likelihood estimates of $\underset{\sim}{\mu}$ and Σ are $\bar{\underset{\sim}{x}} = [4,6]'$ and

$$\frac{1}{n}\sum_{j=1}^{n}(\underset{\sim}{x}_j-\bar{\underset{\sim}{x}})(\underset{\sim}{x}_j-\bar{\underset{\sim}{x}})' = \frac{1}{4}\left\{\left(\begin{bmatrix}3\\6\end{bmatrix}-\begin{bmatrix}4\\6\end{bmatrix}\right)\left(\begin{bmatrix}3\\6\end{bmatrix}-\begin{bmatrix}4\\6\end{bmatrix}\right)'+\left(\begin{bmatrix}4\\4\end{bmatrix}-\begin{bmatrix}4\\6\end{bmatrix}\right)\left(\begin{bmatrix}4\\4\end{bmatrix}-\begin{bmatrix}4\\6\end{bmatrix}\right)'\right.$$

$$\left.+\left(\begin{bmatrix}5\\7\end{bmatrix}-\begin{bmatrix}4\\6\end{bmatrix}\right)\left(\begin{bmatrix}5\\7\end{bmatrix}-\begin{bmatrix}4\\6\end{bmatrix}\right)'+\left(\begin{bmatrix}4\\7\end{bmatrix}-\begin{bmatrix}4\\6\end{bmatrix}\right)\left(\begin{bmatrix}4\\7\end{bmatrix}-\begin{bmatrix}4\\6\end{bmatrix}\right)'\right\}$$

$$= \frac{1}{4}\left\{\begin{bmatrix}-1\\0\end{bmatrix}\begin{bmatrix}-1 & 0\end{bmatrix}+\begin{bmatrix}0\\-2\end{bmatrix}\begin{bmatrix}0 & -2\end{bmatrix}+\begin{bmatrix}1\\1\end{bmatrix}\begin{bmatrix}1 & 1\end{bmatrix}+\begin{bmatrix}0\\1\end{bmatrix}\begin{bmatrix}0 & 1\end{bmatrix}\right\}$$

$$= \frac{1}{4}\begin{bmatrix}2 & 1\\1 & 6\end{bmatrix}$$

4.19 a) By Result 4.7 we know that $(\underset{\sim}{X}_1-\underset{\sim}{\mu})'\,\Sigma^{-1}(\underset{\sim}{X}_1-\underset{\sim}{\mu}) \sim \chi^2_6$

b) From (4-23), $\bar{\underset{\sim}{X}} \sim N_6(\underset{\sim}{\mu},\frac{1}{20}\Sigma)$. Then $\bar{\underset{\sim}{X}}-\underset{\sim}{\mu} \sim N_6(\underset{\sim}{0},\frac{1}{20}\Sigma)$ and finally $\sqrt{20}\,(\bar{\underset{\sim}{X}}-\underset{\sim}{\mu}) \sim N_6(\underset{\sim}{0},\Sigma)$

c) From (4-23), 19S has a Wishart distribution with 19 d.f.

4.20 B(19S)B' is a 2×2 matrix distributed as $W_{19}(\cdot\,|\,B\Sigma B')$ with 19 d.f. where

a) $B\Sigma B'$ has

(1,1) entry $= \sigma_{11} + \frac{1}{4}\sigma_{22} + \frac{1}{4}\sigma_{33} - \sigma_{12} - \sigma_{13} + \frac{1}{2}\sigma_{23}$

(1,2) entry $= -\frac{1}{2}\sigma_{14} + \frac{1}{4}\sigma_{24} + \frac{1}{4}\sigma_{34} - \frac{1}{2}\sigma_{15} + \frac{1}{4}\sigma_{25} + \frac{1}{4}\sigma_{35} + \sigma_{16} - \frac{1}{2}\sigma_{26} - \frac{1}{2}\sigma_{36}$

(2,2) entry $= \sigma_{66} + \frac{1}{4}\sigma_{55} + \frac{1}{4}\sigma_{44} - \sigma_{46} - \sigma_{56} + \frac{1}{2}\sigma_{45}$

b)

$$B\Sigma B' = \begin{bmatrix}\sigma_{11} & \sigma_{13}\\ \sigma_{31} & \sigma_{33}\end{bmatrix}$$

4.21 (a) $\overline{\mathbf{X}}$ is distributed $N_4(\boldsymbol{\mu}, n^{-1}\boldsymbol{\Sigma})$

(b) $\mathbf{X}_1 - \boldsymbol{\mu}$ is distributed $N_4(\mathbf{0}, \boldsymbol{\Sigma})$ so $(\mathbf{X}_1 - \boldsymbol{\mu})'\boldsymbol{\Sigma}^{-1}(\mathbf{X}_1 - \boldsymbol{\mu})$ is distributed as chi-square with p degrees of freedom.

(c) Using Part a),

$$(\overline{\mathbf{X}} - \boldsymbol{\mu})'(n^{-1}\boldsymbol{\Sigma})^{-1}(\overline{\mathbf{X}} - \boldsymbol{\mu}) = n(\overline{\mathbf{X}} - \boldsymbol{\mu})'\boldsymbol{\Sigma}^{-1}(\overline{\mathbf{X}} - \boldsymbol{\mu})$$

is distributed as chi-square with p degrees of freedom.

(d) Approximately distributed as chi-square with p degrees of freedom. Since the sample size is large, $\boldsymbol{\Sigma}$ can be replaced by $\mathbf{S}$.

4.22 a) We see that $n = 75$ is a sufficiently large sample (compared with p) and apply Result 4.13 to get $\sqrt{n}(\underset{\sim}{X}-\underset{\sim}{\mu})$ is approximately $N_p(\underset{\sim}{0},\Sigma)$ and that $\bar{\underset{\sim}{X}}$ is approximately $N_p(\underset{\sim}{\mu},\frac{1}{n}\Sigma)$.

b) By (4-28) we conclude that $\sqrt{n}(\underset{\sim}{X}-\underset{\sim}{\mu})'S^{-1}(\bar{\underset{\sim}{X}}-\underset{\sim}{\mu})$ is approximately χ_p^2.

4.23 a) The Q-Q plot (shown below) is not straight. Normality is suspect.

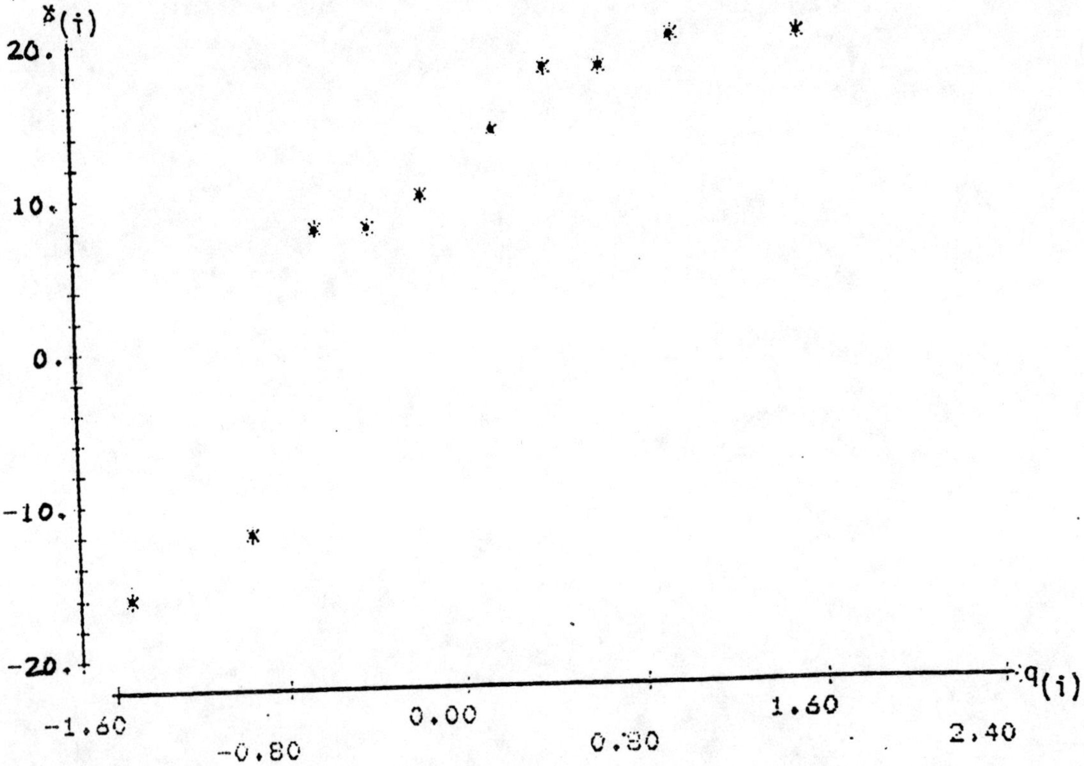

b) Since $r_q = .904 < .935$ (see Table 4.2) we reject the hypothesis of normality of the 10% level.

4.24

(a). Q-Q plots are given below. X_1 has long left hand tails, but X_2 has a long right hand tail.

(b). The critical point for $\alpha = 0.10$ when $n = 10$ is 0.9351. Calculated values of r_Q for X_1, X_2 are 0.93617, 0.94798 respectively. Since they are larger than 0.9351, we do not reject the hypothesis of normality for each variable.

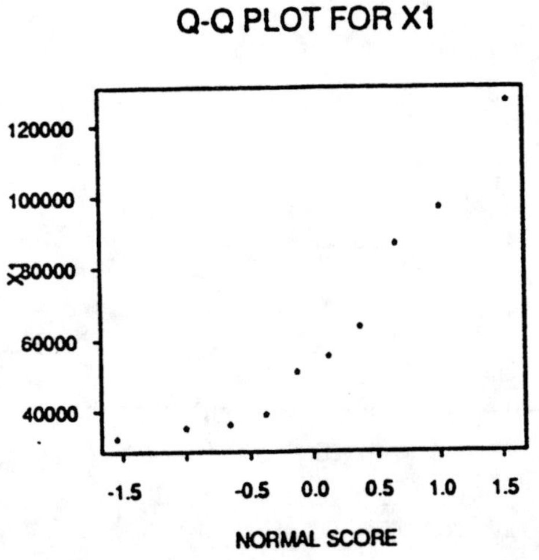

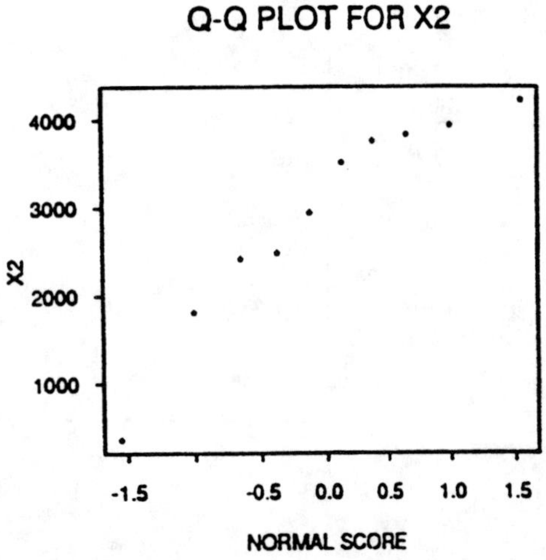

4.25. 10 largest U. S. industrial corporations data

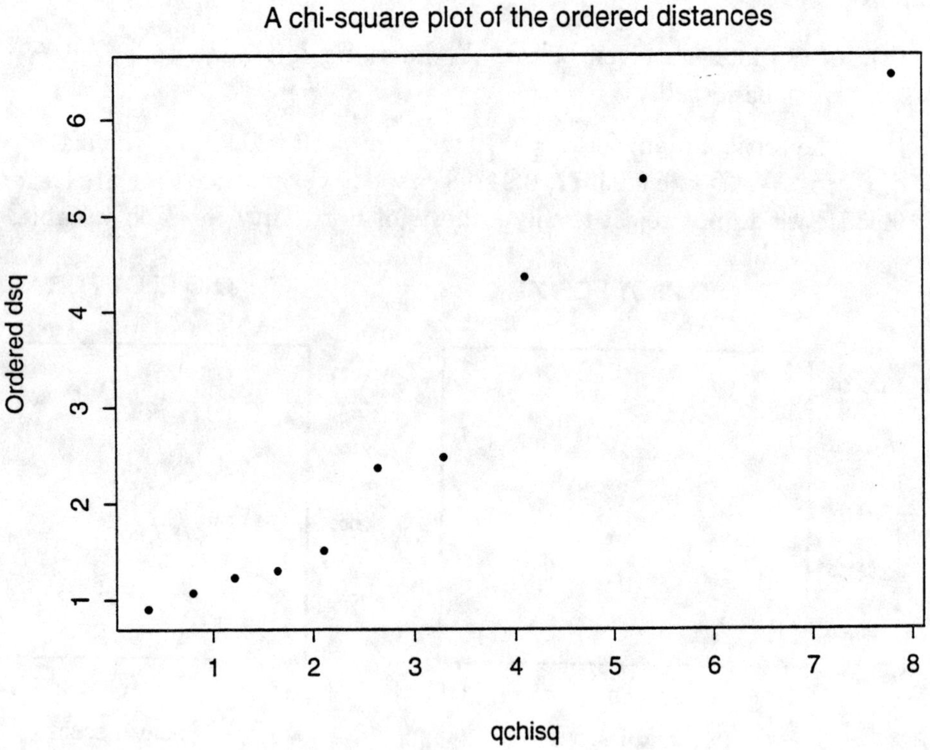

XBAR	S		
	sales	profits	assets
62309.1	1000509114	25575600	1511827230
2927.3	25575600	1430020	45654618
81248.4	1511827230	45654618	2980489782

	Ordered dsq	qchisq
1	0.9078	0.3518
2	1.0749	0.7978
3	1.2284	1.2125
4	1.2963	1.6416
5	1.5003	2.1095
6	2.3614	2.6430
7	2.4713	3.2831
8	4.3524	4.1083
9	5.3677	5.3170
10	6.4396	7.8147

4.26 a) $S = \frac{n}{n-1} S_n = \begin{bmatrix} 5.96 & -1.30 \\ -1.30 & .44 \end{bmatrix}$; $S^{-1} = \begin{bmatrix} .47 & 1.40 \\ 1.40 & 6.40 \end{bmatrix}$; $\bar{\underset{\sim}{x}} = \begin{bmatrix} 7.20 \\ .97 \end{bmatrix}$

Thus d_j^2: 4.06, 2.11, 2.11, .64, 3.27, .01, .52, .65, 2.06, 2.59

b) $\chi_2^2(.5) = 1.39$ thus 4 observations (40%) are within the 50% contour.

c) Q-Q plot is reasonably straight.

d) Given the small number of observations, it is difficult to reject bivariate normality.

4.27 Q-Q plot is shown below.

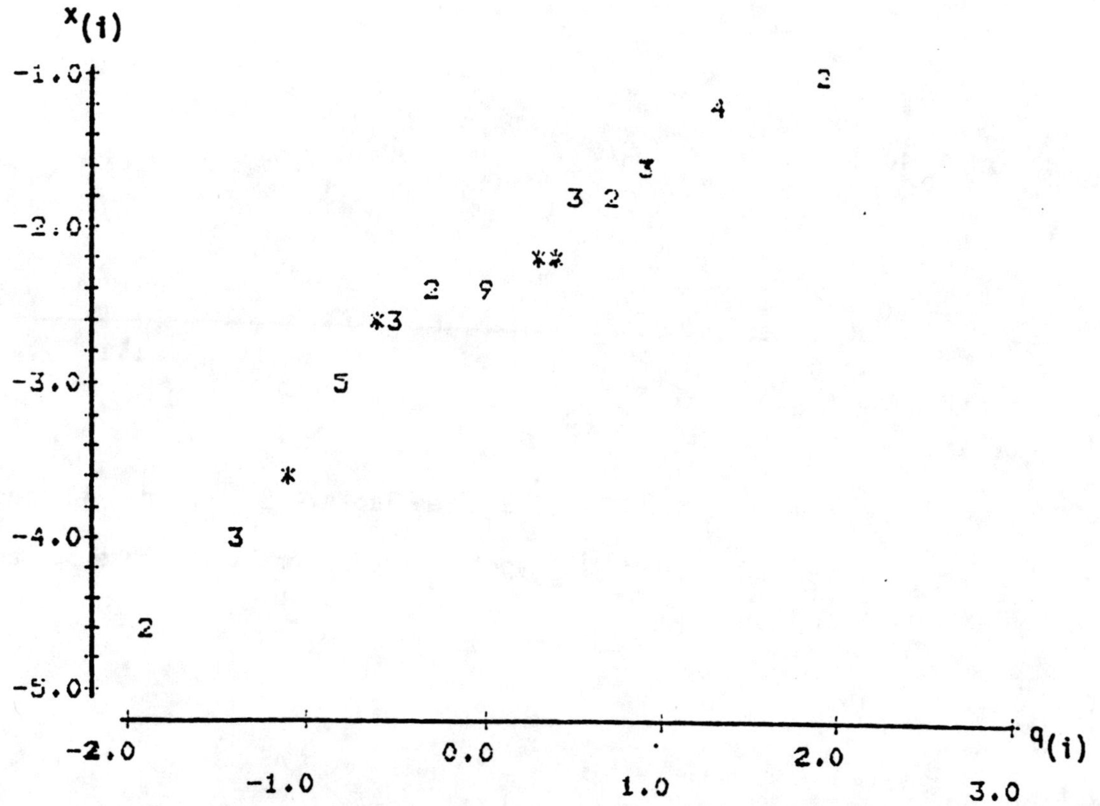

The Q-Q plot is reasonably straight. $r_Q = .971$ ($\lambda=0$)

For $\lambda = 1/4$, $r_Q = .985$ so $\lambda = 1/4$ is a little better choice for the normalizing transformation.

4.28 Q-Q plot is shown below.

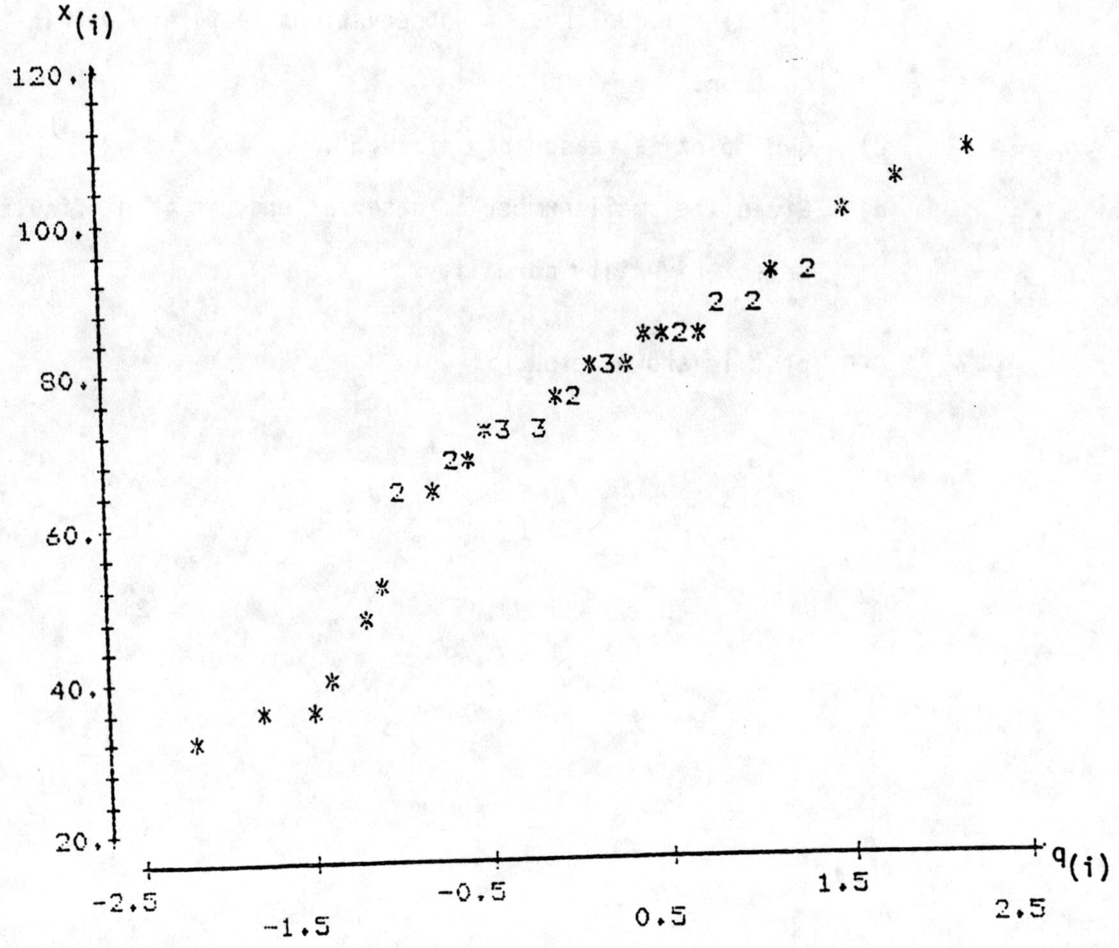

Since $r_Q = .970 < .973$ (See Table 4.2 for $n = 40$ and $\alpha = .05$), we would reject the hypothesis of normality at the 5% level.

4.29

(a).

$$\bar{\mathbf{x}} = \begin{pmatrix} 10.046719 \\ 9.4047619 \end{pmatrix}, \quad \mathbf{S} = \begin{pmatrix} 11.363531 & 3.126597 \\ & 30.978513 \end{pmatrix}.$$

Generalized distances are as follows;

0.4607	0.6592	2.3771	1.6283	0.4135	0.4761	1.1849
10.6392	0.1388	0.8162	1.3566	0.6228	5.6494	0.3159
0.4135	0.1225	0.8988	4.7647	3.0089	0.6592	2.7741
1.0360	0.7874	3.4438	6.1489	1.0360	0.1388	0.8856
0.1380	2.2489	0.1901	0.4607	1.1472	7.0857	1.4584
0.1225	1.8985	2.7783	8.4731	0.6370	0.7032	1.8014

(b). The number of observations whose generalized distances are less than $\chi_2(0.5) = 1.39$ is 26. So the proportion is 26/42=0.6190.

(c). CHI-SQUARE PLOT FOR (X1 X2)

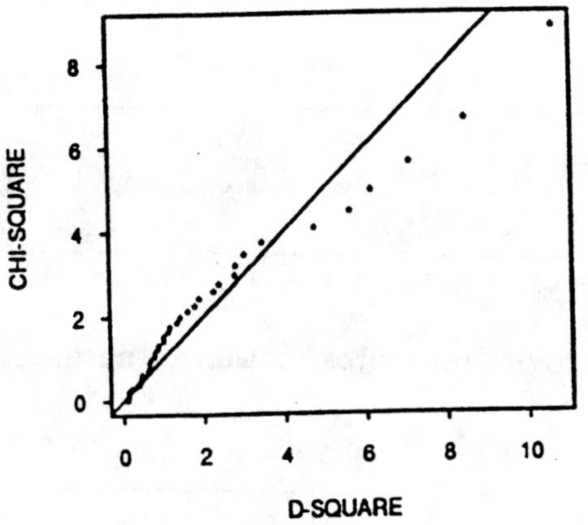

4.30

(a). $\hat{\lambda}_1 = 1.0$. There is no need for transformation. $r_Q = 0.98659 > 0.9198$ where 0.9198 is the critical point for testing normality with $\alpha = 0.05$ when n=10. We accept the null hypothesis of normality.

(b). $\hat{\lambda}_2 = 0.5$. After the transformation, $r_Q = 0.97546 > 0.9198$ where 0.9198 is the critical point for testing normality with $\alpha = 0.05$ when n=10. We accept the null hypothesis of normality.

(c). $(\hat{\lambda}_1, \hat{\lambda}_2) = (0.5, 0.4)$. They are different from results obtained in (a), (b) because the likelihood is flat due to the small sample size.

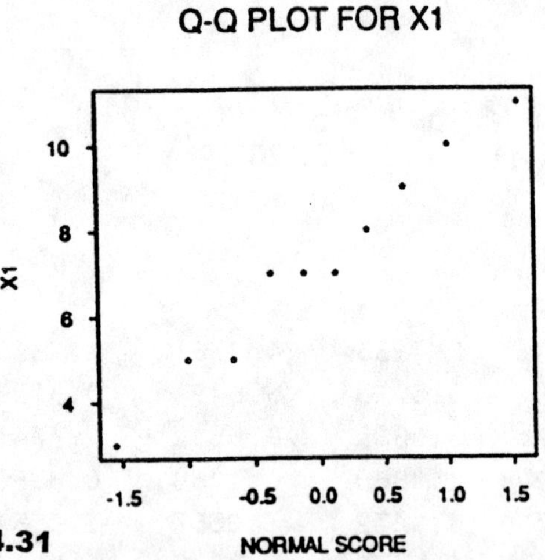

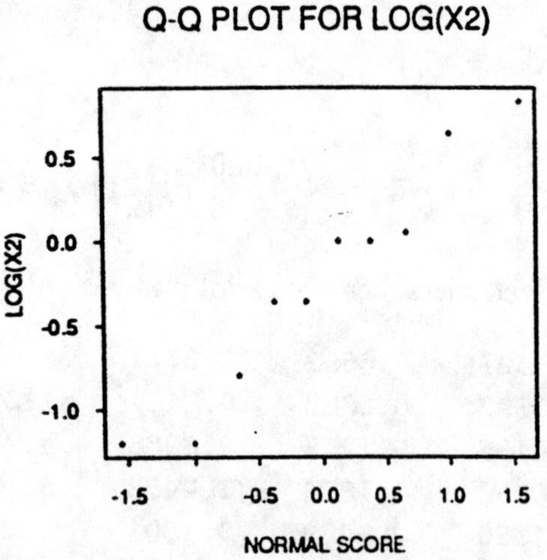

4.31

The non-multiple-sclerosis group:

	X_1	X_2	X_3	X_4	X_5
r_Q	0.94482*	0.96133*	0.95585*	0.97574*	0.94446*
Transformation	$X_1^{-0.5}$	$X_2^{-3.5}$	$(X_3+0.005)^{0.4}$	$X_4^{-3.4}$	$(X_5+0.005)^{0.32}$

*: significant at 5 % level (the critical point = 0.9826 for n=69).

The multiple-sclerosis group:

	X_1	X_2	X_3	X_4	X_5
r_Q	0.97137	0.97209	0.79523*	0.97869	0.84135*
Transformation	–	–	$(X_3+0.005)^{0.26}$	–	$(X_5+0.005)^{0.21}$

*: significant at 5 % level (the critical point = 0.9640 for n=29).

Transformations of X_3 and X_4 do not improve the approximation to normality very much because there are too many zeros.

4.32

	X_1	X_2	X_3	X_4	X_5	X_6
r_Q	0.98464*	0.94526*	0.9970	0.98098*	0.99057	0.92779*
Transformation	$(X_1+0.005)^{-0.59}$	$X_2^{-0.49}$	–	$X_4^{0.25}$	–	$(X_5+0.005)^{0.51}$

*: significant at 5 % level (the critical point = 0.9870 for n=98).

4.33

Marginal Normality:

	X_1	X_2	X_3	X_4
r_Q	0.95986*	0.95039*	0.96341	0.98079

*: significant at 5 % level (the critical point = 0.9652 for n=30).

Bivariate Normality: the χ^2 plots are given in the next page. Those for (X_1, X_2), (X_1, X_3), (X_3, X_4) appear reasonably straight.

CHI-SQUARE PLOT FOR (X1,X2)

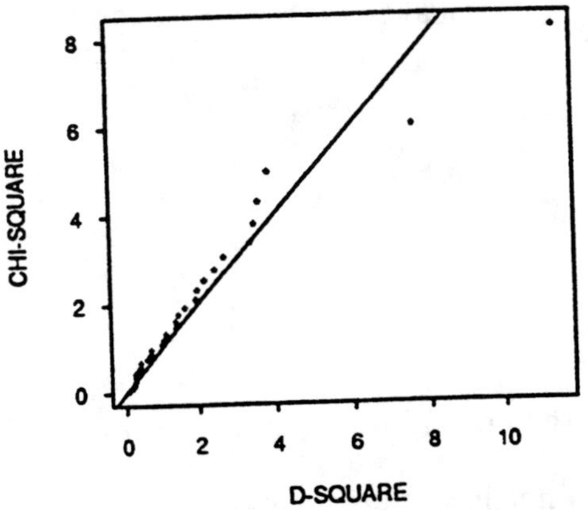

CHI-SQUARE PLOT FOR (X1,X3)

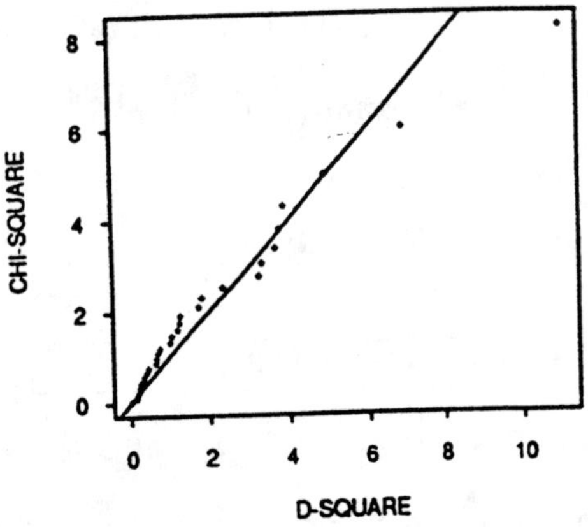

CHI-SQUARE PLOT FOR (X1,X4)

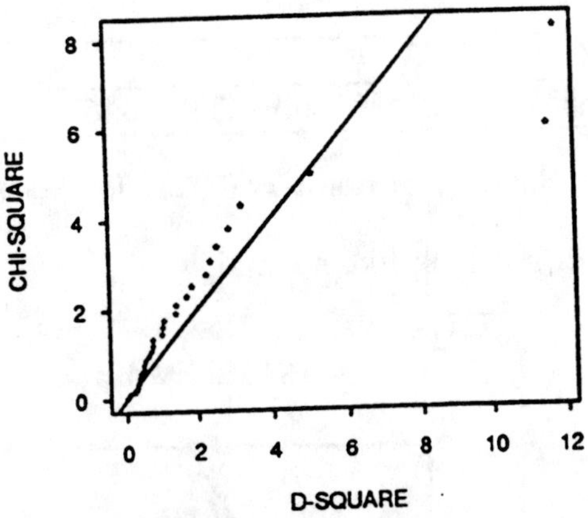

CHI-SQUARE PLOT FOR (X2,X3)

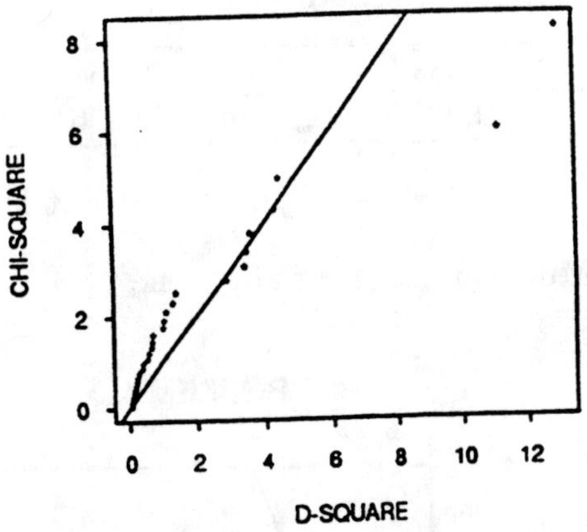

CHI-SQUARE PLOT FOR (X2,X4)

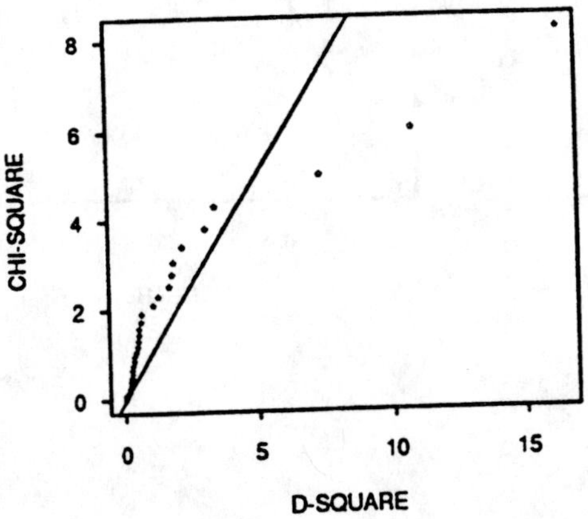

CHI-SQUARE PLOT FOR (X3,X4)

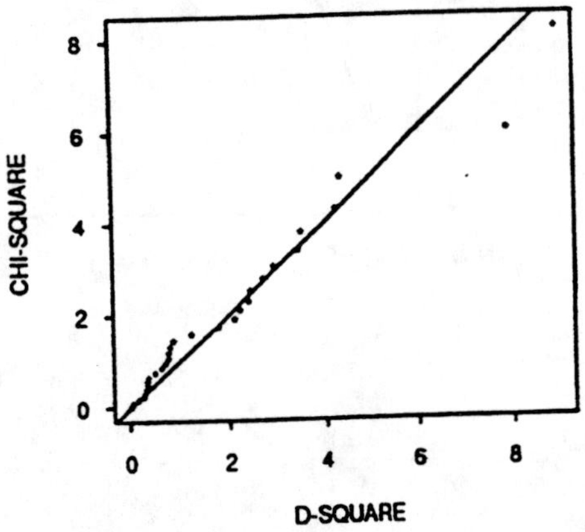

4.34

Marginal Normality:

	X_1	X_2	X_3	X_4	X_5	X_6
r_Q	0.95162*	0.97209	0.98421	0.99011	0.98124	0.99404

*: significant at 5 % level (the critical point = 0.9591 for n=25).

Bivariate Normalitv: Omitted.

4.35

Marginal Normality:

	X_1	X_2	X_3
r_Q	0.89947*	0.99049	0.92270*

*: significant at 5 % level (the critical point = 0.9744 for n=41).

Multivariate Normality: the χ^2 plot does not look like normal.

4.36

Marginal Normality:

	X_1	X_2	X_3	X_4	X_5	X_6	X_7
r_Q	0.98674	0.98496	0.99561	0.98554	0.92155*	0.92558*	0.87316*

*: significant at 5 % level (the critical point = 0.9787 for n=55).

Multivariate Normality: the χ^2 plot does not look like normal.

PROBLEM 4.35

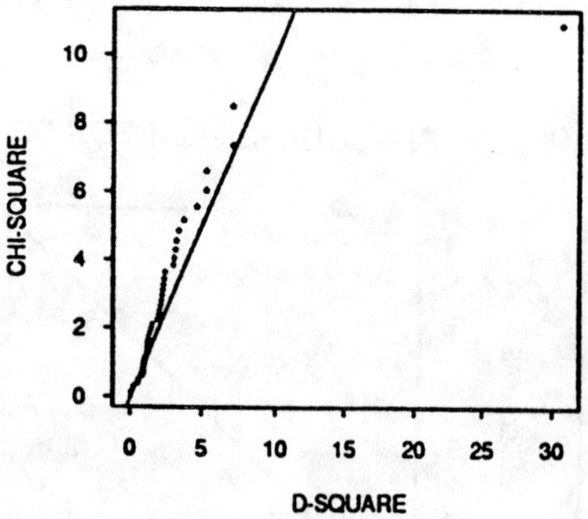

PROBLEM 4.36

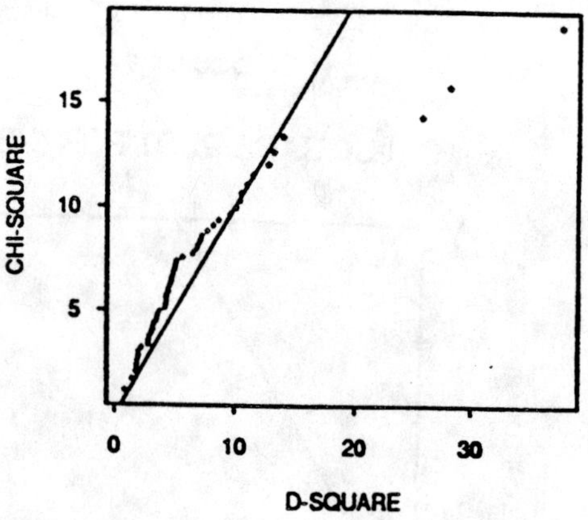

4.37. Marginal and multivariate normality of national track records for women

Normality of National Track Records Data for Women

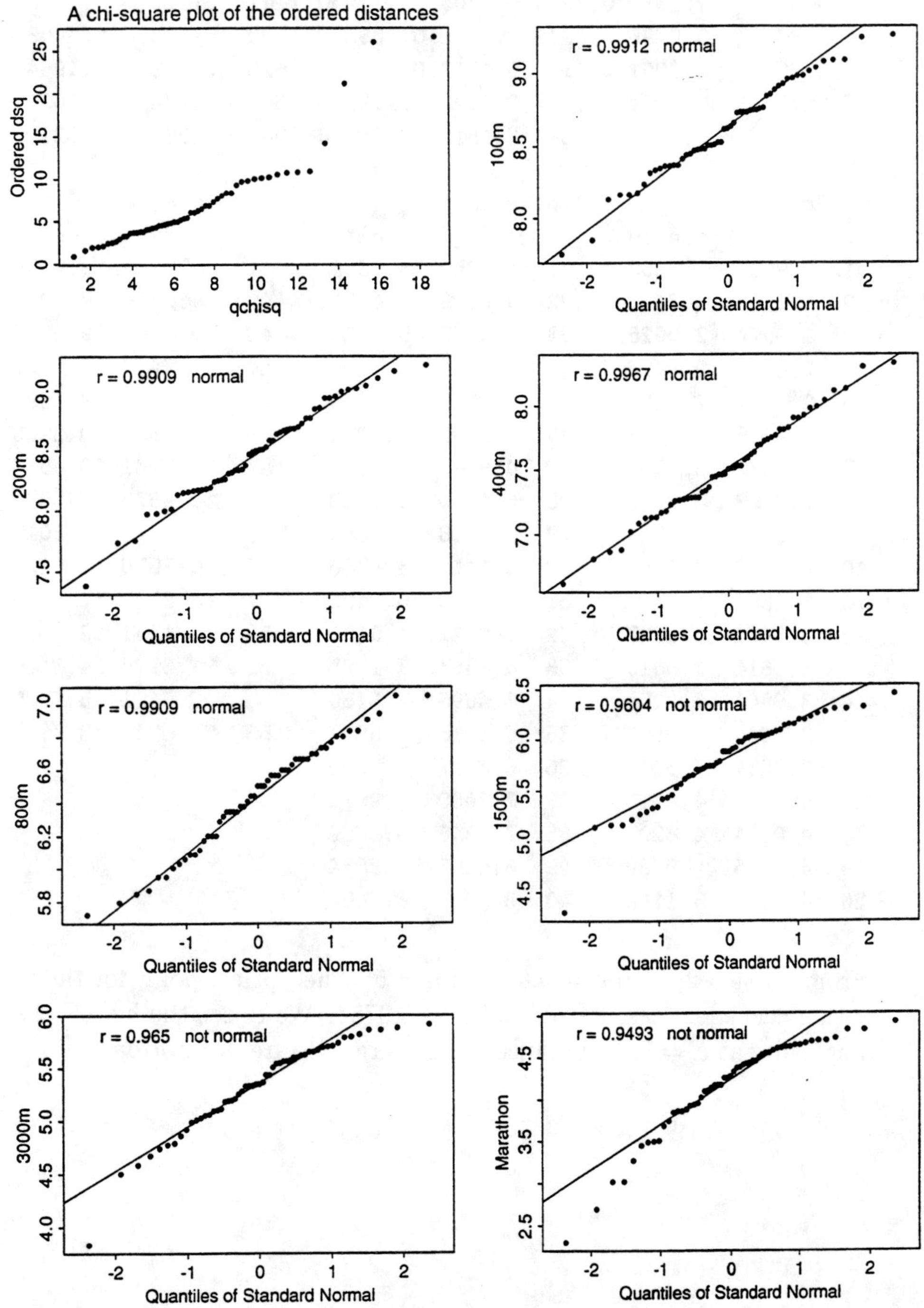

XBAR	S						
	100m	200m	400m	800m	1500m	3000m	Marathon
8.6196	0.1096	0.1238	0.1039	0.0795	0.0991	0.1032	0.1348
8.4777	0.1238	0.1533	0.1265	0.0940	0.1137	0.1174	0.1583
7.5083	0.1039	0.1265	0.1408	0.1112	0.1217	0.1222	0.1518
6.4383	0.0795	0.0940	0.1112	0.1085	0.1220	0.1199	0.1468
5.8099	0.0991	0.1137	0.1217	0.1220	0.1625	0.1618	0.1964
5.3277	0.1032	0.1174	0.1222	0.1199	0.1618	0.1734	0.2097
4.1543	0.1348	0.1583	0.1518	0.1468	0.1964	0.2097	0.3216

	Ordered dsq	qchisq		Ordered dsq	qchisq		Ordered dsq	qchisq
1	0.9290	1.2008	21	4.5050	5.2682	41	8.3382	8.8496
2	1.6133	1.7421	22	4.5519	5.4180	42	9.2314	9.1014
3	1.9977	2.0925	23	4.6375	5.5687	43	9.6604	9.3685
4	2.0342	2.3728	24	4.7574	5.7206	44	9.7473	9.6532
5	2.1271	2.6151	25	4.8349	5.8739	45	9.9785	9.9590
6	2.4705	2.8331	26	4.9553	6.0290	46	10.0686	10.2900
7	2.5632	3.0342	27	4.9569	6.1862	47	10.2061	10.6518
8	2.6799	3.2230	28	5.1769	6.3458	48	10.4673	11.0521
9	2.9658	3.4024	29	5.3518	6.5082	49	10.7073	11.5017
10	3.2640	3.5745	30	5.4411	6.6736	50	10.7679	12.0170
11	3.2654	3.7409	31	6.0510	6.8426	51	10.8518	12.6240
12	3.6450	3.9027	32	6.0621	7.0155	52	14.1541	13.3679
13	3.6614	4.0610	33	6.2487	7.1927	53	21.0996	14.3400
14	3.6945	4.2164	34	6.4695	7.3750	54	26.0421	15.7730
15	3.7644	4.3695	35	6.8118	7.5627	55	26.6244	18.7259
16	3.7894	4.5210	36	6.8217	7.7566			
17	4.0643	4.6713	37	7.3000	7.9574			
18	4.0734	4.8208	38	7.6345	8.1660			
19	4.2555	4.9699	39	8.0307	8.3834			
20	4.2904	5.1189	40	8.3077	8.6108			

From Table 4.2, with $\alpha = 0.05$ and $n = 55$, the critical point for the $Q - Q$ plot correlation coefficient test for normality is 0.9787. We reject the hypothesis of multivariate normality at $\alpha = 0.05$, because some marginals are not normal.

4.38. Marginal and multivariate normality of bull data

Normality of Bull Data

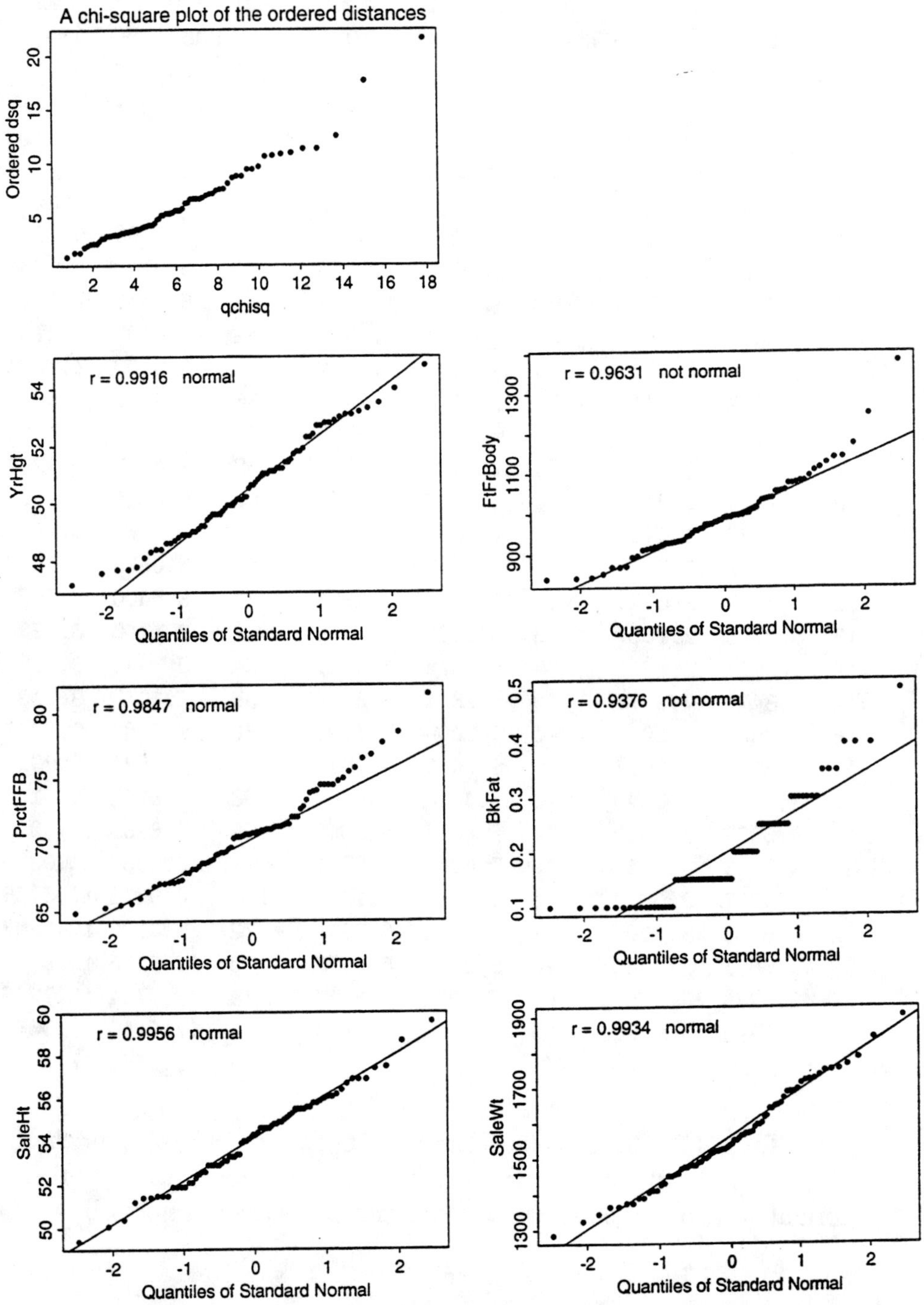

XBAR	S					
	YrHgt	FtFrBody	PrctFFB	BkFat	SaleHt	SaleWt
50.5224	2.9980	100.1305	2.9600	-0.0534	2.9831	82.8108
995.9474	100.1305	8594.3439	209.5044	-1.3982	129.9401	6680.3088
70.8816	2.9600	209.5044	10.6917	-0.1430	3.4142	83.9254
0.1967	-0.0534	-1.3982	-0.1430	0.0080	-0.0506	2.4130
54.1263	2.9831	129.9401	3.4142	-0.0506	4.0180	147.2896
1555.2895	82.8108	6680.3088	83.9254	2.4130	147.2896	16850.6618

	Ordered dsq	qchisq		Ordered dsq	qchisq		Ordered dsq	qchisq
1	1.3396	0.7470	26	3.8618	4.0902	51	6.6693	6.8439
2	1.7751	1.1286	27	3.8667	4.1875	52	6.6748	6.9836
3	1.7762	1.3793	28	3.9078	4.2851	53	6.6751	7.1276
4	2.2021	1.5808	29	4.0413	4.3830	54	6.8168	7.2763
5	2.3870	1.7551	30	4.1213	4.4812	55	6.9863	7.4301
6	2.5512	1.9118	31	4.1445	4.5801	56	7.1405	7.5896
7	2.5743	2.0560	32	4.2244	4.6795	57	7.1763	7.7554
8	2.5906	2.1911	33	4.2522	4.7797	58	7.4577	7.9281
9	2.7604	2.3189	34	4.2828	4.8806	59	7.5816	8.1085
10	3.0189	2.4411	35	4.4599	4.9826	60	7.6287	8.2975
11	3.0495	2.5587	36	4.7603	5.0855	61	8.0873	8.4963
12	3.2679	2.6725	37	4.8587	5.1896	62	8.6430	8.7062
13	3.2766	2.7832	38	5.1129	5.2949	63	8.7748	8.9286
14	3.3115	2.8912	39	5.1876	5.4017	64	8.7940	9.1657
15	3.3470	2.9971	40	5.2891	5.5099	65	9.3973	9.4197
16	3.3669	3.1011	41	5.3004	5.6197	66	9.3989	9.6937
17	3.3721	3.2036	42	5.3518	5.7313	67	9.6524	9.9917
18	3.4141	3.3048	43	5.4024	5.8449	68	10.6254	10.3191
19	3.5279	3.4049	44	5.5938	5.9605	69	10.6958	10.6829
20	3.5453	3.5041	45	5.6060	6.0783	70	10.8037	11.0936
21	3.6097	3.6027	46	5.6333	6.1986	71	10.9273	11.5665
22	3.6485	3.7007	47	5.7754	6.3215	72	11.3006	12.1263
23	3.6681	3.7983	48	6.2524	6.4472	73	11.3216	12.8160
24	3.7236	3.8957	49	6.3264	6.5760	74	12.4744	13.7225
25	3.7395	3.9929	50	6.6491	6.7081	75	17.6149	15.0677
						76	21.5751	17.8649

From Table 4.2, with $\alpha = 0.05$ and $n = 76$, the critical point for the $Q - Q$ plot correlation coefficient test for normality is 0.9839. We reject the hypothesis of multivariate normality at $\alpha = 0.05$, because some marginals are not normal.

Chapter 5

5.1 a) $\underset{\sim}{\bar{x}} = \begin{bmatrix} 6 \\ 10 \end{bmatrix}$; $S = \begin{bmatrix} 8 & -10/3 \\ -10/3 & 2 \end{bmatrix}$

$T^2 = 150/11 = 13.64$

b) T^2 is $3F_{2,2}$ (see (5-5))

c) $H_0: \underset{\sim}{\mu}' = [7,11]$

$\alpha = .05$ so $F_{2,2}(.05) = 19.00$

Since $T^2 = 13.64 < 3F_{2,2}(.05) = 3(19) = 57$; do not reject H_0 at the $\alpha = .05$ level

5.3 a) $$T^2 = \frac{(n-1)\left|\sum_{j=1}^{n}(\underset{\sim}{x}_j-\underset{\sim}{\mu}_0)(\underset{\sim}{x}_j-\underset{\sim}{\mu}_0)'\right|}{\left|\sum_{j=1}^{n}(\underset{\sim}{x}_j-\underset{\sim}{\bar{x}})(\underset{\sim}{x}_j-\underset{\sim}{\bar{x}})'\right|} - (n-1) = \frac{3(244)}{44} - 3 = 13.64$$

b) $$\Lambda = \left(\frac{\left|\sum_{j=1}^{n}(\underset{\sim}{x}_j-\underset{\sim}{\bar{x}})(\underset{\sim}{x}_j-\underset{\sim}{\bar{x}})'\right|}{\left|\sum_{j=1}^{n}(\underset{\sim}{x}_j-\underset{\sim}{\mu}_0)(\underset{\sim}{x}_j-\underset{\sim}{\mu}_0)'\right|}\right)^{n/2} = \left(\frac{44}{244}\right)^2 = .0325$$

Wilks' lambda = $\Lambda^{2/n} = \Lambda^{1/2} = \sqrt{.0325} = .1803$

5.5 $H_0: \underset{\sim}{\mu}' = [.55,.60]$; $T^2 = 1.17$

$\alpha = .05$; $F_{2,40}(.05) = 3.23$

Since $T^2 = 1.17 < \frac{2(41)}{40} F_{2,40}(.05) = 2.05(3.23) = 6.62$, we do not reject H_0 at the $\alpha = .05$ level. The result is consistent with the 95% confidence ellipse for $\underset{\sim}{\mu}$ pictured in Figure 5.1 since $\underset{\sim}{\mu}' = [.55,.60]$ is <u>inside</u> the ellipse.

5.8 $\underset{\sim}{\ell} = S^{-1}(\underset{\sim}{\bar{x}} - \underset{\sim}{\mu}_0) = \begin{bmatrix} 227.273 & -181.818 \\ -181.818 & 212.121 \end{bmatrix} \left(\begin{bmatrix} .564 \\ .603 \end{bmatrix} - \begin{bmatrix} .55 \\ .60 \end{bmatrix} \right)$

$= \begin{bmatrix} 2.636 \\ -1.909 \end{bmatrix}$

$$t^2 = \frac{n(\underset{\sim}{\ell}'(\underset{\sim}{\bar{x}} - \underset{\sim}{\mu}_0))^2}{\underset{\sim}{\ell}' S \underset{\sim}{\ell}} = \frac{42\left([2.636\ -1.909] \begin{bmatrix} .014 \\ .003 \end{bmatrix}\right)^2}{[2.636\ -1.909] \begin{bmatrix} .0144 & .0117 \\ .0117 & .0146 \end{bmatrix} \begin{bmatrix} 2.636 \\ -1.909 \end{bmatrix}} = 1.31 = T^2$$

5.9 a) $\underset{\sim}{\bar{x}}' = [5.1856,\ 16.0700]$

$S = \begin{bmatrix} 176.0042 & 287.2412 \\ 287.2412 & 527.8493 \end{bmatrix}; \quad S^{-1} = \begin{bmatrix} .0508 & -.0276 \\ -.0276 & .0169 \end{bmatrix}$

$$\frac{(n-1)p}{(n-p)} F_{p,n-p}(.10) = \frac{8(2)}{7} F_{2,7}(.10) = \frac{16}{7}(3.26) = 7.45$$

b) 90% simultaneous T2 intervals for the full data set:
Cr: (-6.88, 17.25) Sr: (-4.83, 36.97)

d) With data point (40.53, 73.68) removed,

$\underset{\sim}{\bar{x}}' = [.7675,\ 8.8688]; \quad S = \begin{bmatrix} .3786 & 1.0303 \\ 1.0303 & 69.8598 \end{bmatrix}$

$S^{-1} = \begin{bmatrix} 2.7518 & -.0406 \\ -.0406 & .0149 \end{bmatrix}$

$$\frac{(n-1)p}{(n-p)} F_{p,n-p}(.10) = \frac{7(2)}{6} F_{2,6}(.10) = \frac{14}{6}(3.46) = 8.07$$

90% simultaneous T2 intervals:
Cr: (.15, 1.39) Sr: (.47, 17.27)

5.10 Initial estimates are

$$\tilde{\mu} = \begin{bmatrix} 4 \\ 6 \\ 2 \end{bmatrix}, \quad \tilde{\Sigma} = \begin{bmatrix} 0.5 & 0.0 & 0.5 \\ & 2.0 & 0.0 \\ & & 1.5 \end{bmatrix}.$$

The first revised estimates are

$$\tilde{\mu} = \begin{bmatrix} 4.0833 \\ 6.0000 \\ 2.2500 \end{bmatrix}, \quad \tilde{\Sigma} = \begin{bmatrix} 0.6042 & 0.1667 & 0.8125 \\ & 2.500 & 0.0 \\ & & 1.9375 \end{bmatrix}.$$

5.11 The χ^2 distribution with 3 degrees of freedom.

5.12 Length of one-at-a time t-interval / Length of Bonferroni interval $= t_{n-1}(\alpha/2)/t_{n-1}(\alpha/2m)$.

	m		
n	2	4	10
15	0.8546	0.7489	0.6449
25	0.8632	0.7644	0.6678
50	0.8691	0.7749	0.6836
100	0.8718	0.7799	0.6910
∞	0.8745	0.7847	0.6983

5.13

(a).

$$\begin{aligned} E(X_{ij}) &= (1)p_i + (0)(1-p_i) = p_i. \\ Var(X_{ij}) &= (1-p_i)^2 p_i + (0-p_i)^2(1-p_i) = p_i(1-p_i) \end{aligned}$$

(b). $Cov(X_{ij}, X_{kj}) = E(X_{ij}X_{ik}) - E(X_{ij})E(X_{kj}) = 0 - p_i p_k = -p_i p_k.$

5.14

(a). Using $\hat{p}_j \pm \sqrt{\chi^2_4(0.05)}\sqrt{\hat{p}_j(1-\hat{p}_j)/n}$, the 95 % confidence intervals for p_1, p_2, p_3, p_4, p_5 are
(0.221, 0.370), (0.258, 0.412), (0.098, 0.217), (0.029, 0.112), (0.084, 0.198) respectively.
(b). Using $\hat{p}_1 - \hat{p}_2 \pm \sqrt{\chi^2_4(0.05)}\sqrt{(\hat{p}_1(1-\hat{p}_1) + \hat{p}_2(1-\hat{p}_2) - 2\hat{p}_1\hat{p}_2)/n}$, the 95 % confidence interval for $p_1 - p_2$ is (−0.118, 0.0394). There is no significant difference in two proportions.

5.15

$\hat{p}_1 = 0.585$, $\hat{p}_2 = 0.310$, $\hat{p}_3 = 0.105$. Using $\hat{p}_j \pm \sqrt{\chi^2_3(0.05)}\sqrt{\hat{p}_j(1-\hat{p}_j)/n}$, the 95 % confidence intervals for p_1, p_2, p_3 are (0.488, 0.682), (0.219, 0.401), (0.044, 0.166), respectively.

5.16

(a). Hotelling's $T^2 = 223.31$. The critical point for the statistic ($\alpha = 0.05$) is 8.33. We reject $H_0 : \mu = (500, 50, 30)'$. That is, The group of students represented by scores are significantly different from average college students.

(b). The lengths of three axes are 23.730, 2.473, 1.183. And directions of corresponding axes are

$$\begin{pmatrix} 0.994 \\ 0.103 \\ 0.038 \end{pmatrix}, \begin{pmatrix} -0.104 \\ 0.995 \\ 0.006 \end{pmatrix}, \begin{pmatrix} -0.037 \\ -0.010 \\ 0.999 \end{pmatrix}.$$

(c). Data look fairly normal.

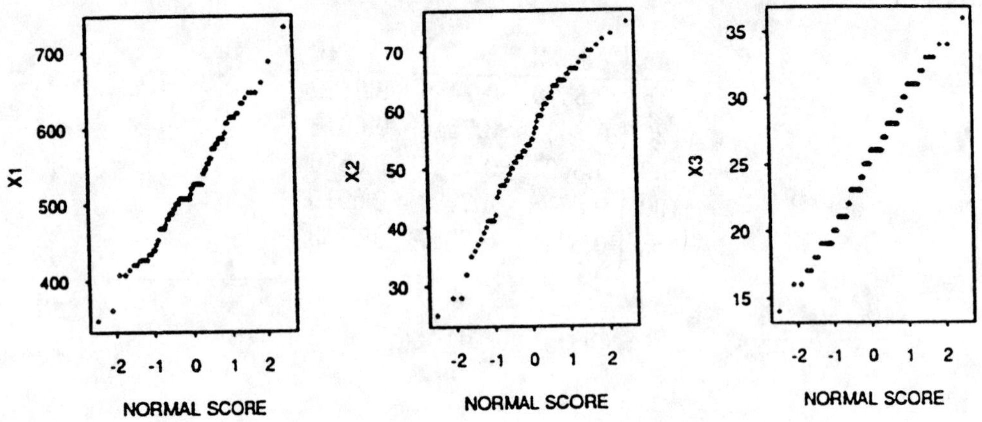

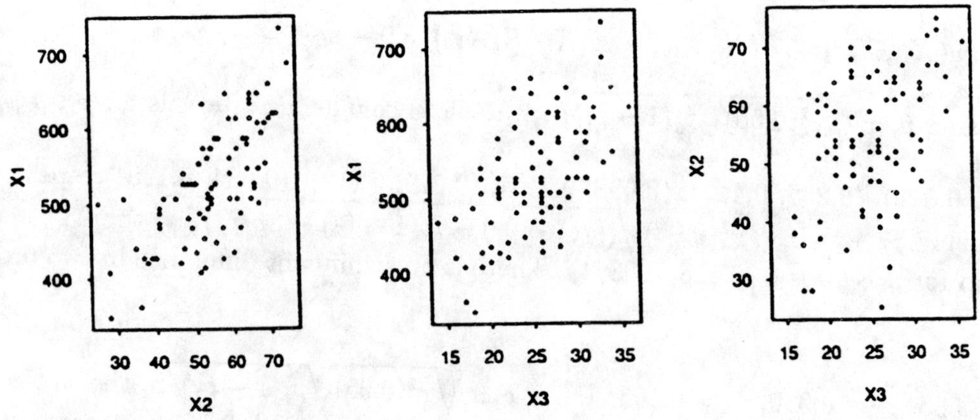

5.17 a) The summary statistics are:

$$n = 30, \quad \bar{\mathbf{x}} = \begin{bmatrix} 1860.50 \\ 8354.13 \end{bmatrix} \quad \text{and} \quad S = \begin{bmatrix} 124055.17 & 361621.03 \\ 361621.03 & 3486330.90 \end{bmatrix}$$

where S has eigenvalues and eigenvectors

$\lambda_1 = 3407292 \qquad \underset{\sim}{e}_1' = [.105740,\ .994394]$

$\lambda_2 = 82748 \qquad \underset{\sim}{e}_2' = [.994394, -.105740]$

Then, since $\frac{1}{n}\frac{p(n-1)}{n-p} F_{p,n-p}(\alpha) = \frac{1}{30}\ \frac{2(29)}{28}\ F_{2,28}(.05) = .2306$,

a 95% confidence region for $\underset{\sim}{\mu}$ is given by the set of $\underset{\sim}{\mu}$

$$[1860.50-\mu_1,\ 8354.13-\mu_2] \begin{bmatrix} 124055.17 & 361621.03 \\ 361621.03 & 3486330.90 \end{bmatrix}^{-1} \begin{bmatrix} 1860.50-\mu_1 \\ 8354.13-\mu_2 \end{bmatrix}$$

$$\leq .2306$$

The half lengths of the axes of this ellipse are $\sqrt{.2306}\,\sqrt{\lambda_1} = 886.4$ and $\sqrt{.2306}\,\sqrt{\lambda_2} = 138.1$. Therefore the ellipse has the form

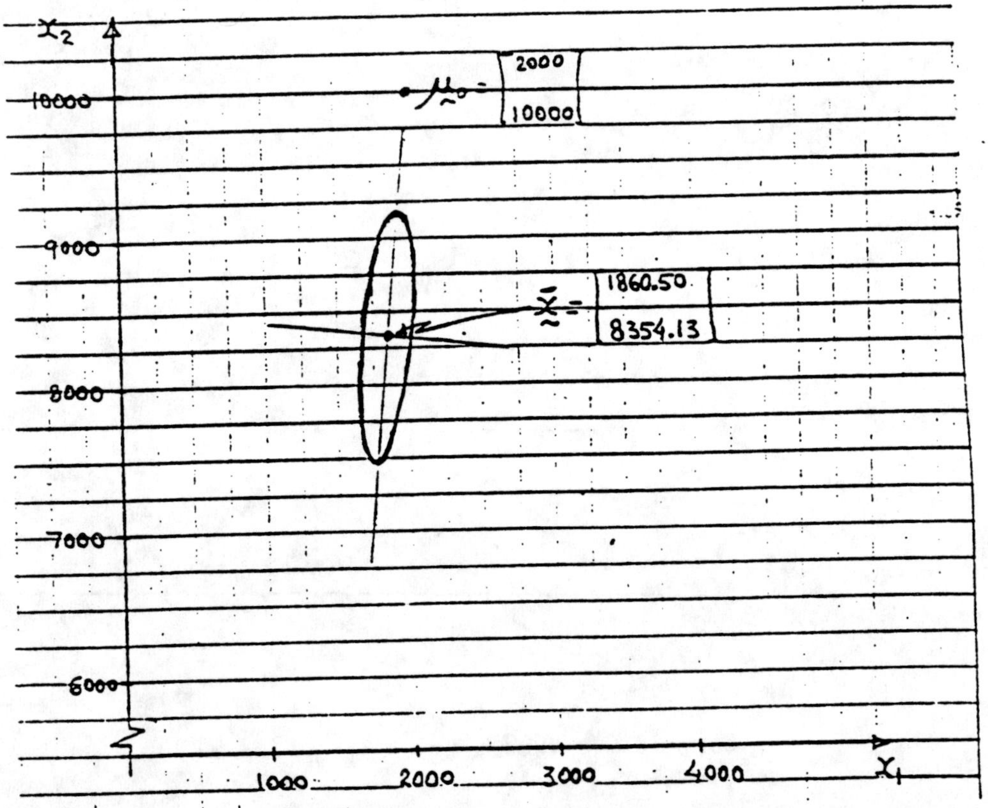

b) Since $\underset{\sim}{\mu}_0$ = [2000, 10000]' does not fall within the 95% confidence ellipse, we would reject the hypothesis $H_0: \underset{\sim}{\mu} = \underset{\sim}{\mu}_0$ at the 5% level. Thus, the data analyzed are <u>not</u> consistent with these values.

c) The Q-Q plots for both stiffness and bending strength (see below) show that the marginal normality is not seriously violated. Also the correlation coefficients for the test of normality are .989 and .990 respectively so that we fail to reject even at the 1% significance level. Finally, the scatter diagram (see below) does not indicate departure from bivariate normality. So, the bivariate normal distribution is a plausible probability model for these data.

Q-Q Plot-Bending Strength

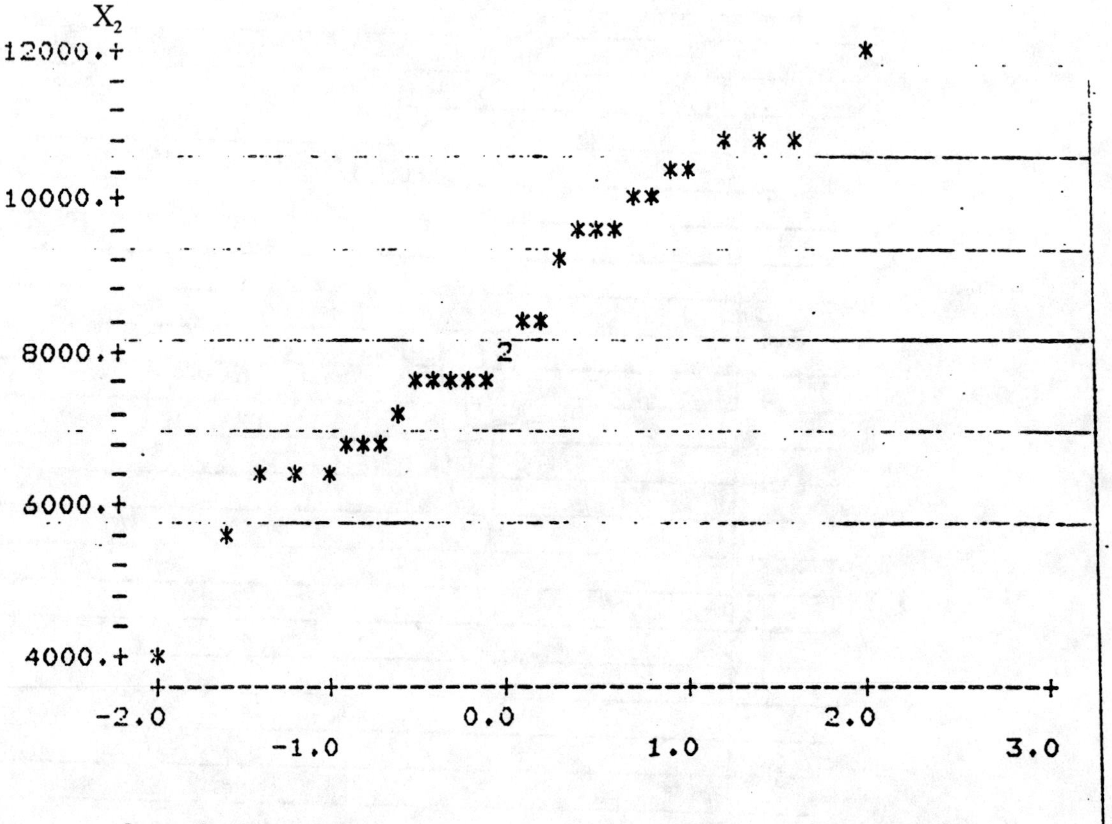

Correlation .989

Q-Q Plot-Stiffness

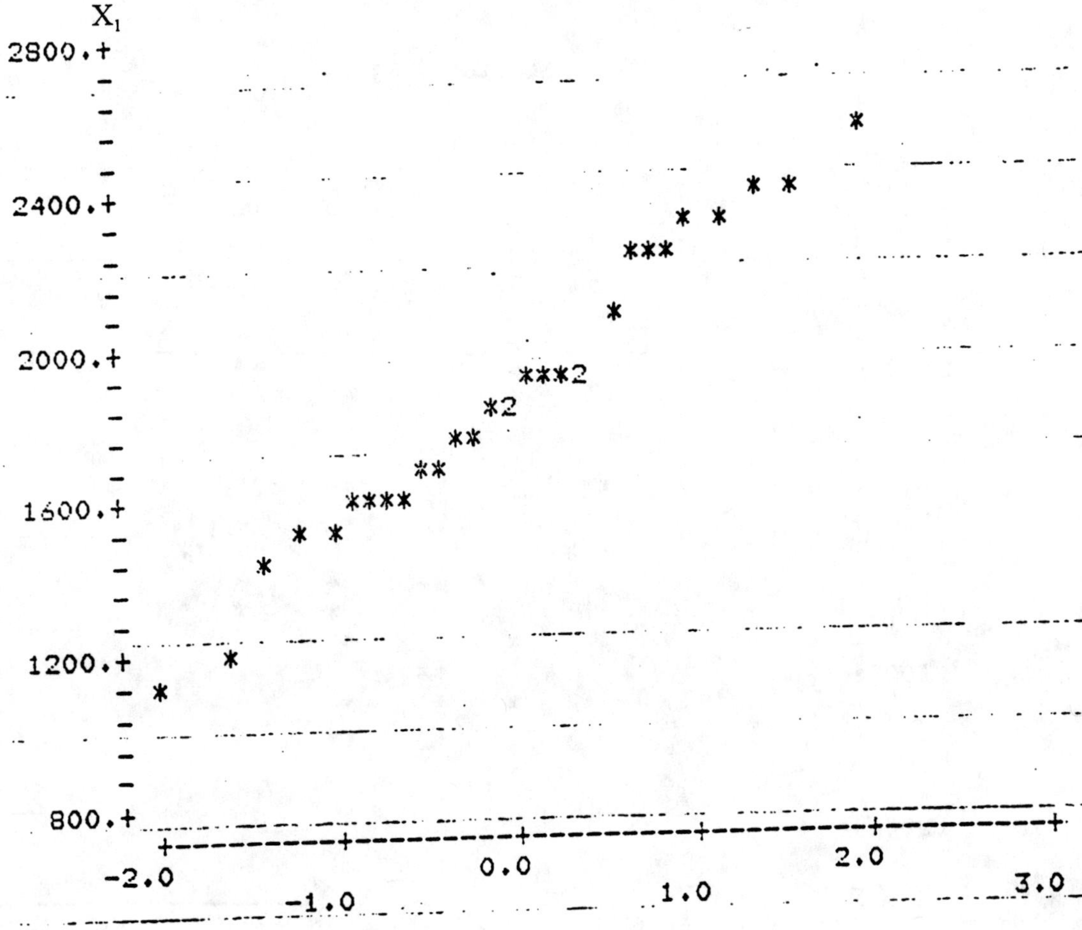

Correlation = .990

Scatter Diagram

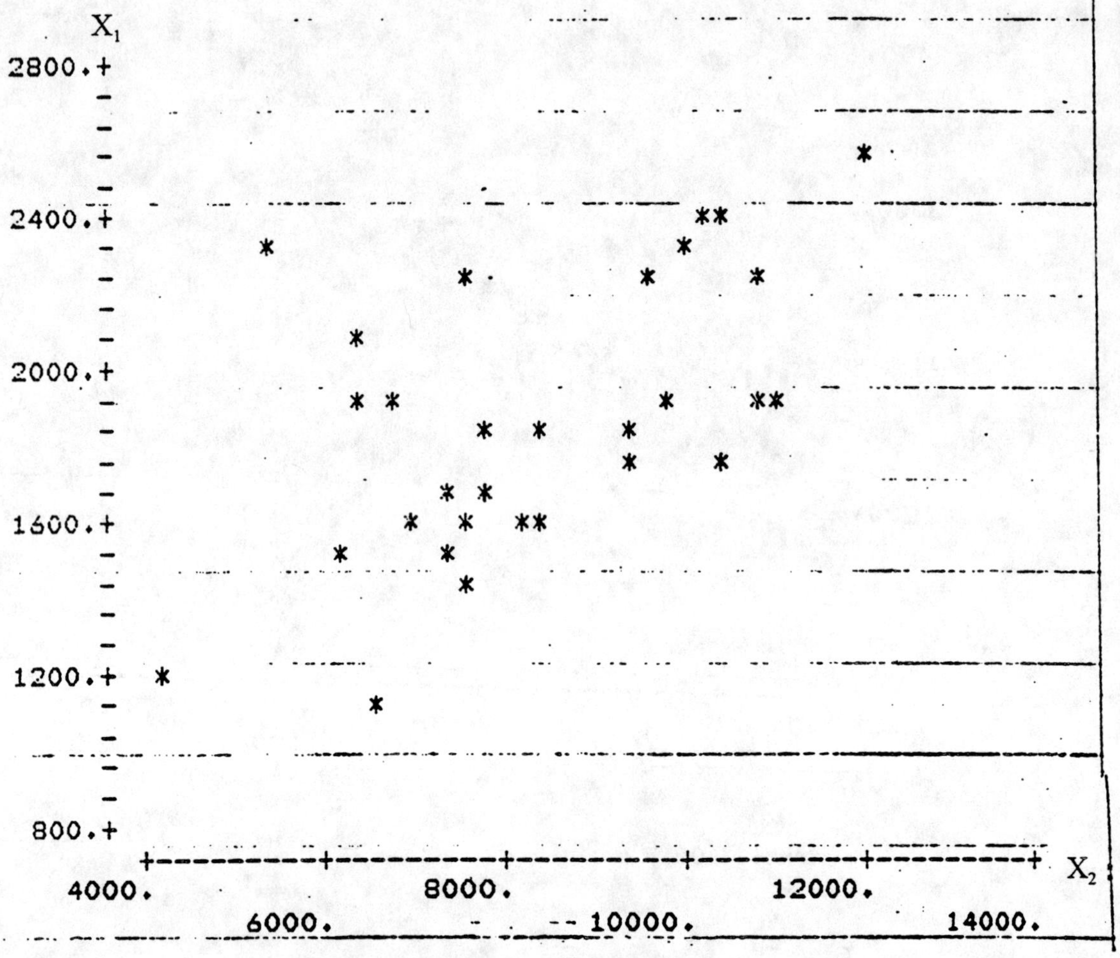

5.18 (a). Yes, they are plausible since the hypothesized vector μ_0 (denoted as $*$ in the plot) is inside the 95% confidence region.

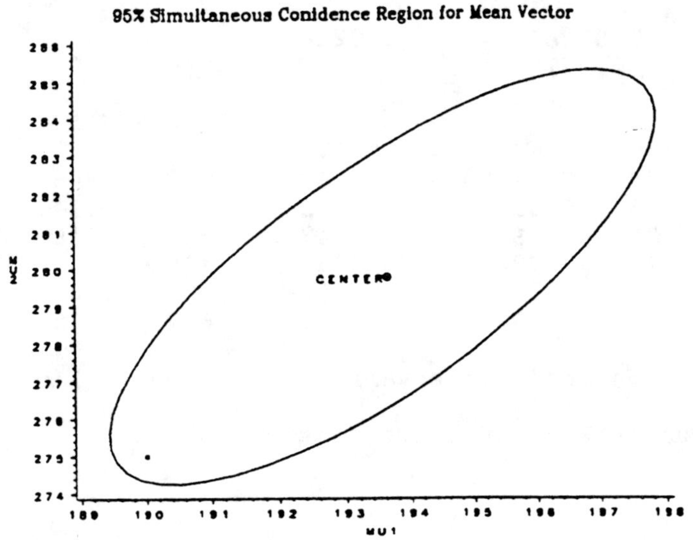

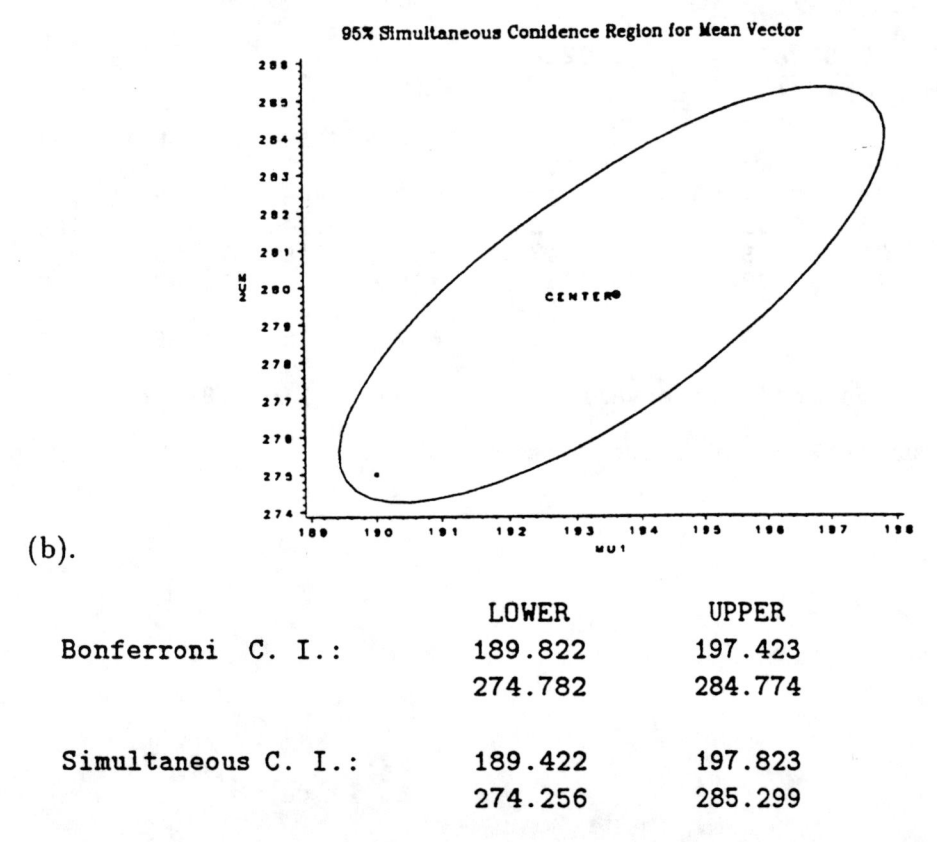

(b).

	LOWER	UPPER
Bonferroni C. I.:	189.822	197.423
	274.782	284.774
Simultaneous C. I.:	189.422	197.823
	274.256	285.299

Simultaneous confidence intervals are larger than Bonferroni's confidence intervals. Simultaneous confidence intervals will touch the simultaneous confidence region from outside.
(c). Q-Q plots suggests non-normality of (X_1, X_2). Could try transforming X_1.

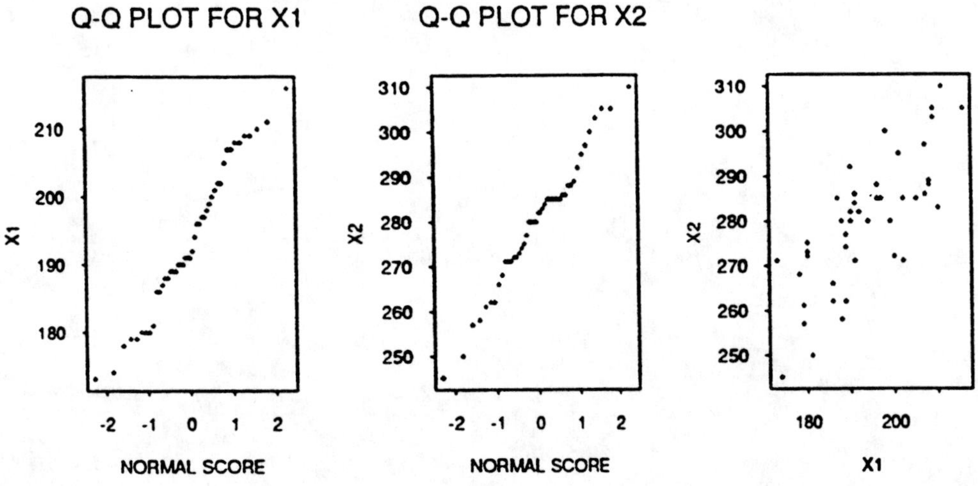

5.19

```
HOTELLING T SQUARE   =   9.0218
P-VALUE     0.3616
```

	N	MEAN	STDEV	T2 INTERVAL	TO	BONFERRONI	TO
x1	25	0.84380	0.11402	.742	.946	.778	.909
x2	25	0.81832	0.10685	.723	.914	.757	.880
x3	25	1.79268	0.28347	1.540	2.046	1.629	1.956
x4	25	1.73484	0.26360	1.499	1.970	1.583	1.887
x5	25	0.70440	0.10756	.608	.800	.642	.766
x6	25	0.69384	0.10295	.602	.786	.635	.753

The Bonferroni intervals use t (.00417) = 2.88 and

the T2 intevals use the constant 4.465.

5.20

(a). After eliminating outliers, the approximation to normality is improved.

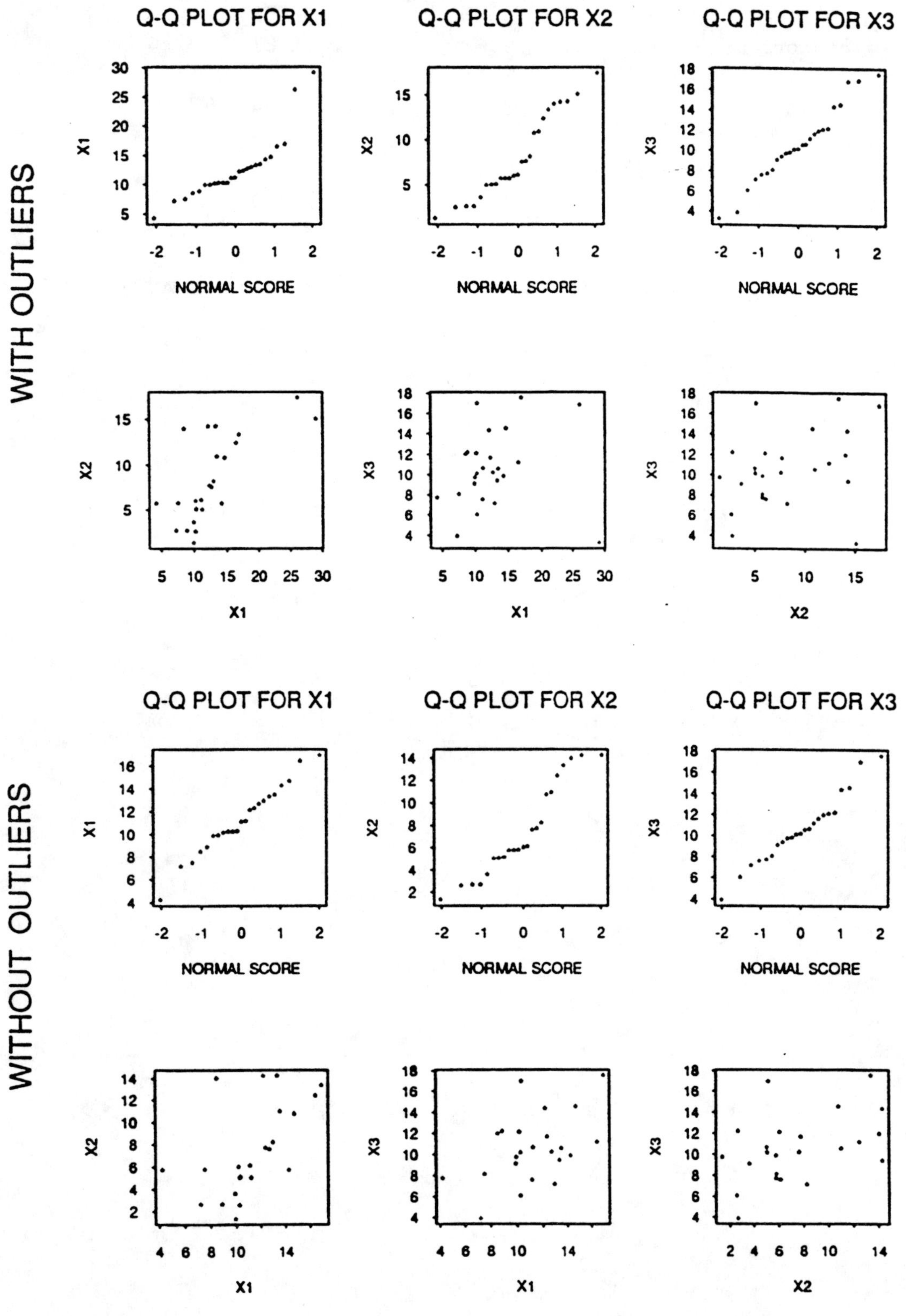

(b) Outliers removed.

	LOWER	UPPER
Bonferroni C. I.:	9.63	12.87
	5.24	9.67
	8.82	12.34
Simultaneous C. I.:	9.25	13.24
	4.72	10.19
	8.41	12.76

Simultaneous confidence intervals are larger than Bonferroni's confidence intervals.

(b) Full data set:

	Lower	Upper
Bonferroni C. I.:	9.79	15.33
	5.78	10.55
	8.65	12.44
Simultaneous C. I.:	9.16	15.96
	5.23	11.09
	8.21	12.87

5.21. Individual $\overline{X}$ charts for the Madison, Wisconsin, Police Department data

	xbar	s	LCL	UCL	
LegalOT	3557.8	606.5	1738.1	5377.4	
ExtraOT	1478.4	1182.8	-2070.0	5026.9	use LCL = 0
Holdover	2676.9	1207.7	-946.2	6300.0	use LCL = 0
COA	13563.6	1303.2	9654.0	17473.2	
MeetOT	800.0	474.0	-622.1	2222.1	use LCL = 0

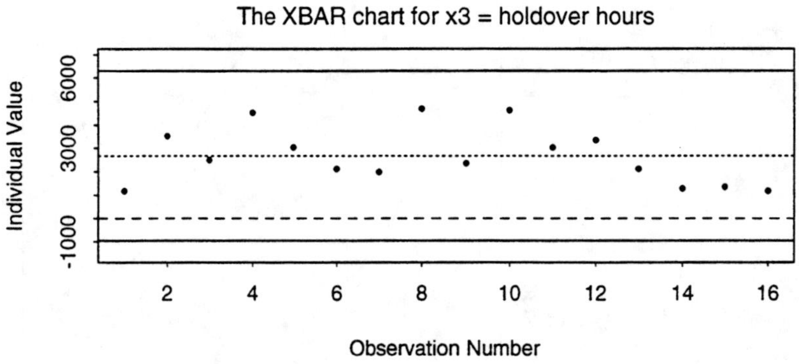

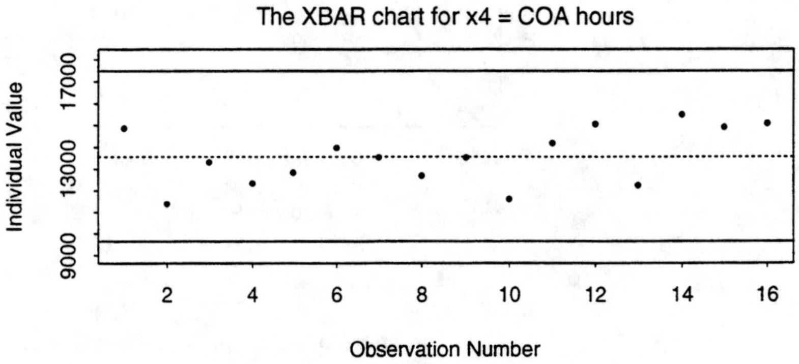

Both holdover and COA hours are stable and in control.

5.22 Quality ellipse and T^2 chart for the holdover and COA overtime hours. All points are in control. The quality control 95% ellipse is

$$1.37 \times 10^{-6}(x_3 - 2677)^2 + 1.18 \times 10^{-6}(x_4 - 13564)^2$$
$$+1.80 \times 10^{-6}(x_3 - 2677)(x_4 - 13564) = 5.99.$$

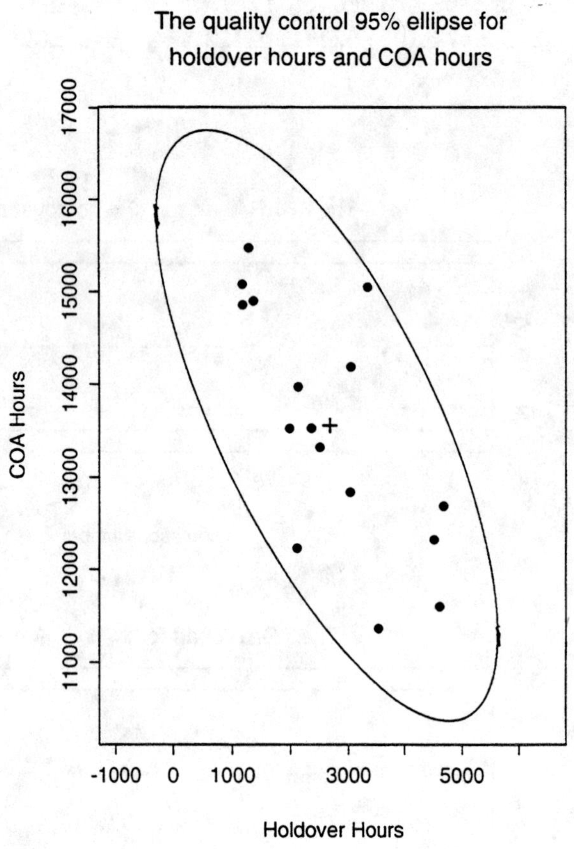

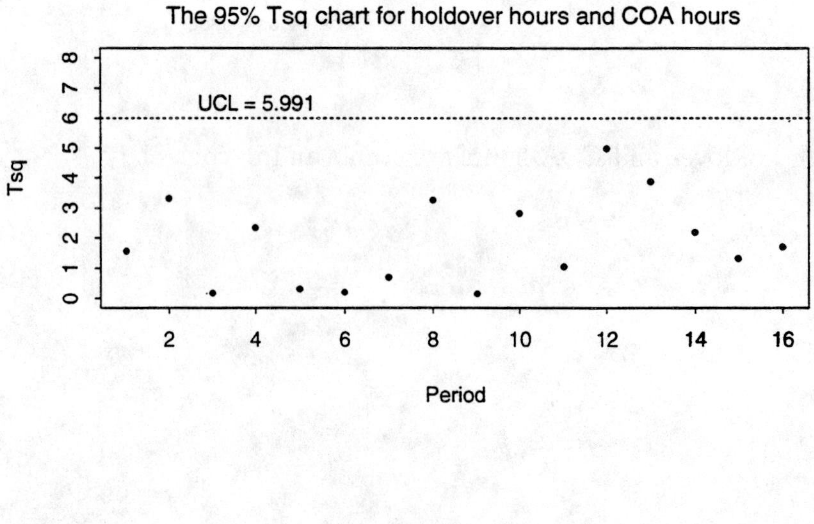

5.23. T^2 chart using the data on x_1 = legal appearances overtime hours, x_2 = extraordinary event overtime hours, and x_3 = holdover overtime hours. All points are in control.

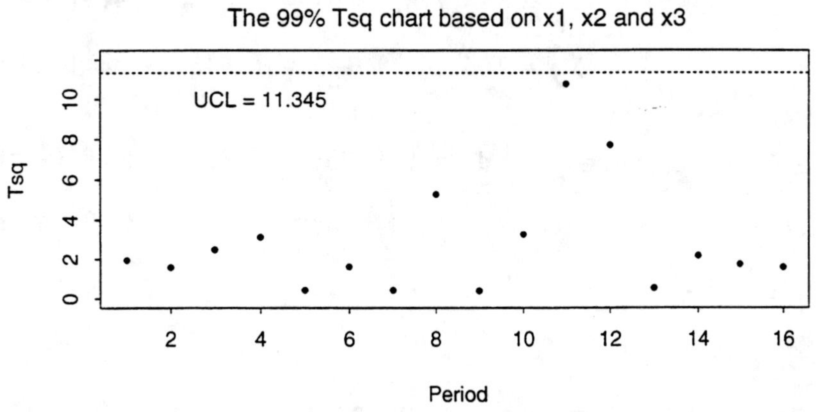

5.24. The 95% prediction ellipse for x_3 = holdover hours and x_4 = COA hours is

$$1.37\times10^{-6}(x_3 - 2677)^2 + 1.18\times10^{-6}(x_4 - 13564)^2$$
$$+1.80\times10^{-6}(x_3 - 2677)(x_4 - 13564) = 8.51.$$

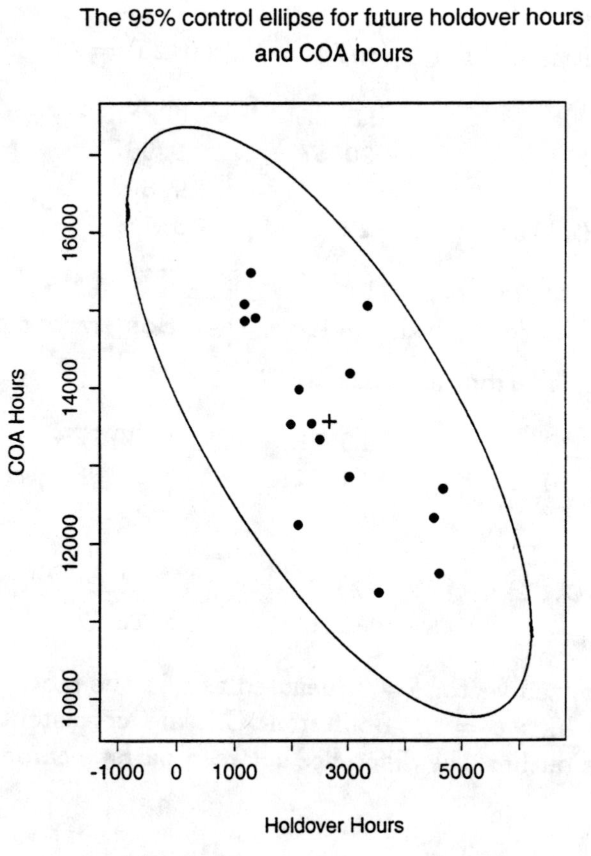

Chapter 6

6.1 Eigenvalues and eigenvectors of S_d are:

$$\lambda_1 = 449.778, \quad \underset{\sim}{e}_1' = [.333, \; .943]$$
$$\lambda_2 = 168.082, \quad \underset{\sim}{e}_2' = [.943, \; -.333]$$

Ellipse centered at $\bar{\underset{\sim}{d}}' = [-9.36, \; 13.27]$. Half length of major axis is 20.57 units. Half length of minor axis is 12.58 units. Major and minor axes lie in $\underset{\sim}{e}_1$ and $\underset{\sim}{e}_2$ directions, respectively.

Yes, the test answers the question: Is $\underset{\sim}{\delta} = \underset{\sim}{0}$ inside the 95% confidence ellipse?

6.2 Using a critical value $t_{n-1}(\alpha/2p) = t_{10}(0.0125) = 2.6338$,

	LOWER	UPPER
Bonferroni C. I.:	-20.57	1.85
	-2.97	29.52
Simultaneous C. I.:	-22.45	3.73
	-5.70	32.25

Simultaneous confidence intervals are larger than Bonferroni's confidence intervals.

6.3 The 95% Bonferroni intervals are

	LOWER	UPPER
Bonferroni C. I.:	-21.92	-2.08
	-3.36	20.56
Simultaneous C. I.:	-23.70	-0.30
	-5.50	22.70

Since the hypothesized vector $\delta = 0$ (denoted as $*$ in the plot) is outside the joint confidence region, we reject $H_0 : \delta = 0$. Bonferroni C.I. are consistent with this result. After the elimination of the outlier, the difference between pairs became significant.

95% Simultaneous Conidence Region for Delta Vector

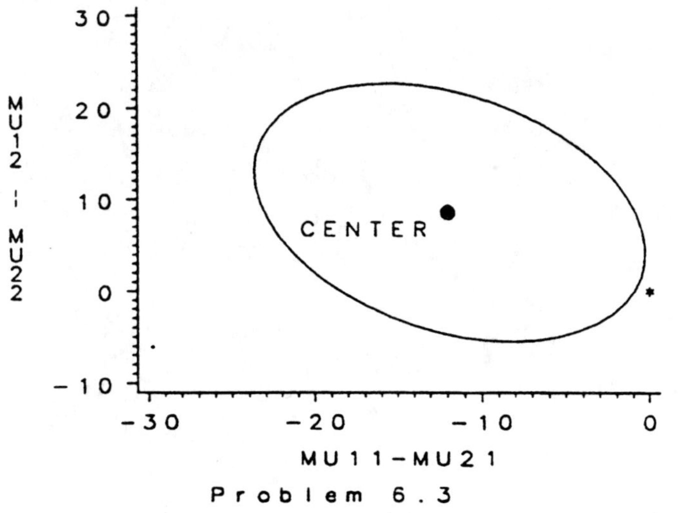

6.4

(a). Hotelling's $T^2 = 10.215$. Since the critical point with $\alpha = 0.05$ is 9.459, we reject $H_0 : \delta = 0$.
(b).

	LOWER	UPPER
Bonferroni C. I.:	-1.09	-0.02
	-0.04	0.64
Simultaneous C. I.:		0.07
		0.69

(c). The χ^2 plot does not look like a straight line. That is, there seems to be a violation of normality assumption.

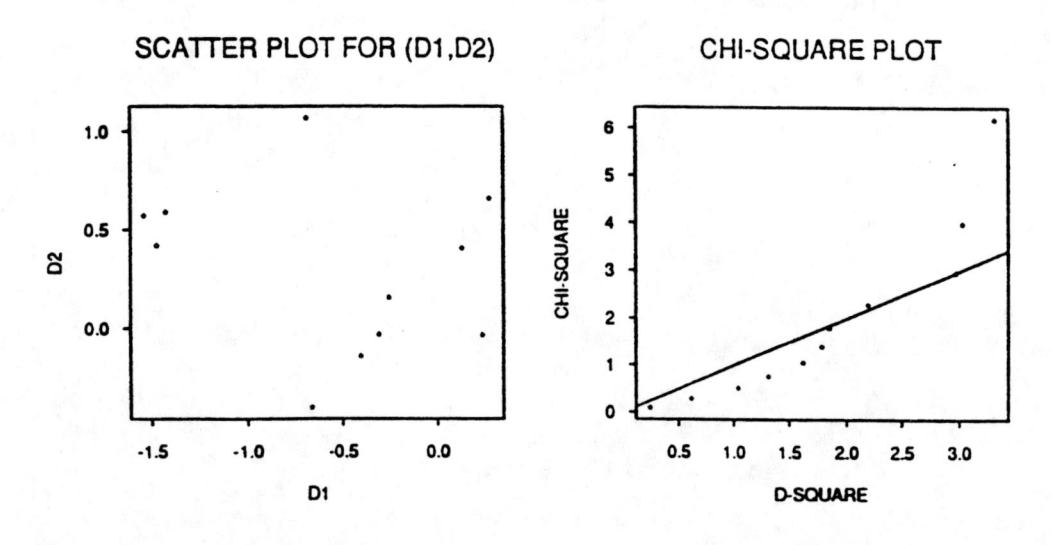

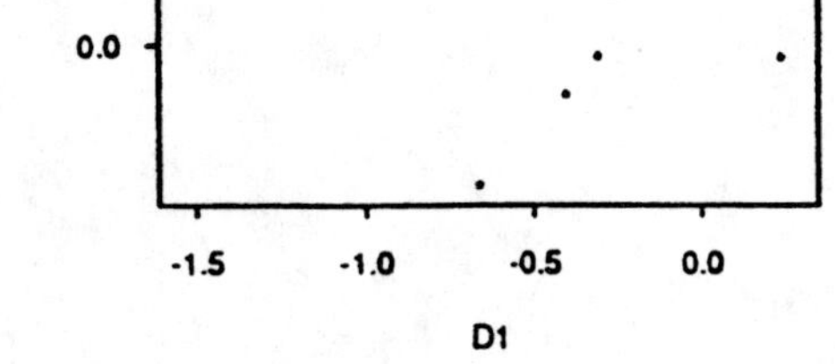

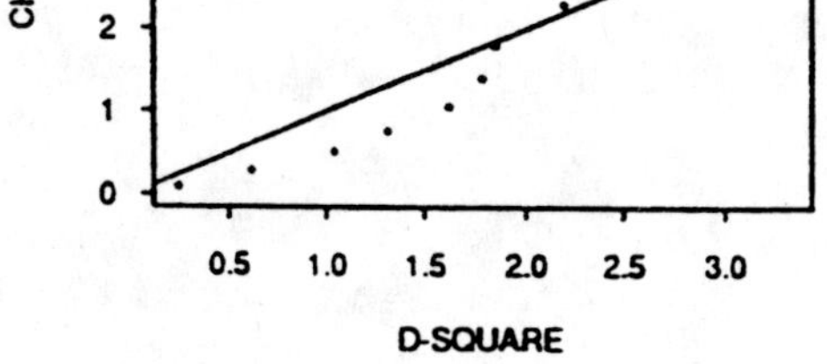

6.5 a) $H_0: C\underset{\sim}{\mu} = \underset{\sim}{0}$ where $C = \begin{bmatrix} 1 & -1 & 0 \\ 0 & 1 & -1 \end{bmatrix}$, $\underset{\sim}{\mu}' = [\mu_1, \mu_2, \mu_3]$.

$$C\bar{\underset{\sim}{x}} = \begin{bmatrix} -11.2 \\ 6.9 \end{bmatrix}, \quad CSC' = \begin{bmatrix} 55.5 & -32.6 \\ -32.6 & 66.4 \end{bmatrix}$$

$$T^2 = n(C\bar{\underset{\sim}{x}})'(CSC')^{-1}(C\bar{\underset{\sim}{x}}) = 90.4; \qquad n = 40; \quad q = 3$$

$$\frac{(n-1)(q-1)}{(n-q+1)} F_{q-1,n-q+1}(.05) = \frac{(39)2}{38}(3.25) = 6.67$$

Since $T^2 = 90.4 > 6.67$ reject $H_0: C\underset{\sim}{\mu} = \underset{\sim}{0}$

b) 95% simultaneous confidence intervals:

$\mu_1 - \mu_2$: $(46.1 - 57.3) \pm \sqrt{6.67}\ \sqrt{\frac{55.5}{40}} = -11.2 \pm 3.0$

$\mu_2 - \mu_3$: 6.9 ± 3.3

$\mu_1 - \mu_3$: -4.3 ± 3.3

The means are all different from one another.

6.6 a) Treatment 2: Sample mean vector $\begin{bmatrix} 2 \\ 4 \end{bmatrix}$; sample covariance matrix

$$\begin{bmatrix} 1 & -3/2 \\ -3/2 & 3 \end{bmatrix}$$

Treatment 3: Sample mean vector $\begin{bmatrix} 3 \\ 2 \end{bmatrix}$; sample covariance matrix

$$\begin{bmatrix} 2 & -4/3 \\ -4/3 & 4/3 \end{bmatrix}$$

$$S_{pooled} = \begin{bmatrix} 1.6 & \\ -1.4 & \end{bmatrix}$$

b) $$T^2 = [2-3,\ 4-2] \left[\left(\frac{1}{3} + \frac{1}{4}\right) \begin{bmatrix} 1.6 & -1.4 \\ -1.4 & 2 \end{bmatrix} \right]^{-1} \begin{bmatrix} 2-3 \\ 4-2 \end{bmatrix} = 3.88$$

$$\frac{(n_1+n_2-2)p}{(n_1+n_2-p-1)} F_{p,n_1+n_2-p-1}(.01) = \frac{(5)2}{4}(18) = 45$$

Since $T^2 = 3.88 < 45$ do not reject $H_0: \underset{\sim}{\mu}_2 - \underset{\sim}{\mu}_3 = \underset{\sim}{0}$ at the $\alpha = .01$ level.

c) 99% simultaneous confidence intervals:

$$\mu_{21} - \mu_{31}:\ (2-3) \pm \sqrt{45}\ \sqrt{\left(\frac{1}{3}+\frac{1}{4}\right)1.6} = -1 \pm 6.5$$

$$\mu_{22} - \mu_{32}:\ 2 \pm 7.2$$

6.7 $$T^2 = [74.4\ \ 201.6] \left[\left(\frac{1}{45} + \frac{1}{55}\right) \begin{bmatrix} 10963.7 & 21505.5 \\ 21505.5 & 63661.3 \end{bmatrix} \right]^{-1} \begin{bmatrix} 74.4 \\ 201.6 \end{bmatrix} = 16.1$$

$$\frac{(n_1+n_2-2)p}{n_1+n_2-p-1} F_{p,n_1+n_2-p-1}(.05) = 6.26$$

Since $T^2 = 16.1 > 6.26$ reject $H_0: \underset{\sim}{\mu}_1 - \underset{\sim}{\mu}_2 = \underset{\sim}{0}$ at the $\alpha = .05$ level.

$$\hat{\underset{\sim}{\ell}} \propto S^{-1}_{pooled}(\bar{\underset{\sim}{x}}_1 - \bar{\underset{\sim}{x}}_2) = \begin{bmatrix} .0017 \\ .0026 \end{bmatrix}$$

6.8 a) For first variable:

observation = mean + treatment effect + residual

$$\begin{bmatrix} 6 & 5 & 8 & 4 & 7 \\ 3 & 1 & 2 & & \\ 2 & 5 & 3 & 2 & \end{bmatrix} = \begin{bmatrix} 4 & 4 & 4 & 4 & 4 \\ 4 & 4 & 4 & & \\ 4 & 4 & 4 & 4 & \end{bmatrix} + \begin{bmatrix} 2 & 2 & 2 & 2 & 2 \\ -2 & -2 & -2 & & \\ -1 & -1 & -1 & -1 & \end{bmatrix} + \begin{bmatrix} 0 & -1 & 2 & -2 & 1 \\ 1 & -1 & 0 & & \\ -1 & 2 & 0 & -1 & \end{bmatrix}$$

$SS_{obs} = 246$ $SS_{mean} = 192$ $SS_{tr} = 36$ $SS_{res} = 18$

For second variable:

$$\begin{bmatrix} 7 & 9 & 6 & 9 & 9 \\ 3 & 6 & 3 & & \\ 3 & 1 & 1 & 3 & \end{bmatrix} = \begin{bmatrix} 5 & 5 & 5 & 5 & 5 \\ 5 & 5 & 5 & & \\ 5 & 5 & 5 & 5 & \end{bmatrix} + \begin{bmatrix} 3 & 3 & 3 & 3 & 3 \\ -1 & -1 & -1 & & \\ -3 & -3 & -3 & -3 & \end{bmatrix} + \begin{bmatrix} -1 & 1 & -2 & 1 & 1 \\ -1 & 2 & -1 & & \\ 1 & -1 & -1 & 1 & \end{bmatrix}$$

$SS_{obs} = 402$ $SS_{mean} = 300$ $SS_{tr} = 84$ $SS_{res} = 18$

Cross product contributions:

275 240 48 -13

b) MANOVA table:

Source of Variation	SSP	d.f.
Treatment	$B = \begin{bmatrix} 36 & 48 \\ 48 & 84 \end{bmatrix}$	3 - 1 = 2
Residual	$W = \begin{bmatrix} 18 & -13 \\ -13 & 18 \end{bmatrix}$	5 + 3 + 4 - 3 = 9
Total (corrected)	$\begin{bmatrix} 54 & 35 \\ 35 & 102 \end{bmatrix}$	11

c) $\Lambda^* = \frac{|W|}{|B+W|} = \frac{155}{4283} = .0362$

Using Table 6.3 with $p = 2$ and $g = 3$

$$\left(\frac{1-\sqrt{\Lambda^*}}{\sqrt{\Lambda^*}}\right)\left(\frac{\Sigma n_\ell - g - 1}{g-1}\right) = 17.02\,.$$

Since $F_{4,16}(.01) = 4.77$ we conclude that treatment differences exist at $\alpha = .01$ level.

Alternatively, using Bartlett's procedure,

$$-\left(n-1-\frac{(p+g)}{2}\right)\ell n\,\Lambda^* = -(12-1-\tfrac{5}{2})\ell n(.0362) = 28.209$$

Since $\chi_4^2(.01) = 13.28$ we again conclude treatment differences exist at $\alpha = .01$ level.

6.9 For <u>any</u> matrix C

$$\underset{\sim}{\bar{d}} = \frac{1}{n}\Sigma\,\underset{\sim}{d}_j = C(\frac{1}{n}\Sigma\,\underset{\sim}{x}_j) = C\,\underset{\sim}{\bar{x}}$$

and $\underset{\sim}{d}_j - \underset{\sim}{\bar{d}} = C(\underset{\sim}{x}_j - \underset{\sim}{\bar{x}})$

so $S_d = \frac{1}{n-1}\Sigma(\underset{\sim}{d}_j-\underset{\sim}{\bar{d}})(\underset{\sim}{d}_j-\underset{\sim}{\bar{d}})' = C(\frac{1}{n-1}\Sigma(\underset{\sim}{x}_j-\underset{\sim}{\bar{x}})(\underset{\sim}{x}_j-\underset{\sim}{\bar{x}})')C' = CSC'$

6.10 $(\bar{x}\,\underset{\sim}{1})'[(\bar{x}_1-\bar{x})\underset{\sim}{u}_1 + \ldots + (\bar{x}_g-\bar{x})\underset{\sim}{u}_g]$

$$= \bar{x}[(\bar{x}_1-\bar{x})n_1 + \ldots + (\bar{x}_g-\bar{x})n_g]$$

$$= \bar{x}[n_1\bar{x}_1 + \ldots + n_g\bar{x}_g - \bar{x}(n_1 + \ldots + n_g)]$$

$$= \bar{x}[(n_1 + \ldots + n_g)\bar{x} - \bar{x}(n_1 + \ldots + n_g)] = 0$$

6.11 $L(\underset{\sim}{\mu}_1, \underset{\sim}{\mu}_2, \Sigma) = L(\underset{\sim}{\mu}_1, \Sigma) L(\underset{\sim}{\mu}_2, \Sigma)$

$$= \left[\frac{1}{(2\pi)^{\frac{(n_1+n_2)p}{2}} |\Sigma|^{\frac{n_1+n_2}{2}}} \right] \exp\left\{ -\frac{1}{2} \left(\text{tr}\, \Sigma^{-1} [(n_1-1)S_1 + (n_2-1)S_2] \right. \right.$$

$$\left. \left. + n_1(\bar{\underset{\sim}{x}}_1 - \underset{\sim}{\mu}_1)' \Sigma^{-1} (\bar{\underset{\sim}{x}} - \underset{\sim}{\mu}_1) + n_2(\bar{\underset{\sim}{x}}_2 - \underset{\sim}{\mu}_2)' \Sigma^{-1} (\bar{\underset{\sim}{x}}_2 - \underset{\sim}{\mu}_2) \right) \right\}$$

using (4-16) and (4-17). The likelihood is maximized with respect to $\underset{\sim}{\mu}_1$ and $\underset{\sim}{\mu}_2$ at $\hat{\underset{\sim}{\mu}}_1 = \bar{\underset{\sim}{x}}_1$ and $\hat{\underset{\sim}{\mu}}_2 = \bar{\underset{\sim}{x}}_2$ respectively and with respect to Σ at

$$\hat{\Sigma} = \frac{1}{n_1+n_2} [(n_1 - 1)S_1 + (n_2 - 2)S_2] = \left(\frac{n_1+n_2-2}{n_1+n_2} \right) S_{pooled}$$

(For the maximization with respect to Σ see Result 4.10 with

$b = \frac{n_1+n_2}{2}$ and $B = (n_1 - 1)S_1 + (n_2 - 2)S_2$)

6.13 a) and b) For first variable:

$$\underset{\text{Observation}}{\begin{bmatrix} 6 & 4 & 8 & 2 \\ 3 & -3 & 4 & -4 \\ -3 & -4 & 3 & -4 \end{bmatrix}} = \underset{\text{mean}}{\begin{bmatrix} 1 & 1 & 1 & 1 \\ 1 & 1 & 1 & 1 \\ 1 & 1 & 1 & 1 \end{bmatrix}} + \underset{\text{factor 1 effect}}{\begin{bmatrix} 4 & 4 & 4 & 4 \\ -1 & -1 & -1 & -1 \\ -3 & -3 & -3 & -3 \end{bmatrix}} + \underset{\text{factor 2 effect}}{\begin{bmatrix} 1 & -2 & 4 & -3 \\ 1 & -2 & 4 & -3 \\ 1 & -2 & 4 & -3 \end{bmatrix}} + \underset{\text{residual}}{\begin{bmatrix} 0 & 1 & -1 & 0 \\ 2 & -1 & 0 & -1 \\ -2 & 0 & 1 & 1 \end{bmatrix}}$$

$SS_{tot} = 220$ $SS_{mean} = 12$ $SS_{fac\,1} = 104$ $SS_{fac\,2} = 90$ $SS_{res} = 14$

For second variable:

$$\begin{bmatrix} 8 & 6 & 12 & 6 \\ 8 & 2 & 3 & 3 \\ 2 & -5 & -3 & -6 \end{bmatrix} = \begin{bmatrix} 3 & 3 & 3 & 3 \\ 3 & 3 & 3 & 3 \\ 3 & 3 & 3 & 3 \end{bmatrix} + \begin{bmatrix} 5 & 5 & 5 & 5 \\ 1 & 1 & 1 & 1 \\ -6 & -6 & -6 & -6 \end{bmatrix} + \begin{bmatrix} 3 & -2 & 1 & -2 \\ 3 & -2 & 1 & -2 \\ 3 & -2 & 1 & -2 \end{bmatrix} + \begin{bmatrix} -3 & 0 & 3 & 0 \\ 1 & 0 & -2 & 1 \\ 2 & 0 & -1 & -1 \end{bmatrix}$$

$SS_{tot} = 440$ $SS_{mean} = 108$ $SS_{fac\,1} = 248$ $SS_{fac\,2} = 54$ $SS_{res} = 30$

Sum of cross products:

$$SCP_{tot} = SCP_{mean} + SCP_{fac\,1} + SCP_{fac\,2} + SCP_{res}$$

$$227 = 36 + 148 + 51 - 8$$

c) MANOVA table:

Source of Variation	SSP	d.f.
Factor 1	$\begin{bmatrix} 104 & 148 \\ 148 & 248 \end{bmatrix}$	$g - 1 = 3 - 1 = 2$
Factor 2	$\begin{bmatrix} 90 & 51 \\ 51 & 54 \end{bmatrix}$	$b - 1 = 4 - 1 = 3$
Residual	$\begin{bmatrix} 14 & -8 \\ -8 & 30 \end{bmatrix}$	$(g-1)(b-1) = 6$
Total (Corrected)	$\begin{bmatrix} 208 & 191 \\ 191 & 332 \end{bmatrix}$	$gb - 1 = 11$

d) We reject $H_0: \underset{\sim}{\tau}_1 = \underset{\sim}{\tau}_2 = \underset{\sim}{\tau}_3 = \underset{\sim}{0}$ at $\alpha = .05$ level since

$$-[(g-1)(b-1) - (\frac{p+1-(g-1)}{2})]\ln\Lambda^* = -[6 - \frac{3-2}{2}]\ln\left(\frac{|SS_{res}|}{|SSP_{fac\,1} + SSP_{res}|}\right)$$

$$\doteq -5.5 \ln\left(\frac{356}{13204}\right) = 19.87 > \chi_4^2(.05) = 9.49$$

and conclude there are factor 1 effects.

We also reject $H_0: \underset{\sim}{\beta}_1 = \underset{\sim}{\beta}_2 = \underset{\sim}{\beta}_3 = \underset{\sim}{\beta}_4 = \underset{\sim}{0}$ at the $\alpha = .05$ level since

$$-[(g-1)(b-1) - (\frac{p+1-(b-1)}{2})\ln\Lambda^* = -[6 - \frac{3-3}{2}]\ln\left(\frac{|SSP_{res}|}{|SSP_{fac\ 2} + SSP_{res}|}\right)$$

$$= -6\ln\left(\frac{356}{6887}\right) = 17.77 > \chi_6^2(.05) = 12.59$$

and conclude there are factor 2 effects.

6.14 b) MANOVA Table:

Source of Variation	SSP	d.f.
Factor 1	$\begin{bmatrix} 496 & 184 \\ 184 & 208 \end{bmatrix}$	2
Factor 2	$\begin{bmatrix} 36 & 24 \\ 24 & 36 \end{bmatrix}$	3
Interaction	$\begin{bmatrix} 32 & 0 \\ 0 & 44 \end{bmatrix}$	6
Residual	$\begin{bmatrix} 312 & -84 \\ -84 & 400 \end{bmatrix}$	12
Total (Corrected)	$\begin{bmatrix} 876 & 124 \\ 124 & 688 \end{bmatrix}$	23

c) Since $-[gb(n-1) - (p+1-(g-1)(b-1))/2]\ln\Lambda^* = -13.5\ln\left(\frac{|SSP_{res}|}{|SSP_{int} + SSP_{res}|}\right)$

$= -13.5\ln(.808) = 2.88 < \chi_{12}^2(.05) = 21.03$ we do not reject

$H_0: \underset{\sim}{\gamma}_{11} = \underset{\sim}{\gamma}_{12} = \dots = \underset{\sim}{\gamma}_{34} = \underset{\sim}{0}$ (no interaction effects) at the

$\alpha = .05$ level.

Since

$$-[gb(n-1)-(p+1-(g-1))/2]\ln\Lambda^* = -11.5\ln\left(\frac{|SSP_{res}|}{|SSP_{fac\ 1} + SSP_{res}|}\right)$$

$= -11.5\ln(.2447) = 16.19 > \chi_4^2(.05) = 9.49$ we reject

$H_0: \underset{\sim}{\tau}_1 = \underset{\sim}{\tau}_2 = \underset{\sim}{\tau}_3 = \underset{\sim}{0}$ (no factor 1 effects) at the $\alpha = .05$ level.

Since

$$-[gb(n-1)-(p+1-(b-1))/2]\ln\Lambda^* = -12\ln\left(\frac{|SSP_{res}|}{|SSP_{fac\ 2} + SSP_{res}|}\right)$$

$= -12\ln(.7949) = 2.76 < \chi_6^2(.05) = 12.59$ we do not reject

$H_0: \underset{\sim}{\beta}_1 = \underset{\sim}{\beta}_2 = \underset{\sim}{\beta}_3 = \underset{\sim}{\beta}_4 = \underset{\sim}{0}$ (no factor 2 effects) at the $\alpha = .05$ level.

6.15 Example 6.11, $g = b = 2, n = 5$;

a) For $H_0: \underset{\sim}{\tau}_1 = \underset{\sim}{\tau}_2 = \underset{\sim}{0}$, $\Lambda^* = .3819$

Since

$$-[gb(n-1)-(p+1-(g-1))/2]\ln \Lambda^* = -14.5\ln(.3819) =$$

$$= 13.96 > \chi_3^2(.05) = 7.81 ,$$

we reject H_0 at $\alpha = .05$ level. For $H_0: \underset{\sim}{\beta}_1 = \underset{\sim}{\beta}_2 = \underset{\sim}{0}$. $\Lambda^* = .5230$ and $-14.5\ln(.5230) = 9.40$. Again we reject H_0 at $\alpha = .05$ level. These results are consistent with the exact F tests.

6.16 $H_0: C\underset{\sim}{\mu} = \underset{\sim}{0}$; $H_1: C\underset{\sim}{\mu} \neq \underset{\sim}{0}$ where $C = \begin{bmatrix} 1 & -1 & 0 & 0 \\ 0 & 1 & -1 & 0 \\ 0 & 0 & 1 & -1 \end{bmatrix}$.

Summary statistics:

$$\bar{\underset{\sim}{x}} = \begin{bmatrix} 1906.1 \\ 1749.5 \\ 1509.1 \\ 1725.0 \end{bmatrix}; \quad S = \begin{bmatrix} 105625 & 94759 & 87249 & 94268 \\ & 101761 & 76166 & 81193 \\ & & 91809 & 90333 \\ & & & 104329 \end{bmatrix}$$

$$T^2 = n(C\bar{\underset{\sim}{x}})'(CSC')^{-1}(C\bar{\underset{\sim}{x}}) = 254.7$$

$$\frac{(n-1)(q-1)}{(n-q+1)} F_{q-1,n-q+1}(\alpha) = \frac{(30-1)(4-1)}{(30-4+1)} F_{3,27}(.05) = 9.54$$

Since $T^2 = 254.7 > 9.54$ we reject H_0 at $\alpha = .05$ level.

95% simultaneous confidence interval for "dynamic" versus "static" means $(\mu_1 + \mu_2) - (\mu_3 + \mu_4)$ is, with $\underset{\sim}{c}' = [1 \quad 1 \quad -1 \quad -1]$,

$$\underset{\sim}{c}'\bar{\underset{\sim}{x}} \pm \sqrt{\frac{(n-1)(q-1)}{(n-q+1)} F_{q-1,n-q+1}(\alpha)} \sqrt{\frac{\underset{\sim}{c}'S\underset{\sim}{c}}{n}}$$

$$= 421.5 \pm 174.5$$

$$(247, 596)$$

6.17

Female turtle

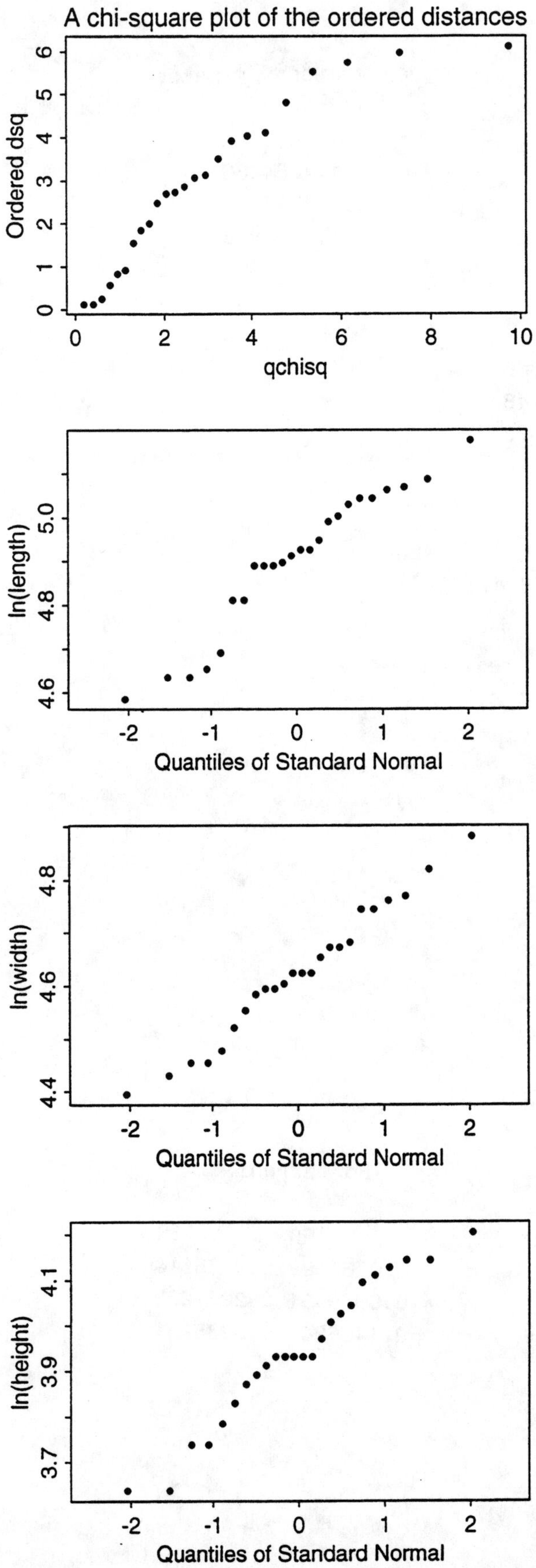

Male turtle

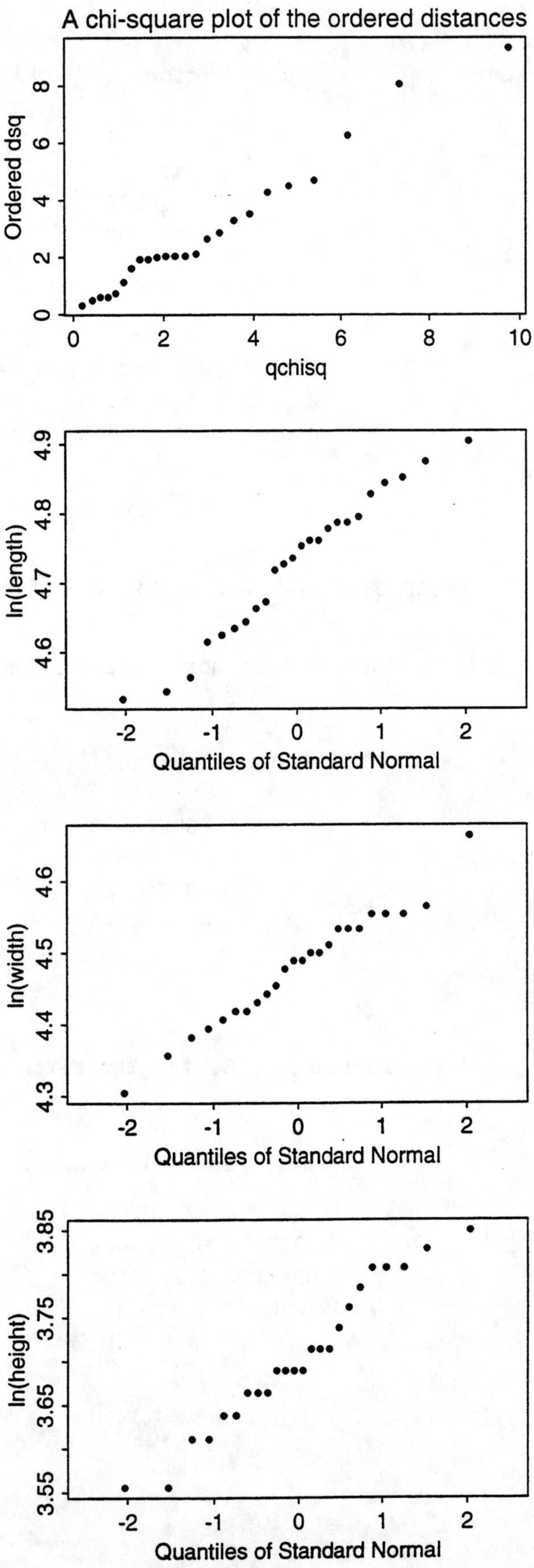

mean vector for females:

X1BAR
4.9006593
4.6229089
3.9402858

mean vector for males:

X2BAR
4.7254436
4.4775738
3.7031858

SPOOLED			
	0.0187388	0.0140655	0.0165386
	0.0140655	0.0113036	0.0127148
	0.0165386	0.0127148	0.0158563

TSQ	CVTSQ	F	CVF	PVALUE
85.052001	8.833461	27.118029	2.8164658	4.355E-10

linear combination most responsible for rejection

of H0 has coefficient vector:

COEFFVEC
-43.72677
-8.710687
67.546415

95% simultaneous CI for the difference

in female and male means

LOWER	UPPER
0.0577676	0.2926638
0.0541167	0.2365537
0.1290622	0.3451377

Bonferroni CI

LOWER	UPPER
0.0768599	0.2735714
0.0689451	0.2217252
0.1466248	0.3275751

6.18

a) $\underset{\sim}{\bar{x}}_1 = \begin{bmatrix} 12.219 \\ 8.113 \\ 9.590 \end{bmatrix}$; $\underset{\sim}{\bar{x}}_2 = \begin{bmatrix} 10.106 \\ 10.762 \\ 18.168 \end{bmatrix}$;

$$S_1 = \begin{bmatrix} 223.0134 & 12.3664 & 2.9066 \\ & 17.5441 & 4.7731 \\ & & 13.9633 \end{bmatrix}$$

$$S_2 = \begin{bmatrix} 4.3623 & .7599 & 2.3621 \\ & 25.8512 & 7.6857 \\ & & 46.6543 \end{bmatrix};$$

$$S_{pooled} = \begin{bmatrix} 15.8112 & 7.8550 & 2.6959 \\ & 20.7458 & 5.8960 \\ & & 26.5750 \end{bmatrix}$$

$$[(\frac{1}{n_1} + \frac{1}{n_2})S_{pooled}]^{-1} = \begin{bmatrix} 1.0939 & -.4084 & -.0203 \\ & .8745 & -.1525 \\ & & .5640 \end{bmatrix}$$

H_0: $\underset{\sim}{\mu}_1 - \underset{\sim}{\mu}_2 = \underset{\sim}{0}$

Since $T^2 = (\underset{\sim}{\bar{x}}_1 - \underset{\sim}{\bar{x}}_2)'[(\frac{1}{n_1} + \frac{1}{n_2})S_{pooled}]^{-1}(\underset{\sim}{\bar{x}}_1 - \underset{\sim}{\bar{x}}_2) = 50.92$

$$> \frac{(n_1+n_2-2)p}{(n_1+n_2-p-1)} F_{p,n_1+n_2-p-1}(.01) = \frac{(57)(3)}{55} F_{3,55}(.01) = 13,$$

we reject H_0 at the $\alpha = .01$ level. There is a difference in the (mean) cost vectors between gasoline trucks and diesel trucks.

b) $$\underset{\sim}{\hat{\ell}} \propto S^{-1}_{pooled}(\underset{\sim}{\bar{x}}_1 - \underset{\sim}{\bar{x}}_2) = \begin{bmatrix} 3.58 \\ -1.88 \\ -4.48 \end{bmatrix}$$

c) 99% simultaneous confidence intervals are:

$$\mu_{11} - \mu_{21}:\ 2.113 \pm 3.790$$

$$\mu_{12} - \mu_{22}:\ -2.650 \pm 4.341$$

$$\mu_{13} - \mu_{23}:\ -8.578 \pm 4.913$$

d) Assumption $\Sigma_1 = \Sigma_2$.

Since S_1 and S_2 are quite different, it may not be reasonable to pool. However, using "large sample" theory ($n_1 = 36$, $n_2 = 23$) we have, by Result 6.4,

$$(\bar{\mathbf{x}}_1 - \bar{\mathbf{x}}_2 - (\boldsymbol{\mu}_1 - \boldsymbol{\mu}_2))'[\frac{1}{n_1} S_1 + \frac{1}{n_2} S_2]^{-1}(\bar{\mathbf{x}}_1 - \bar{\mathbf{x}}_2 - (\boldsymbol{\mu}_1 - \boldsymbol{\mu}_2)) \sim \chi_p^2$$

Since

$$(\bar{\mathbf{x}}_1 - \bar{\mathbf{x}}_2)'[\frac{1}{n_1} S_1 + \frac{1}{n_2} S_2]^{-1}(\bar{\mathbf{x}}_1 - \bar{\mathbf{x}}_2) = 43.15 > \chi_3^2(.01) = 11.34$$

we reject $H_0\colon \boldsymbol{\mu}_1 - \boldsymbol{\mu}_2 = \mathbf{0}$ at the $\alpha = .01$ level. This is consistent with the result in part (a).

6.19 a) Tail length

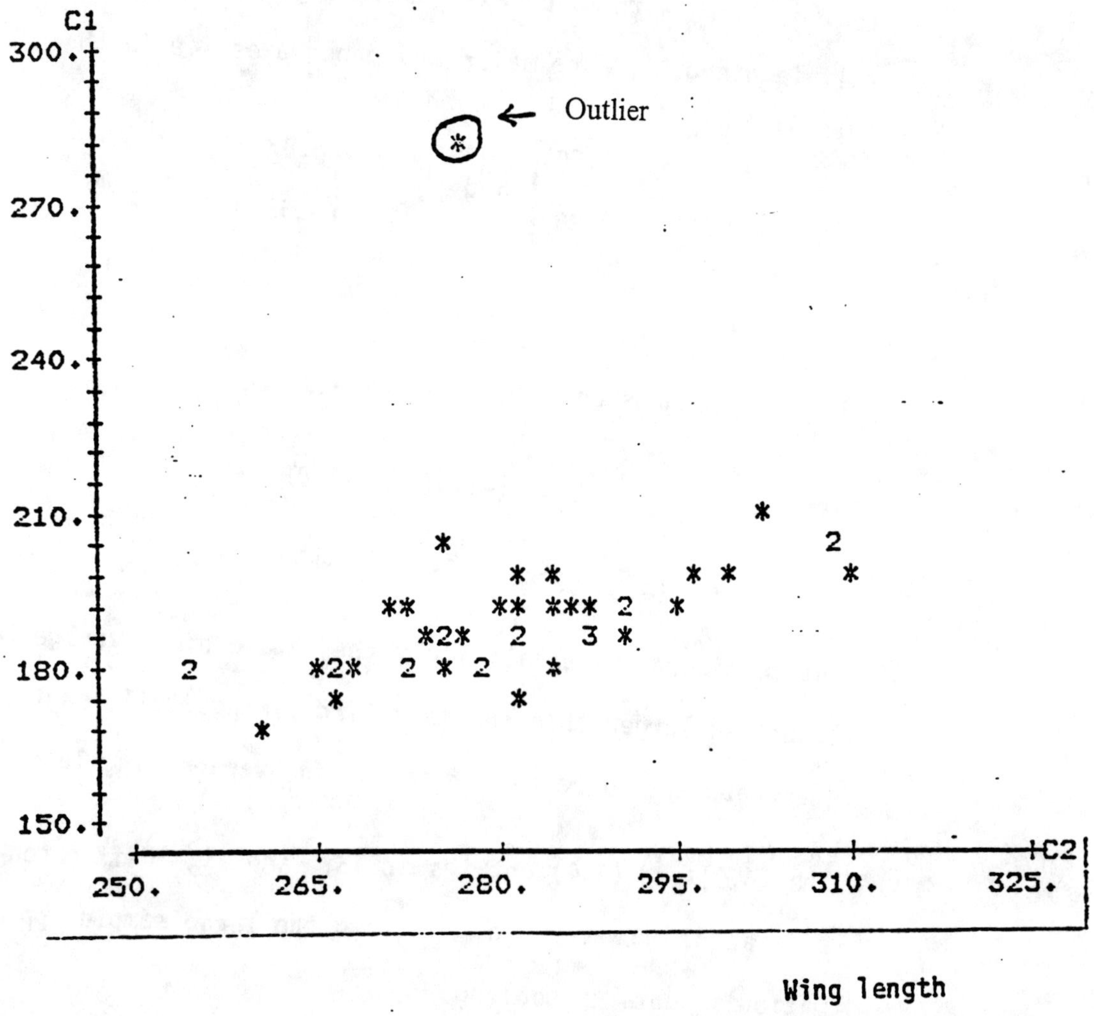

b) After eliminating the outlier,

$$T^2 = \left(\frac{1}{n_1} + \frac{1}{n_2}\right)^{-1} (\bar{\underset{\sim}{x}}_1 - \bar{\underset{\sim}{x}}_2)' \, S^{-1}_{pooled} (\bar{\underset{\sim}{x}}_1 - \bar{\underset{\sim}{x}}_2) = 24.803$$

$$> \frac{(n_1+n_2-2)p}{(n_1+n_2-p-1)} F_{p,n_1+n_2-p-1}(.05) = 6.31,$$

so we reject $H_0: \underset{\sim}{\mu}_1 - \underset{\sim}{\mu}_2 = \underset{\sim}{0}$ at the $\alpha = .05$ level.

c) 95% confidence ellipse for $\mu_1 - \mu_2$ is centered at [-6.46, 1.17] with major axis half-length 8.63 and minor axis half-length 3.06. The major and minor axes lie in the direction of

$$e_1 = \begin{bmatrix} .56 \\ .83 \end{bmatrix} \quad \text{and} \quad e_2 = \begin{bmatrix} .83 \\ -.56 \end{bmatrix}$$

respectively.

95% simultaneous confidence intervals are:

$$\mu_{11} - \mu_{21}: \quad (-11.91, \quad -1.01)$$

$$\mu_{12} - \mu_{22}: \quad (-6.19, \quad 8.53)$$

d) The confidence intervals imply the female bird average tail length is larger than the male bird average tail length. There appears to be no difference in average wing lengths.

6.20 a) The (4,2) and (4,4) entries in S_1 and S_2 differ considerably. However, $n_1 = n_2$ so the large sample approximation amounts to pooling.

b) H_0: $\mu_1 - \mu_2 = 0$ and H_1: $\mu_1 - \mu_2 \neq 0$

$$T^2 = 15.830 > \frac{(38)(4)}{35} F_{4,35}(.05) = 11.47$$

so we reject H_0 at the $\alpha = .05$ level.

c) $$\hat{\ell} \propto S_{\text{pooled}}^{-1}(\bar{x}_1 - \bar{x}_2) = \begin{bmatrix} -.24 \\ .16 \\ -3.74 \\ .01 \end{bmatrix}$$

d) Looking at the coefficients $\hat{\ell}_i\sqrt{s_{ii,pooled}}$, which apply to the standardized variables, we see that X_3: debt to equity ratio has the largest coefficient and therefore might be useful in classifying a bond as "high" or "medium" quality.

6.21

(a) The sample means for female and male are :

$$\bar{x}_F = \begin{bmatrix} 0.3136 \\ 5.1788 \\ 2.3152 \\ 38.1548 \end{bmatrix}, \quad \bar{x}_M = \begin{bmatrix} 0.3972 \\ 5.3296 \\ 3.6876 \\ 49.3404 \end{bmatrix}.$$

The Hotelling's $T^2 = 96.487 > 11.00$ where 11.00 is a critical point corresponding to $\alpha = 0.05$. Therefore, we reject $H_0 : \mu_1 - \mu_2 = 0$. The coefficient of the linear combination of most responsible for rejection is $(-95.600, 6.145, 5.737, -0.762)'$.

(b) The 95% simultaneous C. I. for female mean − male mean:

$$\begin{bmatrix} -0.1697234, & 0.00252336 \\ -1.4650835, & 1.16348346 \\ -1.8760572, & -0.8687428 \\ -17.032834, & -5.3383659 \end{bmatrix}$$

(c) We cannot extend the obtained result to the population of persons in their mid-twenties. Firstly this was a self selected sample of volunteers (friends) and is not even a random sample of graduate students. Further, graduate students are probably more sedentary than the typical persons of their age.

6.22 $n_1 = n_2 = n_3 = 50$; $p = 2$, $g = 3$ (sepal width and petal width responses only!)

$$\bar{\mathbf{x}}_1 = \begin{bmatrix} 3.428 \\ .306 \end{bmatrix}; \quad S_1 = \begin{bmatrix} .14364 & -.00474 \\ & .18576 \end{bmatrix}$$

$$\bar{\mathbf{x}}_2 = \begin{bmatrix} 2.770 \\ 1.326 \end{bmatrix}; \quad S_2 = \begin{bmatrix} .09860 & .04128 \\ & .03920 \end{bmatrix}$$

$$\bar{\mathbf{x}}_3 = \begin{bmatrix} 2.974 \\ 2.026 \end{bmatrix}; \quad S_3 = \begin{bmatrix} .10368 & .04764 \\ & .07563 \end{bmatrix}$$

MANOVA Table:

Source	SSP	d.f.
Treatment	$B = \begin{bmatrix} 11.344 & -21.820 \\ & 75.352 \end{bmatrix}$	2
Residual	$W = \begin{bmatrix} 16.950 & 4.125 \\ & 14.729 \end{bmatrix}$	147
Total	$B+W = \begin{bmatrix} 28.294 & -17.695 \\ & 90.081 \end{bmatrix}$	149

$$\Lambda^* = \frac{|W|}{|B+W|} = \frac{232.64}{2235.64} = .104$$

Since $\left(\frac{\Sigma n_\ell - p - 2}{p}\right)\left(\frac{1 - \sqrt{\Lambda^*}}{\sqrt{\Lambda^*}}\right) = 153.3 > 2.37 \doteq F_{4,292}(.05)$

we reject H_0: $\boldsymbol{\tau}_1 = \boldsymbol{\tau}_2 = \boldsymbol{\tau}_3$ at the $\alpha = .05$ level.

6.23

Without transforming the data, $\Lambda^* = \dfrac{|W|}{|B + W|} = .1159$ and $F = 18.98$.

After transformation, $\Lambda^* = .1198$ and $F = 18.52$.
There is a clear need for transforming the data to make the hypothesis tenable.

6.24 To test for parallelism, consider $H_{01}: C\underset{\sim}{\mu}_1 = C\underset{\sim}{\mu}_2$ with C given by (6-61).

$$C(\bar{\underset{\sim}{x}}_1 - \bar{\underset{\sim}{x}}_2) = \begin{bmatrix} -.413 \\ -.167 \\ -.036 \end{bmatrix}; \quad (CS_{pooles}C')^{-1} = \begin{bmatrix} 1.674 & .947 & .616 \\ & 2.014 & 1.144 \\ & & 2.341 \end{bmatrix}$$

$T^2 = 9.58 > c^2 = 8.0$., we reject H_0 at the $\alpha = .05$ level. The excess electrical usage of the test group was much lower than that of the control group for the 11 A.M., 1 P.M. and 3 P.M. hours. The similar 9 A.M. usage for the two groups contradicts the parallelism hypothesis.

6.25

a) Plots of the husband and wife profiles look similar but seem disparate for the level of "companionate love that you feel for your partner".

b) Parallelism hypothesis $H_0: C\underset{\sim}{\mu}_1 = C\underset{\sim}{\mu}_2$ with C given by (6-43).

$$C(\bar{\underset{\sim}{x}}_1 - \bar{\underset{\sim}{x}}_2) = \begin{bmatrix} -.13 \\ -.17 \\ .33 \end{bmatrix}; \quad CS_{pooled}C' = \begin{bmatrix} .685 & .733 & .029 \\ & .870 & -.028 \\ & & .095 \end{bmatrix}$$

For $\alpha = .05$, $c^2 = 8.7$ (see (6-62)). Since $T^2 = 19.58 > c^2 = 8.7$ we reject H_0 at the $\alpha = .05$ level.

6.26 $T^2 = 106.13 > 16.59$. We reject $H_0 : \mu_1 - \mu_2 = 0$ at 5% significance level. There is a significant difference in the two species.

Sample Mean for L.torrens and L.carteri:

L.torrens	L.carteri	Difference
96.457	99.343	-2.886
42.914	43.743	-0.829
35.371	39.314	-3.943
14.514	14.657	-0.143
25.629	30.000	-4.371
9.571	9.657	-0.086
9.714	9.371	0.343

Pooled Sample Covariance Matrix:

36.008	14.595	6.078	3.675	9.573	2.426	2.649
	16.639	2.764	2.992	6.101	1.053	0.934
		6.437	0.692	1.615	0.211	0.671
			3.039	2.407	0.274	0.229
				13.767	0.565	0.637
					1.213	0.914
						0.990

Linear Combination of most responsible for rejection
of Ho: L.torrens mean - L.carteri mean = 0 is :
(0.006, 0.151, -0.854, 0.268, -0.383, -2.187, 2.971)'

95% Simultaneous C. I. for L.torrens mean - L.carteri mean:

LOWER	UPPER
-8.73	2.96
-4.80	3.14
-6.41	-1.47
-1.84	1.55
-7.98	-0.76
-1.16	0.99
-0.63	1.31

The third and fifth components are most responsible for rejecting H_0. The χ^2 plots look fairy straight.

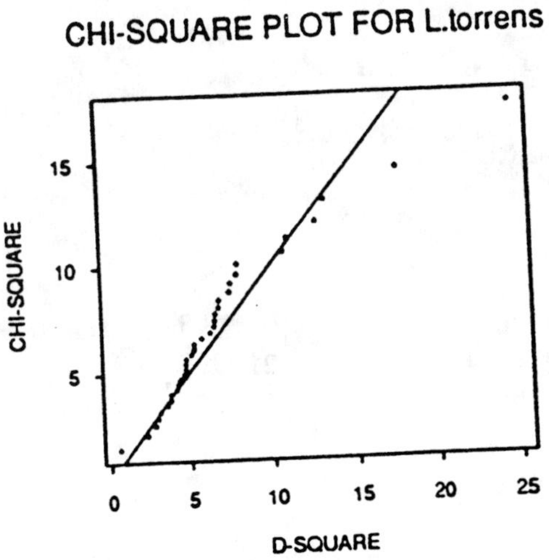

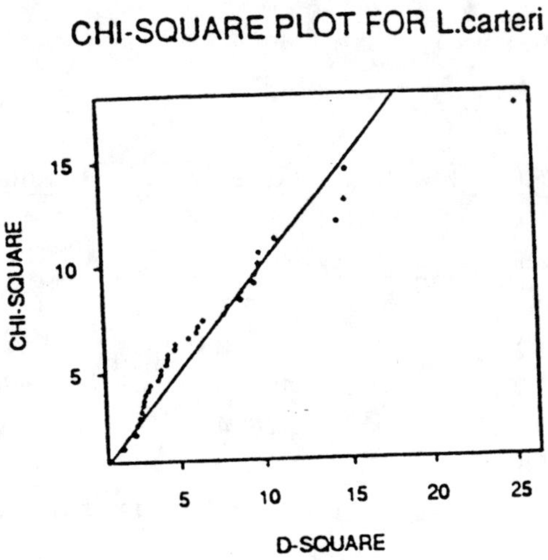

6.27

(a).

	XBAR	S		
Summary Statistics:	0.02548	0.00366259	0.00482862	0.00154159
	0.05784	0.00482862	0.01628931	0.00304801
	0.01056	0.00154159	0.00304801	0.00602526

Hotelling's $T^2 = 5.946$. The critical point is 9.979 and we fail to reject $H_0 : \mu_1 - \mu_2 = 0$ at 5% significance level.

(b). (c).

	LOWER	UPPER
Bonferroni C. I.:	-0.0057	0.0566
	-0.0079	0.1235
	-0.0294	0.0505
Simultaneous C. I.:	-0.0128	0.0637
	-0.0228	0.1385
	-0.0385	0.0596

6.28

```
HOTELLING T SQUARE   =   9.0218
P-VALUE      0.3616
```

	N	MEAN	STDEV	T2 INTERVAL	TO	BONFERRONI	TO
x1	24	0.00012	0.04817	-.0443	.0445	-.0283	.0285
x2	24	-0.00325	0.02751	-.0286	.0221	-.0195	.0130
x3	24	-0.0072	0.1030	-.1020	.0876	-.0679	.0535
x4	24	-0.0123	0.0625	-.0701	.0455	-.0493	.0247
x5	24	0.01513	0.03074	-.0130	.0436	-.0030	.0333
x6	24	0.00017	0.04689	-.0430	.0434	-.0275	.0278

The Bonferroni intervals use t (.00417) = 2.89 and

the T intevals use the constant 4.516.

6.29. **(a)** Two-factor MANOVA of peanuts data

E = Error SS&CP Matrix

	X1	X2	X3
X1	104.205	49.365	76.48
X2	49.365	352.105	121.995
X3	76.48	121.995	94.835

H = Type III SS&CP Matrix for FACTOR1 (Location)

	X1	X2	X3
X1	0.7008333333	-10.6575	7.1291666667
X2	-10.6575	162.0675	-108.4125
X3	7.1291666667	-108.4125	72.520833333

Manova Test Criteria and Exact F Statistics for
the Hypothesis of no Overall FACTOR1 Effect
H = Type III SS&CP Matrix for FACTOR1   E = Error SS&CP Matrix

S=1   M=0.5   N=1

Statistic	Value	F	Num DF	Den DF	Pr > F
Wilks' Lambda	0.10651620	11.1843	3	4	0.0205
Pillai's Trace	0.89348380	11.1843	3	4	0.0205
Hotelling-Lawley Trace	8.38824348	11.1843	3	4	0.0205
Roy's Greatest Root	8.38824348	11.1843	3	4	0.0205

H = Type III SS&CP Matrix for FACTOR2 (Variety)

	X1	X2	X3
X1	196.115	365.1825	42.6275
X2	365.1825	1089.015	414.655
X3	42.6275	414.655	284.10166667

Manova Test Criteria and F Approximations for
the Hypothesis of no Overall FACTOR2 Effect
H = Type III SS&CP Matrix for FACTOR2   E = Error SS&CP Matrix

S=2   M=0   N=1

Statistic	Value	F	Num DF	Den DF	Pr > F
Wilks' Lambda	0.01244417	10.6191	6	8	0.0019
Pillai's Trace	1.70910921	9.7924	6	10	0.0011
Hotelling-Lawley Trace	21.37567504	10.6878	6	6	0.0055
Roy's Greatest Root	18.18761127	30.3127	3	5	0.0012

H = Type III SS&CP Matrix for FACTOR1*FACTOR2

	X1	X2	X3
X1	205.10166667	363.6675	107.78583333
X2	363.6675	780.695	254.22
X3	107.78583333	254.22	85.951666667

```
Manova Test Criteria and F Approximations for
the Hypothesis of no Overall FACTOR1*FACTOR2 Effect
H = Type III SS&CP Matrix for FACTOR1*FACTOR2   E = Error SS&CP Matrix

S=2    M=0     N=1
Statistic                        Value          F      Num DF     Den DF  Pr > F
Wilks' Lambda               0.07429984     3.5582           6          8  0.0508
Pillai's Trace              1.29086073     3.0339           6         10  0.0587
Hotelling-Lawley Trace      7.54429038     3.7721           6          6  0.0655
Roy's Greatest Root         6.82409388    11.3735           3          5  0.0113
```

(b) Residual analysis. The residuals for X_2 at location 2 and variety 5 look large.

CODE	FACTOR1	FACTOR2	PRED1	RES1	PRED2	RES2	PRED3	RES3
a	1	5	194.80	0.50	160.40	-7.30	52.55	-1.15
a	1	5	194.80	-0.50	160.40	7.30	52.55	1.15
b	2	5	185.05	4.65	130.30	9.20	49.95	5.55
b	2	5	185.05	-4.65	130.30	-9.20	49.95	-5.55
c	1	6	199.45	3.55	161.40	-4.60	47.80	2.00
c	1	6	199.45	-3.55	161.40	4.60	47.80	-2.00
d	2	6	200.15	2.55	163.95	2.15	57.25	3.15
d	2	6	200.15	-2.55	163.95	-2.15	57.25	-3.15
e	1	8	190.25	3.25	164.80	-0.30	58.20	-0.40
e	1	8	190.25	-3.25	164.80	0.30	58.20	0.40
f	2	8	200.75	0.75	170.30	-3.50	66.10	-1.10
f	2	8	200.75	-0.75	170.30	3.50	66.10	1.10

Plot of residuals versus fitted values

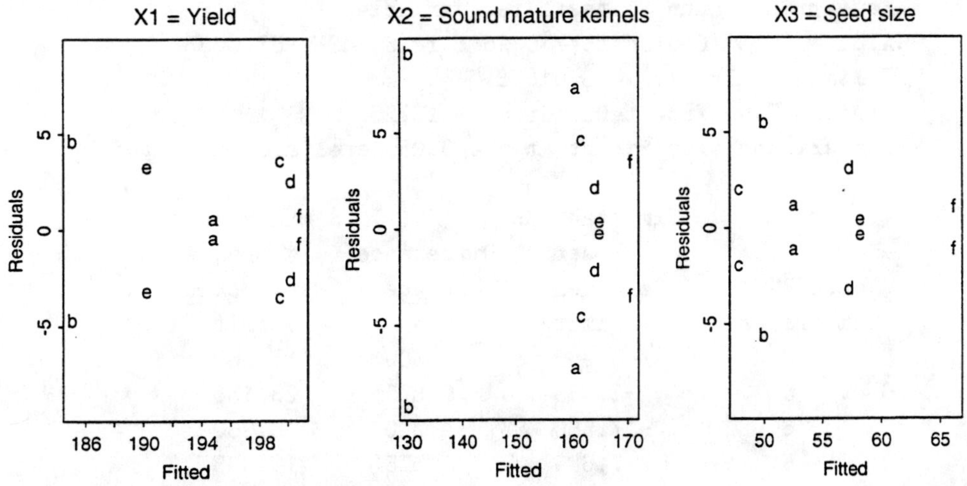

(c) Univariate two-factor ANOVA. With $\alpha = 0.05$, from Wilk's lambda test in part (a), the interaction term can be dropped.

```
Dependent Variable: X1
                    R-Square           C.V.      Root MSE              X1 Mean
                    0.388870       3.187484       6.21798              195.075

Source                    DF   Type III SS   Mean Square   F Value      Pr > F
FACTOR1                    1      0.700833      0.700833      0.02      0.8962
FACTOR2                    2    196.115000     98.057500      2.54      0.1403

Dependent Variable: X2
                    R-Square           C.V.      Root MSE              X2 Mean
                    0.524809       7.506437       11.8996              158.525

Source                    DF   Type III SS   Mean Square   F Value      Pr > F
FACTOR1                    1     162.06750     162.06750      1.14      0.3159
FACTOR2                    2    1089.01500     544.50750      3.85      0.0676

Dependent Variable: X3
                    R-Square           C.V.      Root MSE              X3 Mean
                    0.663596       8.595035       4.75377              55.3083

Source                    DF   Type III SS   Mean Square   F Value      Pr > F
FACTOR1                    1     72.520833     72.520833      3.21      0.1110
FACTOR2                    2    284.101667    142.050833      6.29      0.0229
```

(d) Bonferroni simultaneous comparisons of variety.
Only varieties 5 and 8 differ, and they differ only on X_3.

```
Bonferroni (Dunn) T tests for variable: X1
Alpha= 0.05  Confidence= 0.95  df= 8  MSE= 38.66333
Critical Value of T= 3.01576
Minimum Significant Difference= 13.26
Comparisons significant at the 0.05 level are indicated by '***'.

                Simultaneous             Simultaneous
                    Lower    Difference     Upper
  FACTOR2      Confidence     Between    Confidence
 Comparison         Limit       Means       Limit
6     - 8          -8.960       4.300      17.560
6     - 5          -3.385       9.875      23.135
8     - 6         -17.560      -4.300       8.960
8     - 5          -7.685       5.575      18.835
5     - 6         -23.135      -9.875       3.385
5     - 8         -18.835      -5.575       7.685
```

```
Bonferroni (Dunn) T tests for variable: X2
Alpha= 0.05  Confidence= 0.95  df= 8  MSE= 141.6
Critical Value of T= 3.01576
Minimum Significant Difference= 25.375
Comparisons significant at the 0.05 level are indicated by '***'.

                 Simultaneous              Simultaneous
                    Lower     Difference      Upper
  FACTOR2        Confidence     Between    Confidence
 Comparison         Limit        Means        Limit
8      - 6        -20.500        4.875       30.250
8      - 5         -3.175       22.200       47.575
6      - 8        -30.250       -4.875       20.500
6      - 5         -8.050       17.325       42.700
5      - 8        -47.575      -22.200        3.175
5      - 6        -42.700      -17.325        8.050

Bonferroni (Dunn) T tests for variable: X3
Alpha= 0.05  Confidence= 0.95  df= 8  MSE= 22.59833
Critical Value of T= 3.01576
Minimum Significant Difference= 10.137
Comparisons significant at the 0.05 level are indicated by '***'.

                 Simultaneous              Simultaneous
                    Lower     Difference      Upper
  FACTOR2        Confidence     Between    Confidence
 Comparison         Limit        Means        Limit
8      - 6         -0.512        9.625       19.762
8      - 5          0.763       10.900       21.037    ***
6      - 8        -19.762       -9.625        0.512
6      - 5         -8.862        1.275       11.412
5      - 8        -21.037      -10.900       -0.763    ***
5      - 6        -11.412       -1.275        8.862
```

6.30. Fitting a linear growth curve to calcium measurements on the dominant ulna

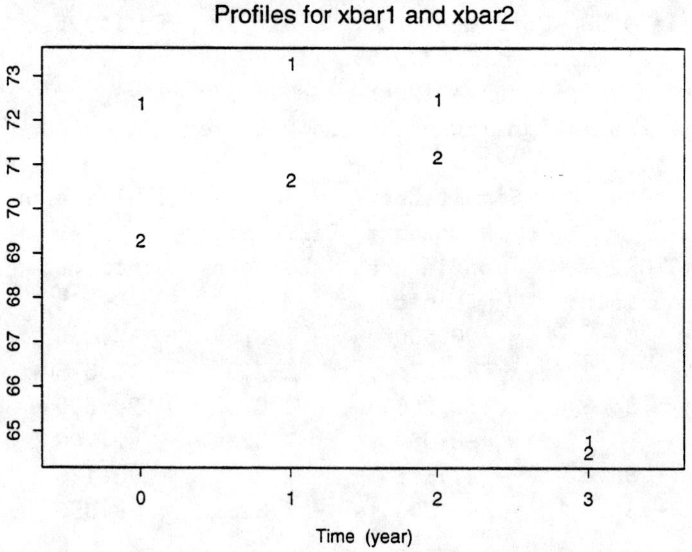


```
XBAR                 Grand mean
72.3800 69.2875      71.1939
73.2933 70.6562      71.8273
72.4733 71.1812      72.1848
64.7867 64.5312      65.2667

MLE of beta          [B'Sp^(-1)B]^(-1)
73.4707 70.5049      93.1313 -5.2393
-1.9035 -0.9818      -5.2393  1.2948

S1
92.1189 86.1106 73.3623 74.5890
86.1106 89.0764 72.9555 71.7728
73.3623 72.9555 71.8907 63.5918
74.5890 71.7728 63.5918 75.4441

S2
98.1745  97.0134 89.4824 86.1111
97.0134 100.5960 88.1425 88.2095
89.4824  88.1425 86.3496 80.5506
86.1111  88.2095 80.5506 81.4156

Spooled
95.2511 91.7500 81.7003 80.5487
91.7500 95.0348 80.8108 80.2745
81.7003 80.8108 79.3694 72.3636
80.5487 80.2745 72.3636 78.5328

W = (N-g)*Spooled
2762.282 2660.749 2369.308 2335.912
2660.749 2756.009 2343.514 2327.961
2369.308 2343.514 2301.714 2098.544
2335.912 2327.961 2098.544 2277.452

Estimated covariance matrix
 7.1816 -0.4040  0.0000  0.0000
-0.4040  0.0998  0.0000  0.0000
 0.0000  0.0000  6.7328 -0.3788
 0.0000  0.0000 -0.3788  0.0936

W1
2803.839 2610.438 2271.920 2443.549
2610.438 2821.243 2464.120 2196.065
2271.920 2464.120 2531.625 1845.313
2443.549 2196.065 1845.313 2556.818

Lambda = |W|/|W1| = 0.201
```

Since, with $\alpha = 0.01$, $-\left[N - \frac{1}{2}(p-q+g)\right]\log(\Lambda) = 45.72 > \chi^2_{(4-1-1)2}(0.01) = 13.28$, we reject the null hypothesis of a linear fit at $\alpha = 0.01$.

6.31. Fitting a quadratic growth curve to calcium measurements on the dominant ulna, treating all 31 subjects as a single group.

```
XBAR              MLE of beta          [B'Sp^(-1)B]^(-1)
70.7839           71.6039              92.2789 -5.9783  0.0799
71.9323            3.8673              -5.9783  9.3020 -2.9033
71.8065           -1.9404               0.0799 -2.9033  1.0760
64.6548

S                                       W = (n-1)*S
94.5441 90.7962 80.0081 78.0676         2836.322 2723.886 2400.243 2342.027
90.7962 93.6616 78.9965 77.7725         2723.886 2809.848 2369.894 2333.175
80.0081 78.9965 77.1546 70.0366         2400.243 2369.894 2314.639 2101.099
78.0676 77.7725 70.0366 75.9319         2342.027 2333.175 2101.099 2277.957

Estimated covariance matrix             W2
 3.1894 -0.2066  0.0028                 2857.167 2764.522 2394.410 2369.674
-0.2066  0.3215 -0.1003                 2764.522 2889.063 2358.522 2387.070
 0.0028 -0.1003  0.0372                 2394.410 2358.522 2316.271 2093.362
                                        2369.674 2387.070 2093.362 2314.625

Lambda = |W|/|W2| = 0.7653
```

Since, with $\alpha = 0.01$, $-\left[n - \frac{1}{2}(p - q + 1)\right] \log(\Lambda) = 7.893 > \chi^2_{4-2-1}(0.01) = 6.635$, we reject the null hypothesis of a quadratic fit at $\alpha = 0.01$.

Chapter 7

7.1 $\hat{\beta} = (Z'Z)^{-1}Z'y = \frac{1}{120}\begin{bmatrix}120 & -10\\ -10 & 1\end{bmatrix}\begin{bmatrix}72\\ 872\end{bmatrix} = \frac{1}{15}\begin{bmatrix}-10\\ 19\end{bmatrix} = \begin{bmatrix}-.667\\ 1.267\end{bmatrix}$

$$\hat{y} = Z\hat{\beta} = \frac{1}{15}\begin{bmatrix}180\\ 85\\ 123\\ 351\\ 199\\ 142\end{bmatrix} = \begin{bmatrix}12.000\\ 5.667\\ 8.200\\ 23.400\\ 13.267\\ 9.467\end{bmatrix} \;;\quad \hat{\varepsilon} = y - \hat{y} = \begin{bmatrix}15\\ 9\\ 3\\ 25\\ 9\\ 13\end{bmatrix} - \begin{bmatrix}12.000\\ 5.667\\ 8.200\\ 23.400\\ 13.267\\ 9.467\end{bmatrix} = \begin{bmatrix}3.000\\ 3.333\\ -5.200\\ 1.600\\ -6.267\\ 3.533\end{bmatrix}$$

Residual sum of squares: $\hat{\varepsilon}'\hat{\varepsilon} = 101.467$

Fitted equation: $\hat{y} = -.667 + 1.267\, z_1$

7.2 Standardized variables

z_1	z_2	y
-.292	-1.088	.391
-1.166	-.726	-.391
-.817	-.726	-1.174
1.283	.363	1.695
-.117	.726	-.652
1.108	1.451	.130

Fitted equation:

$\hat{y} = 1.33z_1 - .79z_2$

Also, prior to standardizing the variables, $\bar{z}_1 = 11.667$, $\bar{z}_2 = 5.000$ and $\bar{y} = 12.000$; $\sqrt{s_{z_1z_1}} = 5.716$, $\sqrt{s_{z_2z_2}} = 2.757$ and $\sqrt{s_{yy}} = 7.667$.

The fitted equation for the original variables is

$$\frac{\hat{y} - 12}{7.667} = 1.33\left(\frac{z_1 - 11.667}{5.716}\right) - .79\left(\frac{z_2 - 5}{2.757}\right)$$

$$\hat{y} = .43 + 1.78z_1 - 2.19z_2$$

7.3 Follow hint and note that $\hat{\varepsilon}^* = Y^* - \hat{Y}^* = V^{-1/2}Y - V^{-1/2}Z\beta_W$ and $(n-r-1)\sigma^2 = \hat{\varepsilon}^{*\prime}\hat{\varepsilon}^*$ is distributed as χ^2_{n-r-1}.

7.4 a) $V = I$ so $\hat{\beta}_W = (\underset{\sim}{z}'\underset{\sim}{z})^{-1}\underset{\sim}{z}'\underset{\sim}{y} = (\sum_{j=1}^{n} z_j y_j)/(\sum_{j=1}^{n} z_j^2)$.

b) V^{-1} is diagonal with $j^{\underline{th}}$ diagonal element $1/z_j$ so

$$\hat{\beta}_W = (\underset{\sim}{z}'V^{-1}\underset{\sim}{z})^{-1}\ \underset{\sim}{z}'V^{-1}\underset{\sim}{y} = (\sum_{j=1}^{n} y_j)/(\sum_{j=1}^{n} z_j)$$

c) V^{-1} is diagonal with $j^{\underline{th}}$ diagonal element $1/z_j^2$ so

$$\hat{\beta}_W = (\underset{\sim}{z}'V^{-1}\underset{\sim}{z})^{-1}\underset{\sim}{z}'V^{-1}\underset{\sim}{y} = (\sum_{j=1}^{n} (y_j/z_j))/n$$

7.5 Solution follows from Hint.

7.6 a) First note that $\Lambda^- = \text{diag}[\lambda_1^{-1},\ldots,\lambda_{r_1+1}^{-1},0,\ldots 0]$ is a generalized inverse of Λ since

$$\Lambda\Lambda^- = \begin{bmatrix} I_{r_1+1} & 0 \\ 0 & 0 \end{bmatrix} \quad \text{so} \quad \Lambda\Lambda^-\Lambda = \begin{bmatrix} \lambda_1 & & & & & 0 \\ & \ddots & & & & \\ & & \lambda_{r_1+1} & & & \\ & & & 0 & & \\ & & & & \ddots & \\ 0 & & & & & 0 \end{bmatrix} = \Lambda$$

Since $$Z'Z = \sum_{i=1}^{p} \lambda_i \underset{\sim}{e}_i \underset{\sim}{e}_i' = P\Lambda P'$$

$$(Z'Z)^- = \sum_{i=1}^{r_1+1} \lambda_i^{-1}\ \underset{\sim}{e}_i \underset{\sim}{e}_i' = P\Lambda^- P'$$

with $PP' = P'P = I_p$, we check that the defining relation holds

$$\begin{aligned}(Z'Z)(Z'Z)^-(Z'Z) &= P\Lambda \underbrace{P'(P}\Lambda^- \underbrace{P')P}\Lambda P' \\ &= P\Lambda\Lambda^-\ \Lambda P' \\ &= P\Lambda P' = Z'Z\end{aligned}$$

b) By the hint, if $Z\hat{\underset{\sim}{\beta}}$ is the projection, $\underset{\sim}{0} = Z'(\underset{\sim}{y} - Z\hat{\underset{\sim}{\beta}})$ or $Z'Z\hat{\underset{\sim}{\beta}} = Z'\underset{\sim}{y}$. In c), we show that $Z\hat{\underset{\sim}{\beta}}$ is the projection of $\underset{\sim}{y}$.

c) Consider $\underset{\sim}{g}_i = \lambda_i^{-1/2} Z\underset{\sim}{e}_i$ for $i = 1,2,\ldots,r_1+1$. Then

$$Z(Z'Z)^- Z' = Z\left(\sum_{i=1}^{r_1+1} \lambda_i^{-1} \underset{\sim}{e}_i \underset{\sim}{e}_i'\right)Z' = \sum_{i=1}^{r_1+1} \underset{\sim}{g}_i \underset{\sim}{g}_i'$$

The $\{\underset{\sim}{g}_i\}$ are r_1+1 mutually perpendicular unit length vectors that span the space of all linear combinations of the columns of Z. The projection of $\underset{\sim}{y}$ is then (see Result 2A.2 and Definition 2A.12)

$$\sum_{i=1}^{r_1+1} (\underset{\sim}{g}_i'\underset{\sim}{y})\underset{\sim}{g}_i = \sum_{i=1}^{r_1+1} \underset{\sim}{g}_i(\underset{\sim}{g}_i'\underset{\sim}{y}) = \left(\sum_{i=1}^{r_1+1} \underset{\sim}{g}_i\underset{\sim}{g}_i'\right)\underset{\sim}{y} = Z(Z'Z)^- Z'\underset{\sim}{y}$$

d) See Hint.

7.7 Write $\underset{\sim}{\beta} = \begin{bmatrix} \underset{\sim}{\beta}_{(1)} \\ \hline \underset{\sim}{\beta}_{(2)} \end{bmatrix}$ and $Z = \left[Z_1 \mid Z_2 \right]$.

Recall from Result 7.4 that $\hat{\underset{\sim}{\beta}} = \begin{bmatrix} \hat{\underset{\sim}{\beta}}_{(1)} \\ \hline \hat{\underset{\sim}{\beta}}_{(2)} \end{bmatrix} = (Z'Z)^{-1}Z'\underset{\sim}{y}$ is distributed as $N_{r+1}(\underset{\sim}{\beta},\sigma^2(Z'Z)^{-1})$ independently of $n\hat{\sigma}^2 = (n-r-1)s^2$ which is distributed as $\sigma^2 \chi^2_{n-r-1}$. From the Hint, $(\hat{\underset{\sim}{\beta}}_{(2)}-\underset{\sim}{\beta}_{(2)})'(C^{22})^{-1}(\hat{\underset{\sim}{\beta}}_{(2)}-\underset{\sim}{\beta}_{(2)})$ is $\sigma^2\chi^2_{r-q}$ and this is distributed independently of s^2. (The latter follows because the full random vector $\hat{\underset{\sim}{\beta}}$ is distributed independently of s^2). The result follows from the definition of a F random variable as the ratio of two independent χ^2 random variables divided by their degrees of freedom.

7.8 (a) $H^2 = Z(Z'Z)^{-1}Z'Z(Z'Z)^{-1}Z' = Z(Z'Z)^{-1}Z' = H$.

(b) Since $I - H$ is an idempotent matrix, it is positive semidefinite. Let a be an $n \times 1$ unit vector with j th element 1. Then $0 \le a'(I-H)a = (1-h_{jj})$. That is, $h_{jj} \le 1$. On the other hand, $(Z'Z)^{-1}$ is positive definite. Hence $h_{jj} = b_j'(Z'Z)^{-1}b_j > 0$ where b_j is the j th row of Z.
$\sum_{i=1}^{r+1} h_{jj} = tr(Z(Z'Z)^{-1}Z') = tr((Z'Z)^{-1}Z'Z) = tr(I_{r+1}) = r+1$.

(c) Using

$$(Z'Z)^{-1} = \frac{1}{n\sum_{i=1}^{n}(z_j-\bar{z})^2}\begin{bmatrix} \sum_{i=1}^{n} z_i^2 & -\sum_{i=1}^{n} z_i \\ -\sum_{i=1}^{n} z_i & n \end{bmatrix},$$

we obtain

$$\begin{aligned} h_{jj} &= (1\ z_j)(Z'Z)^{-1}\begin{pmatrix} 1 \\ z_j \end{pmatrix} \\ &= \frac{1}{n\sum_{i=1}^{n}(z_j-\bar{z})^2}\left(\sum_{i=1}^{n} z_i^2 - 2z_j\sum_{i=1}^{n} z_i + nz_j^2\right) \\ &= \frac{1}{n} + \frac{(z_j-\bar{z})^2}{\sum_{i=1}^{n}(z_j-\bar{z})^2} \end{aligned}$$

7.9

$$Z' = \begin{bmatrix} 1 & 1 & 1 & 1 & 1 \\ -2 & -1 & 0 & 1 & 2 \end{bmatrix}; \quad (Z'Z)^{-1} = \begin{bmatrix} 1/5 & 0 \\ 0 & 1/10 \end{bmatrix}$$

$$\hat{\underset{\sim}{\beta}}_{(1)} = (Z'Z)^{-1}Z'\underset{\sim}{y}_{(1)} = \begin{bmatrix} 3 \\ -.9 \end{bmatrix}; \quad \hat{\underset{\sim}{\beta}}_{(2)} = (Z'Z)^{-1}Z'\underset{\sim}{y}_{(2)} = \begin{bmatrix} 0 \\ 1.5 \end{bmatrix}$$

$$\hat{\boldsymbol{\beta}} = \left[\hat{\underset{\sim}{\beta}}_{(1)} \;\middle|\; \hat{\underset{\sim}{\beta}}_{(2)}\right] = \begin{bmatrix} 3 & 0 \\ -.9 & 1.5 \end{bmatrix}.$$

Hence

$$\hat{\mathbf{Y}} = Z\hat{\boldsymbol{\beta}} = \begin{bmatrix} 4.8 & -3.0 \\ 3.9 & -1.5 \\ 3.0 & 0 \\ 2.1 & 1.5 \\ 1.2 & 3.0 \end{bmatrix};$$

$$\hat{\boldsymbol{\varepsilon}} = \mathbf{Y} - \hat{\mathbf{Y}} = \begin{bmatrix} 5 & -3 \\ 3 & -1 \\ 4 & -1 \\ 2 & 2 \\ 1 & 3 \end{bmatrix} - \begin{bmatrix} 4.8 & -3.0 \\ 3.9 & -1.5 \\ 3.0 & 0 \\ 2.1 & 1.5 \\ 1.2 & 3.0 \end{bmatrix} = \begin{bmatrix} .2 & 0 \\ -.9 & .5 \\ 1.0 & -1.0 \\ -.1 & .5 \\ -.2 & 0 \end{bmatrix}$$

$$\mathbf{Y}'\mathbf{Y} = \hat{\mathbf{Y}}'\hat{\mathbf{Y}} + \hat{\boldsymbol{\varepsilon}}'\hat{\boldsymbol{\varepsilon}}$$

$$\begin{bmatrix} 55 & -15 \\ -15 & 24 \end{bmatrix} = \begin{bmatrix} 53.1 & -13.5 \\ -13.5 & 22.5 \end{bmatrix} + \begin{bmatrix} 1.9 & -1.5 \\ -1.5 & 1.5 \end{bmatrix}$$

7.10 a) Using Result 7.7, the 95% confidence interval for the mean reponse is given by

$$[1, .5]\begin{bmatrix}3.0\\-.9\end{bmatrix} \pm 3.18\sqrt{[1, .5]\begin{bmatrix}.2 & 0\\0 & .1\end{bmatrix}\begin{bmatrix}1\\.5\end{bmatrix}\left(\frac{1.9}{3}\right)} \quad \text{or}$$

(1.35, 3.75).

b) Using Result 7.8, the 95% prediction interval for the actual Y is given by

$$[1, -.5]\begin{bmatrix}3.0\\-.9\end{bmatrix} \pm 3.18\sqrt{\left\{1 + [1, .5]\begin{bmatrix}.2 & 0\\0 & .1\end{bmatrix}\begin{bmatrix}1\\.5\end{bmatrix}\right\}\left(\frac{1.9}{3}\right)} \quad \text{or}$$

(-.25, 5.35) .

c) Using (7-48) a 95% prediction ellipse for the actual Y's is given by

$$[y_{01} - 2.55, y_{02} - .75]\begin{bmatrix}7.5 & 7.5\\7.5 & 9.5\end{bmatrix}\begin{bmatrix}y_{01} - 2.55\\y_{02} - .75\end{bmatrix}$$

$$\leq (1 + .225)\left(\frac{(2)(3)}{2}\right)(19) = 69.825$$

7.11 The proof follows along the line of (7-40), (7-41) and material following with Σ^{-1} replaced by A. Note

$$(\underset{\sim}{Y}-ZB)'(\underset{\sim}{Y}-Z'B) = \sum_{j=1}^{n}(\underset{\sim}{Y}_j-B\underset{\sim}{z}_j)(\underset{\sim}{Y}_j-B\underset{\sim}{z}_j)'$$

and, as in (7-40)

$$\Sigma_{j=1}^{n} d_j^2(B) = tr[A^{-1}(\underset{\sim}{Y}-ZB)'(\underset{\sim}{Y}-ZB)] \ .$$

Next, by (7-41)

$$(\underset{\sim}{Y}-ZB)'(\underset{\sim}{Y}-ZB) = (\underset{\sim}{Y}-Z\hat{\underset{\sim}{\beta}}+Z\hat{\underset{\sim}{\beta}}-ZB)'(Y-Z\hat{\underset{\sim}{\beta}}+Z\hat{\underset{\sim}{\beta}}-ZB) = \hat{\varepsilon}'\hat{\varepsilon} + (\hat{\underset{\sim}{\beta}}-B)'Z'Z(\hat{\underset{\sim}{\beta}}-B)]$$

so

$$\Sigma_{j=1}^{n} d_j^2(B) = tr[A^{-1}\hat{\varepsilon}'\hat{\varepsilon}] + tr[A^{-1}(\hat{\underset{\sim}{\beta}}-B)'Z'Z(\hat{\underset{\sim}{\beta}}-B)]$$

The first term does not depend on the choice of B. Using Result 2A.12(c)

$$\begin{aligned} tr[A^{-1}(\hat{\underset{\sim}{\beta}}-B)'Z'Z(\hat{\underset{\sim}{\beta}}-B) &= tr[(\hat{\underset{\sim}{\beta}}-B)'Z'Z(\hat{\underset{\sim}{\beta}}-B)A] \\ &= tr[Z'Z(\hat{\underset{\sim}{\beta}}-B)A(\hat{\underset{\sim}{\beta}}-B)'] \\ &= tr[Z(\hat{\underset{\sim}{\beta}}-B)A(\hat{\underset{\sim}{\beta}}-B)'Z'] \\ &\geq \underset{\sim}{c}'A\underset{\sim}{c} > 0 \end{aligned}$$

where $\underset{\sim}{c}$ is any non-zero row of $\underset{\sim}{Z}(\hat{\underset{\sim}{\beta}}-B)$. Unless $B = \hat{\underset{\sim}{\beta}}$, $Z(\hat{\underset{\sim}{\beta}}-B)$ will have a non-zero row. Thus $\hat{\underset{\sim}{\beta}}$ is the best choice for any positive definite A.

7.12 (a) best linear predictor = $-4 + 2Z_1 - Z_2$

(b) mean square error = $\sigma_{yy} - \underset{\sim}{\sigma}'_{zy} \Sigma_{zz}^{-1} \underset{\sim}{\sigma}_{zy} = 4$

(c) $$\rho_{Y(x)} = \sqrt{\frac{\underset{\sim}{\sigma}'_{zy} \Sigma_{zz}^{-1} \underset{\sim}{\sigma}_{zy}}{\sigma_{yy}}} = \frac{\sqrt{5}}{3} = .745$$

(d) Following equation (7-62), we partition Σ as

$$\Sigma = \left[\begin{array}{cc|c} 9 & 3 & 1 \\ 3 & 2 & 1 \\ \hline 1 & 1 & 1 \end{array}\right]$$

and determine covariance of $\begin{bmatrix} Y \\ Z_1 \end{bmatrix}$ given z_2 to be

$$\begin{bmatrix} 9 & 3 \\ 3 & 2 \end{bmatrix} - \begin{bmatrix} 1 \\ 1 \end{bmatrix} (1)^{-1} [1, 1] = \begin{bmatrix} 8 & 2 \\ 2 & 1 \end{bmatrix}.$$ Therefore

$$\rho_{YZ_1 \cdot Z_2} = \frac{2}{\sqrt{8}\sqrt{1}} = \frac{\sqrt{2}}{2} = .707$$

7.13 (a) By Result 7.13, $\hat{\underset{\sim}{\beta}} = S_{zz}^{-1} \underset{\sim}{s}_{zy} = \begin{bmatrix} 3.73 \\ 5.57 \end{bmatrix}$

(b) Let $\underset{\sim}{Z}'_{(2)} = [Z_2, Z_3]$ $$R_{z_1(z_2 z_3)} = \sqrt{\frac{\underset{\sim}{s}'_{z_{(2)} z_1} S^{-1}_{z_{(2)} z_{(2)}} \underset{\sim}{s}_{z_{(2)} z_1}}{s_{z_1 z_1}}}$$

$$= \sqrt{\frac{3452.33}{5691.34}} = .78$$

(c) Partition $\underset{\sim}{Z} = \begin{bmatrix} \underset{\sim}{Z}_{(1)} \\ \hline Z_3 \end{bmatrix}$ so

$$S = \left[\begin{array}{cc|c} 5691.34 & & \\ 600.51 & 126.05 & \\ \hline 217.25 & 23.37 & 23.11 \end{array}\right] = \left[\begin{array}{c|c} S_{z_{(1)}z_{(1)}} & s'_{\tilde{z}_3 z_{(1)}} \\ \hline s_{\tilde{z}_3 z_{(1)}} & s_{z_3 z_3} \end{array}\right]$$

and

$$S_{z_{(1)}z_{(1)}} - s'_{\tilde{z}_3 z_{(1)}} \, s^{-1}_{z_3 z_3} \, s_{\tilde{z}_3 z_{(1)}} = \begin{bmatrix} 3649.04 & 380.82 \\ 380.82 & 102.42 \end{bmatrix}$$

Thus

$$r_{z_1 z_2 \cdot z_3} = \frac{380.82}{\sqrt{3649.04}\,\sqrt{102.42}} = .62$$

7.14 (a) The large positive correlation between a manager's experience and achieved rate of return on portfolio indicates an apparent advantage for managers with experience. The negative correlation between attitude toward risk and achieved rate of return indicates an apparent advantage for conservative managers.

(b) From (7-63)

$$r_{yz_1 \cdot z_2} = \frac{s_{yz_1 \cdot z_2}}{\sqrt{s_{yy \cdot z_2}}\,\sqrt{s_{z_1 z_1 \cdot z_2}}} = \frac{s_{yz_1} - \dfrac{s_{yz_2}\, s_{z_1 z_2}}{s_{z_2 z_2}}}{\sqrt{s_{yy} - \dfrac{s^2_{yz_2}}{s_{z_2 z_2}}}\,\sqrt{s_{z_1 z_1} - \dfrac{s^2_{z_1 z_2}}{s_{z_2 z_2}}}}$$

$$= \frac{r_{yz_1} - r_{yz_2}\, r_{z_1 z_2}}{\sqrt{1 - r^2_{yz_2}}\,\sqrt{1 - r^2_{z_1 z_2}}} = .31$$

Removing "years of experience" from consideration, we now have a positive correlation between "attitude toward risk" and "achieved return". After adjusting for years of experience, there is an apparent advantage to managers who take risks.

7.15 (a) MINITAB computer output gives: $\hat{y} = 11{,}870 + 2634z_1 + 45.2z_2$; residual sum of squares = 204995012 with 17 degrees of freedom. Thus $s = 3473$. Now for example, the estimated standard deviation of $\hat{\beta}_0$ is $\sqrt{1.9961s^2} = 4906$. Similar calculations give the estimated standard deviations of $\hat{\beta}_1$ and $\hat{\beta}_2$.

(b) An analysis of the residuals indicate there are no apparent model inadequacies.

(c) The 95% prediction interval is ($51,228; $66,239)

(d) Using (7-17), $F = \dfrac{(45.2)(.0067)^{-1}(45.2)}{12058533} = .025$

Since $F_{1,17}(.05) = 4.45$ we cannot reject $H_0{:}\beta_2 = 0$. It appears as if Z_2 is not needed in the model provided Z_1 is included in the model.

7.16

Predictors	p = r + 1	C_p
Z_1	2	1.025
Z_2	2	12.24
Z_1, Z_2	3	3

7.17

(a)

Analysis of Variance

Source	DF	Sum of Squares	Mean Square	F Value	Prob>F
Model	2	6519120.8603	3259560.4302	3.593	0.0844
Error	7	6351059.2397	907294.17709		
C Total	9	12870180.1			

Root MSE	952.51991	R-square	0.5065
Dep Mean	2927.30000	Adj R-sq	0.3655
C.V.	32.53920		

Parameter Estimates

Variable	DF	Parameter Estimate	Standard Error	T for H0: Parameter=0	Prob > |T|
INTERCEP	1	1464.452545	711.36299816	2.059	0.0785
SALES	1	0.010348	0.02077169	0.498	0.6336
ASSETS	1	0.010069	0.01203479	0.837	0.4304

(b). The eighth company is an outlier. We should check whether we can eliminate this observation without losing any information.

Obs	Student Residual	-2-1-0 1 2	Cook's D	Rstudent	Hat Diag H	Cov Ratio
1	-0.479	| | |	0.101	-0.4508	0.5695	3.3379
2	-0.329	| | |	0.020	-0.3074	0.3523	2.3396
3	0.408	| | |	0.031	0.3821	0.3594	2.3062
4	0.948	| |* |	0.035	0.9400	0.1045	1.1743
5	1.065	| |** |	0.665	1.0767	0.6377	2.5802
6	-0.683	| *| |	0.042	-0.6543	0.2144	1.6438
7	0.794	| |* |	0.043	0.7706	0.1692	1.4401
8	-2.325	| ****| |	0.422	-4.5111	0.1898	0.0231
9	0.351	| | |	0.009	0.3280	0.1865	1.8508
10	0.421	| | |	0.016	0.3949	0.2166	1.8770

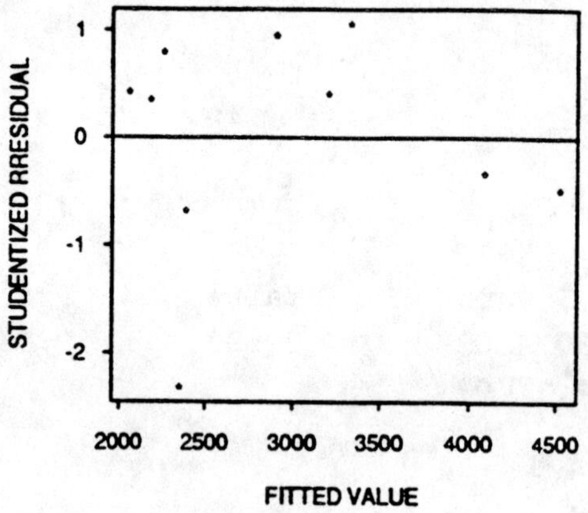

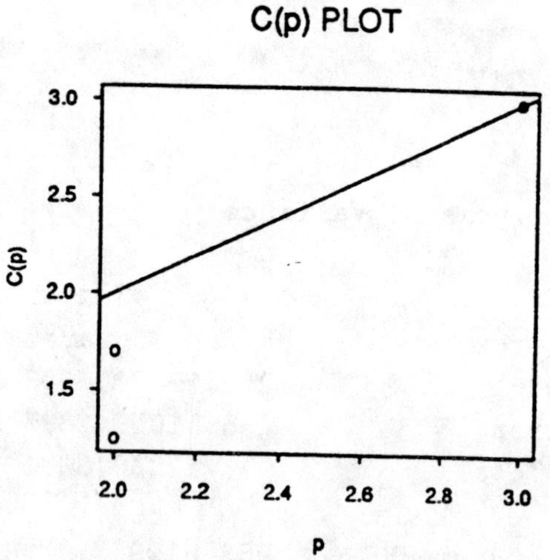

(c). Predicted value is 2583 (standard deviation is 463). The 95% prediction interval is (79, 5088).
(d). P-value for testing $H_0 : \beta_2 = 0$ is $0.4304 > 0.05$. We fail to reject H_0. This could be due to the outlier, so test H_0 again after eliminating eighth observation. If we fail to reject it, then we should drop "assets" term and refit the model.

7.18

C(p)	R-square	In	Variables in Model	P
1.24816	0.48903471	1	ASSETS	2
1.70001	0.45718133	1	SALES	2
3.00000	0.50652911	2	SALES- ASSETS	3

7.19 (a) Using the BMDP9R computer program the "best" regression equation involving $\ell n(y)$ and $z_1, z_2, \ldots, z_5$ is

$$\widehat{\ell n(y)} = 2.756 - .322 z_2 + .114 z_4$$

with $s = 1.058$ and $R^2 = .60$. It may be possible to find a better model using predictors of the form z_1, z_2, z_3^2 and so forth.

(b) A plot of the residuals versus the predicted values indicates no apparent problems. A Q-Q plot of the residuals is a bit wavy.

Perhaps a transformation other than the logarithmic transformation would produce a better model.

7.20 Eigenvalues of the correlation matrix of the predictor variables z_1, $z_2,\ldots,z_5$ are 1.4465, 1.1435, .8940, .8545, .6615. The corresponding eigenvectors give the coefficients of $z_1, z_2,\ldots,z_5$ in the principle component. For example, the first principal component, written in terms of standardized predictor variables, is

$$\hat{x}_1 = .6064z_1^* - .3901z_2^* - .6357z_3^* - .2755z_4^* - .0045z_5^* .$$

A regression of $\ell n(y)$ on the first principle component gives

$$\widehat{\ell n}(y) = 1.7371 - .0701\hat{x}_1$$

with $s = .701$ and $R^2 = .015$.

A regression of $\ell n(y)$ on the fourth principle component produces the best of the one principle component predictor variable regressions. In this case $\widehat{\ell n}(y) = 1.7371 + .3604\hat{x}_4$ and $s = .618$ and $R^2 = .235$.

7.21 This data set doesn't appear to yield a regression relationship which explains a large proportion of the variation in the responses.

(a) (i) One reader, starting with a full quadratic model in the predictors z_1 and z_2, suggested the fitted regression equation:

$$\hat{y}_1 = -7.3808 + .5281z_2 - .0038z_2^2$$

with $s = 3.05$ and $R^2 = .22$. (Can you do better than this?)

(ii) A plot of the residuals versus the fitted values suggests the response may not have constant variance. Also a Q-Q plot of the residuals has the following general appearance:

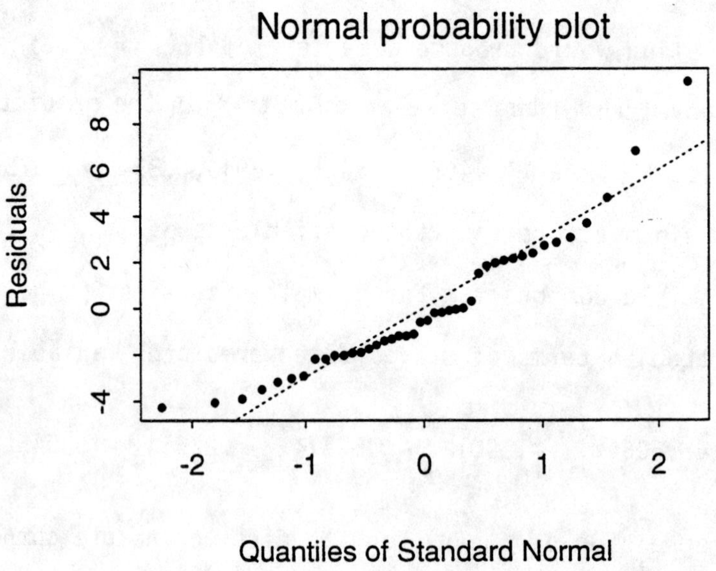

Therefore the normality assumption may also be suspect. Perhaps a better regression can be obtained after the responses have been transformed or re-expressed in a different metric.

(iii) Using the results in (a)(i), a 95% prediction interval of $z_1 = 10$ (not needed) and $z_2 = 80$ is

$$10.84 \pm 2.02\sqrt{7.47} \quad \text{or} \quad (5.32, 16.37).$$

7.23. (a) Regression analysis using the response Y_1 = SalePr.

Summary of Backward Elimination Procedure for Dependent Variable X2

Step	Variable Removed	Number In	Partial R**2	Model R**2	C(p)	F	Prob>F
1	X9	7	0.0041	0.5826	7.6697	0.6697	0.4161
2	X3	6	0.0043	0.5782	6.3735	0.7073	0.4033
3	X5	5	0.0127	0.5655	6.4341	2.0795	0.1538

Dependent Variable: X2 SalePr

Analysis of Variance

Source	DF	Sum of Squares	Mean Square	F Value	Prob>F
Model	5	16462859.832	3292571.9663	18.224	0.0001
Error	70	12647164.839	180673.78342		
C Total	75	29110024.671			

Root MSE 425.05739 R-square 0.5655

Parameter Estimates

Variable	DF	Parameter Estimate	Standard Error	T for H0: Parameter=0	Prob > |T|
INTERCEP	1	-5605.823664	1929.3986440	-2.905	0.0049
X1	1	-77.633612	22.29880197	-3.482	0.0009
X4	1	-2.332721	0.75490590	-3.090	0.0029
X6	1	389.364490	89.17300145	4.366	0.0001
X7	1	1749.420733	701.21819165	2.495	0.0150
X8	1	133.177529	46.66673277	2.854	0.0057

The 95% prediction interval for SalePr for z_0 is

$$z_0'\hat{\beta} \pm t_{70}(0.025)\sqrt{(425.06)^2(1 + z_0'(\mathbf{Z}'\mathbf{Z})^{-1}z_0)}.$$

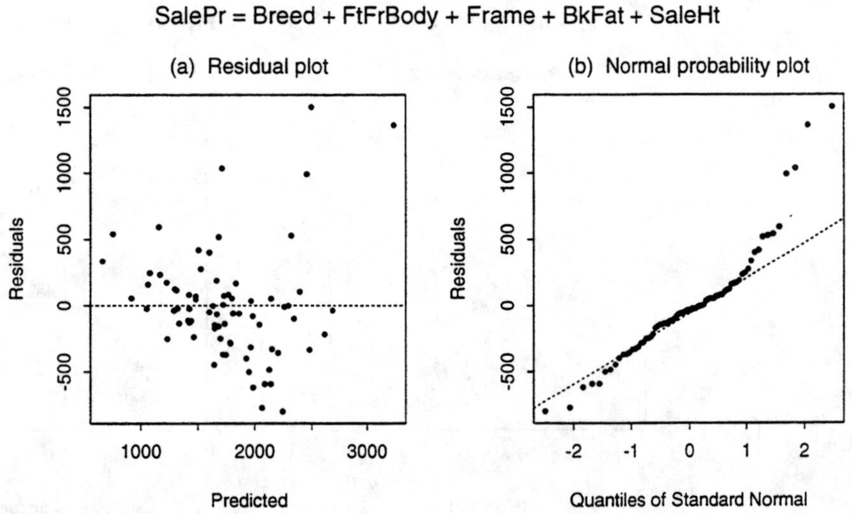

(b) Regression analysis using the response $Y_1 = \ln(\text{SalePr})$.

```
Summary of Backward Elimination Procedure for Dependent Variable LOGX2
        Variable   Number   Partial    Model
Step    Removed        In      R**2     R**2       C(p)         F   Prob>F
   1    X3              7    0.0033   0.6368     7.6121    0.6121   0.4368
   2    X7              6    0.0057   0.6311     6.6655    1.0594   0.3070
   3    X9              5    0.0122   0.6189     6.9445    2.2902   0.1348
   4    X4              4    0.0081   0.6108     6.4537    1.4890   0.2265

Dependent Variable: LOGX2
Analysis of Variance
                           Sum of         Mean
Source           DF       Squares       Square        F Value        Prob>F
Model             4       4.02968      1.00742         27.854        0.0001
Error            71       2.56794      0.03617
C Total          75       6.59762

    Root MSE        0.19018      R-square        0.6108
Parameter Estimates
                    Parameter        Standard    T for H0:
Variable   DF        Estimate           Error    Parameter=0     Prob > |T|
INTERCEP    1        5.235773      0.91286786          5.736         0.0001
X1          1       -0.049418      0.00846029         -5.841         0.0001
X5          1       -0.027613      0.00827438         -3.337         0.0013
X6          1        0.183611      0.03992448          4.599         0.0001
X8          1        0.058996      0.01927655          3.060         0.0031
```

The 95% prediction interval for ln(SalePr) for z_0 is

$$z_0'\hat{\beta} \pm t_{70}(0.025)\sqrt{(0.1902)^2(1 + z_0'(\mathbf{Z}'\mathbf{Z})^{-1}z_0)}.$$

The few outliers among these latter residuals are not so pronounced.

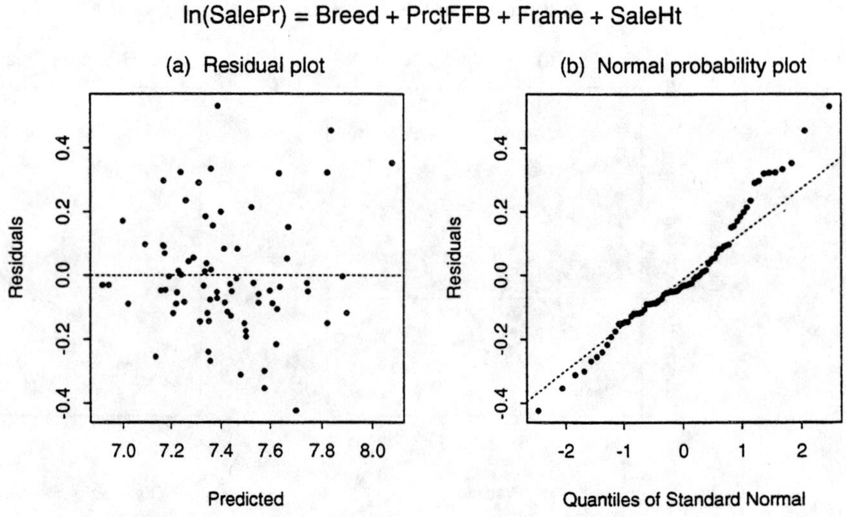

7.24. (a) Regression analysis using the response Y_1 = SaleHt and the predictors Z_1 = YrHgt and Z_2 = FtFrBody.

```
Dependent Variable: X8              SaleHt
Analysis of Variance
                          Sum of           Mean
Source          DF       Squares         Square       F Value        Prob>F
Model            2     235.74533      117.87267       131.165        0.0001
Error           73      65.60204        0.89866
C Total         75     301.34737

     Root MSE        0.94798      R-square         0.7823
Parameter Estimates
                     Parameter       Standard      T for H0:
Variable  DF          Estimate          Error    Parameter=0     Prob > |T|
INTERCEP   1          7.846281     3.36221288          2.334         0.0224
X3         1          0.802235     0.08088562          9.918         0.0001
X4         1          0.005773     0.00151072          3.821         0.0003
```

The 95% prediction interval for SaleHt for $z_0' = (1, 50.5, 970)$ is

$$53.96 \pm t_{73}(0.025)\sqrt{0.8987(1.0148)} \;=\; (52.06, 55.86).$$

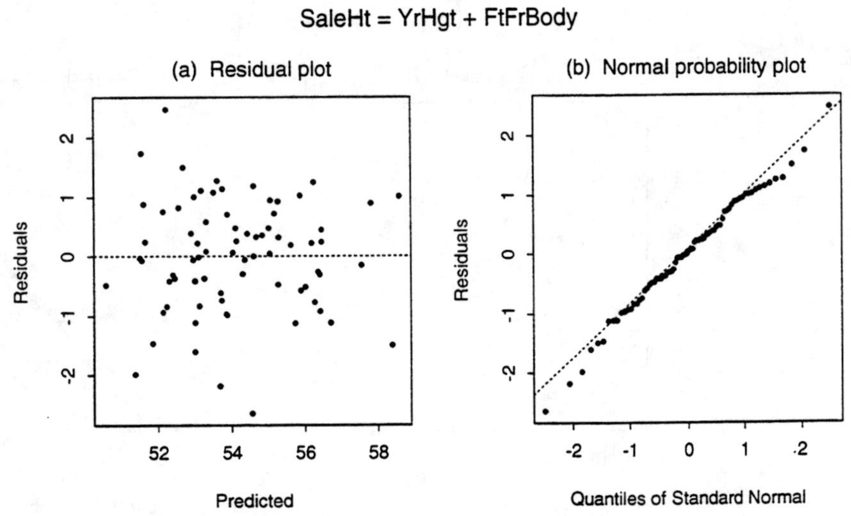

(b) Regression analysis using the response $Y_1 = \text{SaleHt}$ and the predictors $Z_1 = \text{YrHgt}$ and $Z_2 = \text{FtFrBody}$.

```
Dependent Variable: X9          SaleWt
Analysis of Variance
                            Sum of            Mean
Source            DF       Squares          Square          F Value         Prob>F
Model              2 390456.63614 195228.31807           16.319         0.0001
Error             73 873342.99544  11963.60268
C Total           75 1263799.6316

    Root MSE         109.37826       R-square          0.3090
Parameter Estimates
                    Parameter        Standard      T for H0:
Variable  DF         Estimate           Error    Parameter=0       Prob > |T|
INTERCEP   1       675.316794    387.93499836          1.741           0.0859
X3         1         2.719286      9.33265244          0.291           0.7716
X4         1         0.745610      0.17430765          4.278           0.0001
```

The 95% prediction interval for SaleWt for $z_0' = (1, 50.5, 970)$ is

$$1535.9 \pm t_{73}(0.025)\sqrt{11963.6(1.0148)} \;=\; (1316.3, 1755.5).$$

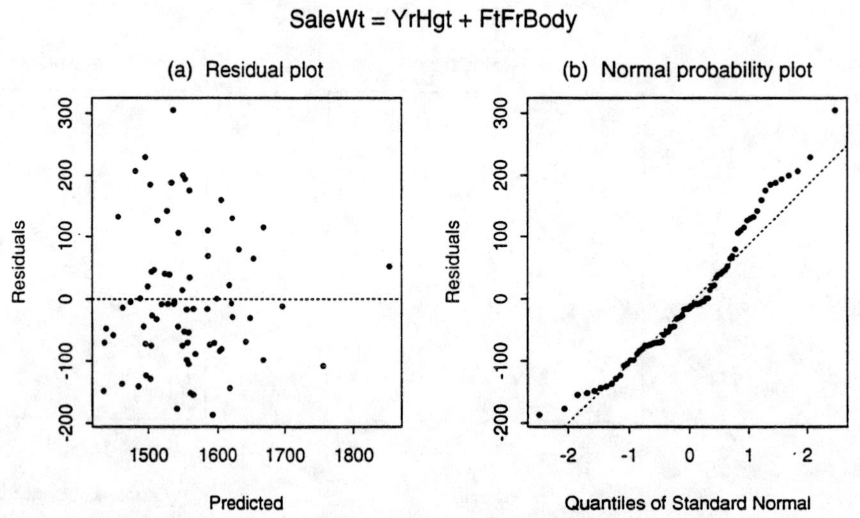

Multivariate regression analysis using the responses Y_1 = SaleHt and Y_2 = SaleWt and the predictors Z_1 = YrHgt and Z_2 = FtFrBody.

```
Multivariate Test:  HO: YrHgt = 0
Multivariate Statistics and Exact F Statistics
S=1     M=0     N=35
```

Statistic	Value	F	Num DF	Den DF	Pr > F
Wilks' Lambda	0.38524567	57.4469	2	72	0.0001
Pillai's Trace	0.61475433	57.4469	2	72	0.0001
Hotelling-Lawley Trace	1.59574625	57.4469	2	72	0.0001
Roy's Greatest Root	1.59574625	57.4469	2	72	0.0001

```
Multivariate Test:  HO: FtFrBody = 0
Multivariate Statistics and Exact F Statistics
S=1     M=0     N=35
```

Statistic	Value	F	Num DF	Den DF	Pr > F
Wilks' Lambda	0.75813396	11.4850	2	72	0.0001
Pillai's Trace	0.24186604	11.4850	2	72	0.0001
Hotelling-Lawley Trace	0.31902811	11.4850	2	72	0.0001
Roy's Greatest Root	0.31902811	11.4850	2	72	0.0001

The theory requires using x_3 (YrHgt) to predict both SaleHt and SaleWt, even though this term could be dropped in the prediction equation for SaleWt. The 95% prediction ellipse for both SaleHt and SaleWt for $z_0' = (1, 50.5, 970)$ is

$$1.3498(Y_{01} - 53.96)^2 + 0.0001(Y_{02} - 1535.9)^2 - 0.0098(Y_{01} - 53.96)(Y_{02} - 1535.9)$$
$$= 1.0148\frac{2(73)}{72}F_{2,72}(0.05) = 6.4282.$$

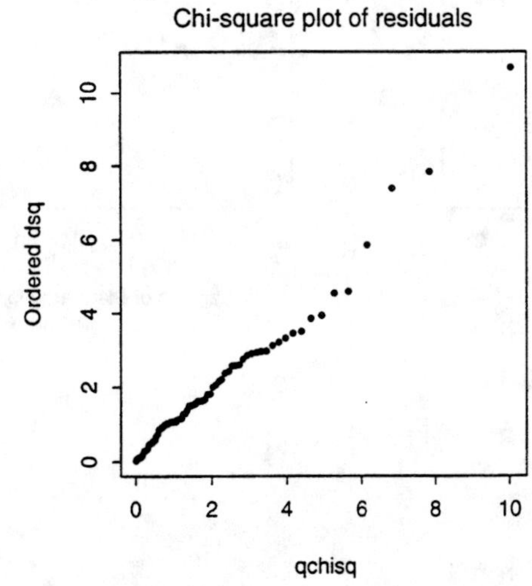

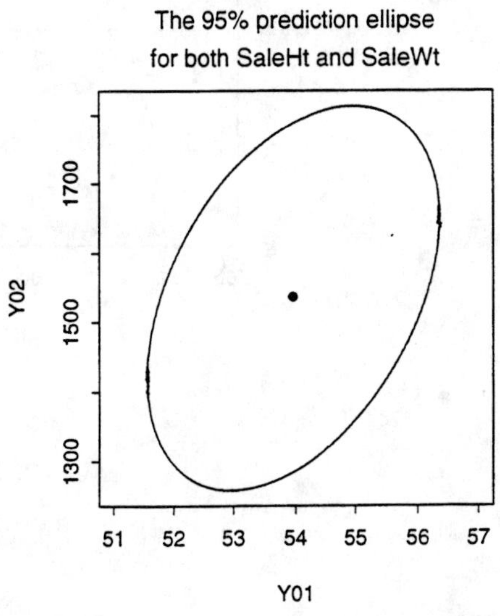

7.25. (a) Regression analysis using the first response Y_1. The backward elimination procedure gives $Y_1 = \beta_{01} + \beta_{11}Z_1 + \beta_{21}Z_2$. All variables left in the model are significant at the 0.05 level. (It is possible to drop the intercept but we retain it.)

```
Dependent Variable: Y1          TOT
Analysis of Variance
                         Sum of          Mean
Source          DF      Squares        Square         F Value          Prob>F
Model            2 5905583.8728 2952791.9364           22.962          0.0001
Error           14 1800356.3625 128596.88303
C Total         16 7705940.2353

    Root MSE        358.60408       R-square           0.7664
Parameter Estimates
                    Parameter         Standard       T for H0:
Variable  DF         Estimate            Error     Parameter=0      Prob > |T|
INTERCEP   1        56.720053     206.70336862           0.274          0.7878
Z1         1       507.073084     193.79082471           2.617          0.0203
Z2         1         0.328962       0.04977501           6.609          0.0001
```

The 95% prediction interval for $Y_1 = \text{TOT}$ for $z_0' = (1, 1, 1200)$ is

$$958.5 \pm t_{14}(0.025)\sqrt{128596.9(1.0941)} \;=\; (154.0, 1763.1).$$

TOT = GEN + AMT

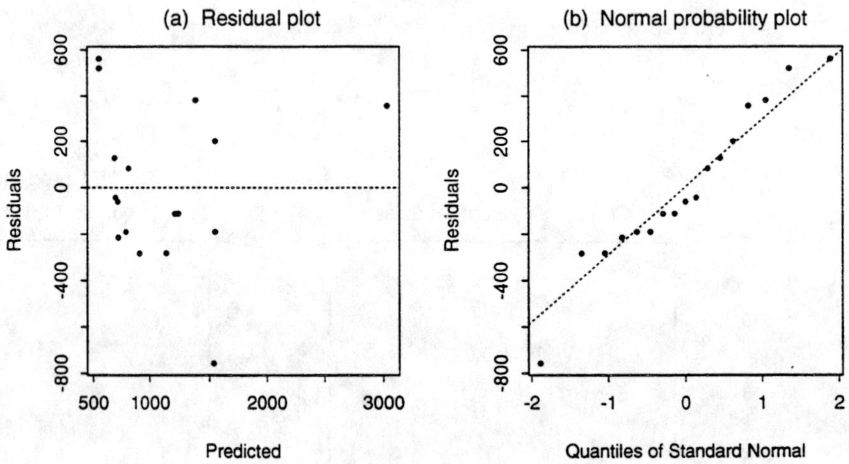

(b) Regression analysis using the second response Y_2. The backward elimination procedure gives $Y_2 = \beta_{02} + \beta_{12}Z_1 + \beta_{22}Z_2$. All variables left in the model are significant at the 0.05 level.

```
Dependent Variable: Y2          AMI
Analysis of Variance
                          Sum of           Mean
Source           DF      Squares         Square        F Value        Prob>F
Model             2 5989720.5384 2994860.2692         25.871         0.0001
Error            14  1620657.344 115761.23886
C Total          16 7610377.8824

    Root MSE       340.23703       R-square        0.7870
Parameter Estimates
                      Parameter         Standard    T for H0:
Variable  DF           Estimate            Error  Parameter=0     Prob > |T|
INTERCEP   1        -241.347910     196.11640164       -1.231         0.2387
Z1         1         606.309666     183.86521452        3.298         0.0053
Z2         1           0.324255       0.04722563        6.866         0.0001
```

The 95% prediction interval for $Y_2 = \text{AMI}$ for $z_0' = (1, 1, 1200)$ is

$$754.1 \pm t_{14}(0.025)\sqrt{115761.2(1.0941)} \; = \; (-9.234, 1517.4).$$

AMI = GEN + AMT

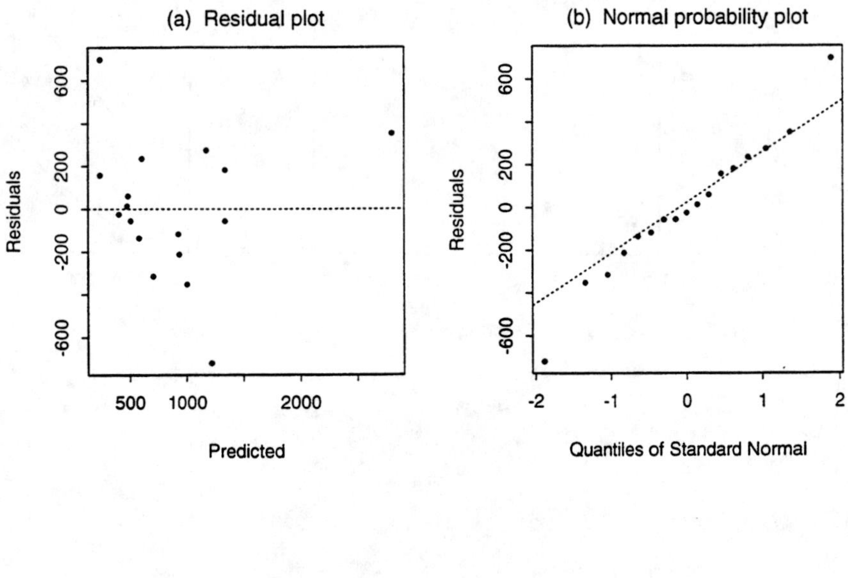

(c) Multivariate regression analysis using Y_1 and Y_2.

```
Multivariate Test:  HO: PR=0, DIAP=0, QRS=0
Multivariate Statistics and F Approximations
S=2    M=0    N=4
```

Statistic	Value	F	Num DF	Den DF	Pr > F
Wilks' Lambda	0.44050214	1.6890	6	20	0.1755
Pillai's Trace	0.60385990	1.5859	6	22	0.1983
Hotelling-Lawley Trace	1.16942861	1.7541	6	18	0.1657
Roy's Greatest Root	1.07581808	3.9447	3	11	0.0391

Based on Wilks' Lambda, the three variables Z_3, Z_4 and Z_5 are not significant. The 95% prediction ellipse for both TOT and AMI for $z_0' = (1, 1, 1200)$ is

$$4.305 \times 10^{-5}(Y_{01} - 958.5)^2 + 4.782 \times 10^{-5}(Y_{02} - 754.1)^2$$
$$- 8.214 \times 10^{-5}(Y_{01} - 958.5)(Y_{02} - 754.1) = 1.0941 \frac{2(14)}{13} F_{2,13}(0.05) = 8.968.$$

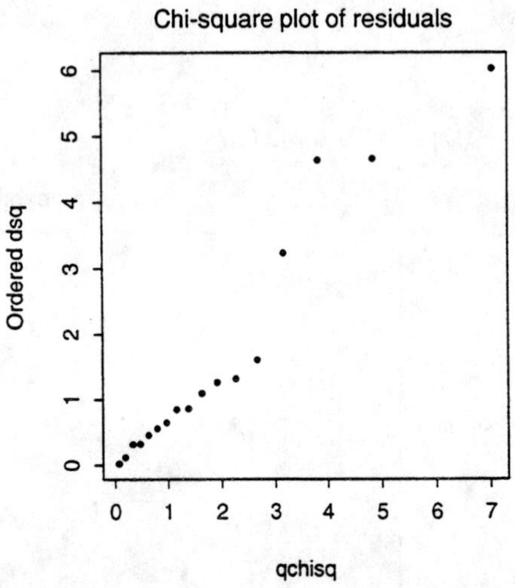

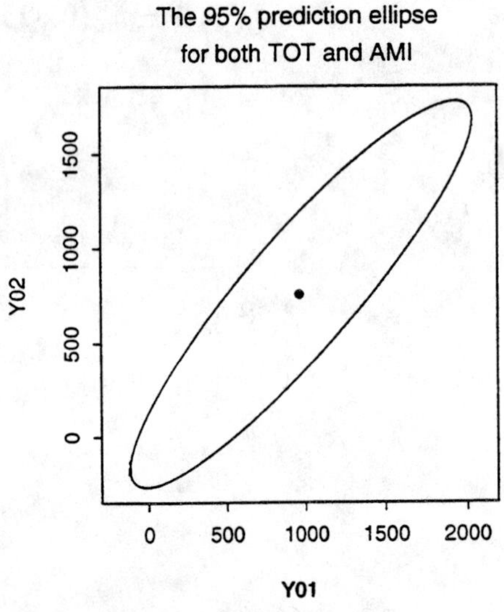

Chapter 8

8.1 Eigenvalues of $\boldsymbol{\Sigma}$ are $\lambda_1 = 6$, $\lambda_2 = 1$. The principal components are

$$Y_1 = .894X_1 + .447X_2$$
$$Y_2 = .447X_1 - .894X_2$$

$Var(Y_1) = \lambda_1 = 6$. Therefore, proportion of total population variance explained by Y_1 is $6/(6+1) = .86$.

8.2

$$\boldsymbol{\rho} = \begin{bmatrix} 1 & .6325 \\ .6325 & 1 \end{bmatrix}$$

(a) $Y_1 = .707Z_1 + .707Z_2$

$Y_2 = .707Z_1 - .707Z_2$

$Var(Y_1) = \lambda_1 = 1.6325$

Proportion of total population variance explained by Y_1 is $1.6325/(1+1) = .816$

(b) No. The two (standardized) variables contribute equally to the principal components in 8.2(a). The two variables contribute unequally to the principal components in 8.1 because of their unequal variances.

(c) $\rho_{Y_1Z_1} = .903$; $\rho_{Y_1Z_2} = .903$; $\rho_{Y_2Z_1} = .429$

8.3 Eigenvalues of $\boldsymbol{\Sigma}$ are 2, 4, 4. Eigenvectors associated with the eigenvalues 4, 4 are not unique. One choice is $\mathbf{e}_2' = [0 \ \ 1 \ \ 0]$ and $\mathbf{e}_3' = [0 \ \ 0 \ \ 1]$. With these assignments the principal components are $Y_1 = X_1$, $Y_2 = X_2$ and $Y_3 = X_3$.

8.4 Eigenvalues of $\boldsymbol{\Sigma}$ are solutions of $|\boldsymbol{\Sigma}-\lambda I| = (\sigma^2-\lambda)^3-2(\sigma^2-\lambda)(\sigma^2\rho)^2 = 0$ Thus $(\sigma^2-\lambda)[(\sigma^2-\lambda)^2-2\sigma^4\rho^2] = 0$ so $\lambda = \sigma^2$ or $\lambda = \sigma^2(1\pm\rho\sqrt{2})$. For $\lambda_1 = \sigma^2$, $\mathbf{e}_1' = [1/\sqrt{2}, 0, -1/\sqrt{2}]$. For $\lambda_2 = \sigma^2(1+\rho\sqrt{2})$; $\mathbf{e}_2' = [1/2, 1/\sqrt{2}, 1/2]$. For $\lambda_3 = \sigma^2(1-\rho\sqrt{2})$, $\mathbf{e}_3' = [1/2, -1/\sqrt{2}, 1/\sqrt{2}]$

Principal Component	Variance	Proportion of Total Variance Explained
$Y_1 = \frac{1}{\sqrt{2}} X_1 - \frac{1}{\sqrt{2}} X_3$	σ^2	1/3
$Y_2 = \frac{1}{2} X_1 + \frac{1}{\sqrt{2}} X_2 + \frac{1}{2} X_3$	$\sigma^2(1+\rho\sqrt{2})$	$\frac{1}{3}(1+\rho\sqrt{2})$
$Y_3 = \frac{1}{2} X_1 - \frac{1}{\sqrt{2}} X_2 + \frac{1}{2} X_3$	$\sigma^2(1-\rho\sqrt{2})$	$\frac{1}{3}(1-\rho\sqrt{2})$

8.5 (a) Eigenvalues of $\underset{\sim}{\rho}$ satisfy

$$|\underset{\sim}{\rho}-\lambda I| = (1-\lambda)^3 + 2\rho^3 - 3(1-\lambda)\rho^2 = 0$$

or $(1+2\rho-\lambda)(1-\rho-\lambda)^2 = 0$. Hence $\lambda_1 = 1+2\rho$; $\lambda_2 = \lambda_3 = 1-\rho$ and results are consistent with (8-16) for $p = 3$.

(b) By direct multiplication

$$\underset{\sim}{\rho}\left(\frac{1}{\sqrt{p}}\underset{\sim}{1}\right) = (1+(p-1)\rho)\left(\frac{1}{\sqrt{p}}\underset{\sim}{1}\right)$$

thus varifying the first eigenvalue-eigenvector pair. Further

$\underset{\sim}{\rho}\,\underset{\sim}{e}_i = (1-\rho)\underset{\sim}{e}_i$, $i = 2,3,\ldots,p$.

8.6

(a) $\hat{y}_1 = 0.999673x_1 + 0.025574x_2$. Sample variance of $\hat{y}_1 = \hat{\lambda}_1 = 1.0012 \times 10^9$.
$\hat{y}_2 = -0.025574x_1 + 0.999673x_2$. Sample variance of $\hat{y}_2 = \hat{\lambda}_2 = 775,734.3$.

(b) Proportion of total sample variance explained by $\hat{y}_1$ is $\hat{\lambda}_1/(\hat{\lambda}_1 + \hat{\lambda}_2) = 0.99923$.

(c)

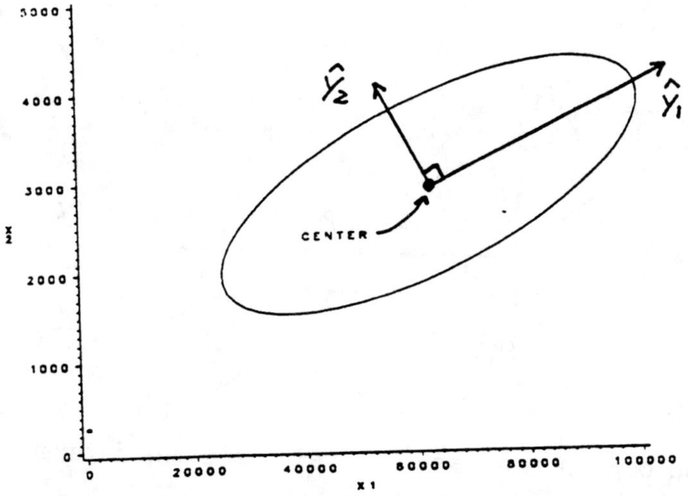

(d) $r_{\hat{y}_1,x_1} = 1.000$ and $r_{\hat{y}_1,x_2} = 0.67668$.
The first component is almost completely determined by x_1 = sales since its variance is approximately 700 times that of x_2 = profits. This is confirmed by the correlation coefficient $r_{\hat{y}_1,x_1}$.

8.7

(a) $\hat{y}_1 = 0.707107z_1 + 0.707107z_2$. Sample variance of $\hat{y}_1 = \hat{\lambda}_1 = 1.67615$.
$\hat{y}_2 = -0.707107z_1 - 0.707107z_2$. Sample variance of $\hat{y}_2 = \hat{\lambda}_2 = 0.32385$.

```
Correlation Matrix
                    SALES        PROFITS
SALES              1.0000         0.6762
PROFITS            0.6762         1.0000
```

(b) Proportion of total sample variance explained by $\hat{y}_1$ is $\hat{\lambda}_1/(\hat{\lambda}_1 + \hat{\lambda}_2) = 0.83808$.

(c) $r_{\hat{y}_1,z_1} = 0.91546$ and $r_{\hat{y}_1,z_2} = 0.91546$.
The standardized "sales" and "profits" contribute equally to the first principal component.

(d) The nature of the principal component is heavily influenced by the relative sizes of the variances of the variables. The correlation coefficients between the components and the variables give some indication of the importance of the variables taking account differences in variances. Standardizing the variables makes the variable units comparable and puts the variables on similar scales.

8.8 (a) $r_{\hat{y}_i z_k} = \hat{e}_{ik}\sqrt{\hat{\lambda}_i}$; $i = 1,2;\ \ k = 1,2,\ldots,5$

i \ k	1	2	3	4	5
1	.784	.773	.794	.712	.712
2	.216	.458	.234	-.473	-.523

The correlations seem to reinforce the interpretations given in Example 8.5.

(b) Using (8-34) and (8-35) we have

k	$\bar{r}_k$
1	.484
2	.472
3	.493
4	.434
5	.433

$\bar{r} = .463$

$\hat{\gamma} = 2.754$

$T = 21.2$ compared to $\chi_9^2(.05) = 16.92$ implies we would reject H_0 at the 5% level. The assumption is that we are dealing with a large random sample from a multivariate normal distribution.

8.9 (a) By (5-10)

$$\max_{\mu, \Sigma} L(\mu, \Sigma) = \frac{e^{-\frac{np}{2}}}{(2\pi)^{\frac{pn}{2}} \left(\frac{n-1}{n}\right)^{\frac{pn}{2}} |S|^{\frac{n}{2}}}$$

The same result applied to each variable independently gives

$$\max_{\mu_i, \sigma_{ii}} L(\mu_i, \sigma_{ii}) = \frac{e^{-\frac{n}{2}}}{(2\pi)^{\frac{n}{2}} \left(\frac{n-1}{n}\right)^{\frac{n}{2}} s_{ii}^{\frac{n}{2}}}$$

Under H_0, $\max_{\mu, \Sigma_0} L(\mu, \Sigma_0) = \prod_{i=1}^{p} L(\mu_i, \sigma_{ii})$

and the likelihood ratio statistic becomes

$$\Lambda = \frac{\max_{\mu, \Sigma_0} L(\mu, \Sigma_0)}{\max_{\mu, \Sigma} L(\mu, \Sigma)} = \frac{|S|^{\frac{n}{2}}}{\prod_{i=1}^{p} s_{ii}^{\frac{n}{2}}}$$

(b) When $\Sigma = \sigma^2 I$, using (4-16) and (4-17) we get

$$\max_{\underset{\sim}{\mu}} L(\underset{\sim}{\mu},\sigma^2 I) = \frac{1}{(2\pi)^{\frac{np}{2}}(\sigma^2)^{\frac{np}{2}}} e^{-\frac{1}{2\sigma^2}\{tr[(n-1)S]\}}$$

so

$$\max_{\underset{\sim}{\mu},\sigma^2} L(\underset{\sim}{\mu},\sigma^2 I) = \frac{(np)^{np/2}\ e^{-np/2}}{(2\pi)^{np/2}(n-1)^{np/2}(tr[S])^{np/2}}$$

$$= \frac{e^{-np/2}}{(2\pi)^{np/2}\ (\frac{n-1}{n})^{np/2}\ (\frac{1}{p}\,tr\,(S))^{np/2}}$$

and the result follows. Under H_0 there are p μ_i's and one variance so the dimension of the parameter space is $\gamma_0 = p + 1$. The unrestricted case has dimension $p + p(p+1)/2$ so the χ^2 has $p(p+1)/2 - 1 = (p+2)(p-1)/2$ d.f.

8.10 (a)

$$S = \begin{bmatrix} .0016 & .0008 & .0008 & .0004 & .0005 \\ & .0012 & .0008 & .0004 & .0003 \\ & & .0016 & .0005 & .0005 \\ & & & .0008 & .0004 \\ \text{(symmetric)} & & & & .0008 \end{bmatrix}$$

Coefficients of Sample Principal Components

Variable \ Component	$\hat{y}_1$	$\hat{y}_2$	$\hat{y}_3$	$\hat{y}_4$	$\hat{y}_5$
x_1 (Allied Ch.)	-.56	.74	.12	.28	.21
x_2 (DuPont)	-.47	-.09	.47	-.69	-.28
x_3 (Un. Carbide)	-.55	-.65	.12	.50	.10
x_4 (Exxon)	-.29	-.11	-.61	-.44	.58
x_5 (Texaco)	-.28	.07	-.62	.06	-.73

(b) $\hat{\lambda}_1 = .00360$; $\hat{\lambda}_2 = .00079$; $\hat{\lambda}_3 = .00074$; $\hat{\lambda}_4 = .00051$; $\hat{\lambda}_5 = .00034$

Total Sample Variance = $\Sigma \hat{\lambda}_i = .00598$

	Component				
	$\hat{y}_1$	$\hat{y}_2$	$\hat{y}_3$	$\hat{y}_4$	$\hat{y}_5$
Sample Variance	.00360	.00079	.00074	.00051	.00034
Percentage of total sample variance explained	60.2%	13.2%	12.4%	8.5%	5.7%
Cumulative percentage of total sample variance explained	60.2%	73.4%	85.8%	94.3%	100%

(c) Using (8-33) Bonferroni 90% simultaneous confidence intervals for λ_1, λ_2 and λ_3 are:

λ_1: (.0028, .0051)

λ_2: (.0006, .0011)

λ_3: (.0006, .0011)

(d) Data can be summarized effectively in 2 or 3 dimensions. In 2 dimensions we can identify a "market component" and an "industry component".

8.11 (a)

$$S = \begin{bmatrix} 4.308 & 1.683 & 1.803 & 2.155 & -2.530 \\ & 1.768 & .588 & .177 & 1.760 \\ & & .801 & 1.065 & -1.580 \\ & & & 1.970 & -3.570 \\ \text{(symmetric)} & & & & 50.400 \end{bmatrix}$$

(b)

$\hat{\lambda}_i$	$\hat{e}_i'$				
50.969	[-.058	.033	-.035	-.076	.994]
6.650	[-.782	-.350	-.327	-.341	-.075]
1.421	[-.023	-.764	.101	.632	.075]
.230	[.542	-.540	.050	-.642	.002]
.014	[-.302	-.010	.938	-.173	.002]

$$\hat{y}_1 = -.058x_1 + .033x_2 - .035x_3 - .076x_4 + .994x_5$$

$$\hat{y}_2 = -.782x_1 - .350x_2 - .327x_3 - .391x_4 - .075x_5$$

(c)

	$\hat{y}_1$	$\hat{y}_2$
	$r_{\hat{y}_1 x_k}$	$r_{\hat{y}_2 x_k}$
x_1	-.20	-.97
x_2	.18	-.68
x_3	-.28	-.94
x_4	-.38	-.72
x_5	.9996	-.03
Sample variance	50.969	6.650
Cumulative percentage of total sample variance explained.	86.0%	97.2%

The first component is dominated by x_5 = median home value ($1,000's). The second component is (essentially) a weighted average of X_1, X_2, X_3 and X_4. Moreover, $\hat{y}_1$ (which is essentially x_5) accounts for a large proportion of the total variability.

These results can be compared to the results in Example 8.3. The scaling of the original variables can greatly affect the nature and interpretation of the principal components.

8.12

$$S = \begin{bmatrix} 2.500 & -2.768 & -.378 & -.464 & -.586 & -2.235 & .171 \\ & 300.516 & 3.914 & -1.395 & 6.779 & 30.779 & .624 \\ & & 1.522 & .673 & 2.316 & 2.822 & .142 \\ & & & 1.182 & 1.089 & -.811 & .177 \\ & & & & 11.364 & 3.133 & 1.045 \\ & & & & & 30.978 & .593 \\ \text{(Symmetric)} & & & & & & .479 \end{bmatrix}$$

$$R = \begin{bmatrix} 1.0 & -.101 & -.194 & -.270 & -.110 & -.254 & .156 \\ & 1.0 & .183 & -.074 & .116 & .319 & .052 \\ & & 1.0 & .502 & .557 & .411 & .166 \\ & & & 1.0 & .297 & -.134 & .235 \\ & & & & 1.0 & .167 & .448 \\ & & & & & 1.0 & .154 \\ \text{(Symmetric)} & & & & & & 1.0 \end{bmatrix}$$

<u>Using S:</u>

$\hat{\lambda}_1 = 304.26;\ \hat{\lambda}_2 = 28.28;\ \hat{\lambda}_3 = 11.46;\ \hat{\lambda}_4 = 2.52;\ \hat{\lambda}_5 = 1.28;$

$\hat{\lambda}_6 = .53;\quad \hat{\lambda}_7 = .21$

The first sample principal component

$$\hat{y}_1 = -.010x_1 + .993x_2 + .014x_3 - .005x_4 + .024x_5 + .112x_6 + .002x_7$$

accounts for 87% of the total sample variance. The first component is essentially "solar radiation". (Note the large sample variance for x_2 in S).

Using R:

$$\hat{\lambda}_1 = 2.34;\ \hat{\lambda}_2 = 1.39;\ \hat{\lambda}_3 = 1.20;\ \hat{\lambda}_4 = .73;\ \hat{\lambda}_5 = .65;$$

$$\hat{\lambda}_6 = .54;\ \hat{\lambda}_7 = .16$$

The first three sample principle components are

$$\hat{y}_1 = .237z_1 -.205z_2 -.551z_3 -.378z_4 -.498z_5 -.324z_6 -.319z_7$$

$$\hat{y}_2 = -.278z_1 +.527z_2 +.007z_3 -.435z_4 -.199z_5 +.567z_6 -.308z_7$$

$$\hat{y}_3 = .644z_1 +.225z_2 -.113z_3 -.407z_4 +.197z_5 +.159z_6 +.541z_7$$

These components account for 70% of the total sample variance.

The first component contrasts "wind" with the remaining variables. It might be some general measure of the pollution level. The second component is largely composed of "solar radiation" and the pollutants "NO" and "O_3". It might represent the effects of solar radiation since solar radiation is involved in the production of NO and O_3 from the other pollutants. The third component is composed largely of "wind" and certain pollutants (e.g. "NO" and "HC"). It might represent a wind transport effect. A "better" interpretation of the components would depend on more extensive subject matter knowledge.

The data can be effectively summarized in three or fewer dimensions. The choice of S or R makes a difference.

8.13

(a) Covariance Matrix

	X1	X2	X3
X1	4.654750889	0.931345370	0.589699088
X2	0.931345370	0.612821160	0.110933412
X3	0.589699088	0.110933412	0.571428861
X4	0.276915309	0.118469052	0.087004959
X5	1.074885659	0.388886434	0.347989910
X6	0.158150852	-0.024851988	0.110131391

	X4	X5	X6
X1	0.276915309	1.074885659	0.158150852
X2	0.118469052	0.388886434	-0.024851988
X3	0.087004959	0.347989910	0.110131391
X4	0.110409072	0.217405649	0.021814433
X5	0.217405649	0.862172372	-0.008817694
X6	0.021814433	-0.008817694	0.861455923

Correlation Matrix

	X1	X2	X3	X4	X5	X6
X1	1.0000	0.5514	0.3616	0.3863	0.5366	0.0790
X2	0.5514	1.0000	0.1875	0.4554	0.5350	-.0342
X3	0.3616	0.1875	1.0000	0.3464	0.4958	0.1570
X4	0.3863	0.4554	0.3464	1.0000	0.7046	0.0707
X5	0.5366	0.5350	0.4958	0.7046	1.0000	-.0102
X6	0.0790	-.0342	0.1570	0.0707	-.0102	1.0000

(b) We will work with **R** since the sample variance of x1 is approximately 40 times larger than that of x4.

Eigenvalues of the Correlation Matrix

	Eigenvalue	Difference	Proportion	Cumulative
PRIN1	2.86431	1.78786	0.477385	0.47738
PRIN2	1.07645	0.29881	0.179408	0.65679
PRIN3	0.77764	0.12733	0.129607	0.78640
PRIN4	0.65031	0.26228	0.108386	0.89479
PRIN5	0.38803	0.14478	0.064672	0.95946
PRIN6	0.24326	.	0.040543	1.00000

Eigenvectors

	PRIN1	PRIN2	PRIN3	PRIN4	PRIN5	PRIN6
X1	0.444858	-.026660	0.339330	-.551149	-.600851	0.146492
X2	0.429300	-.291738	0.498607	-.061367	0.687297	0.076408
X3	0.358773	0.380135	-.628157	-.421060	0.331839	0.211635
X4	0.462854	-.020959	-.124585	0.665604	-.207413	0.532689
X5	0.521276	-.073690	-.203339	0.200526	-.103175	-.794127
X6	0.055877	0.873960	0.429880	0.178715	0.053090	-.116262

(c) It is not possible to summarize the radiotherapy data with a single component. We need the first four components to summarize the data.

(d) Correlations between principal components and $X1 - X6$ are

	PRIN1	PRIN2	PRIN3	PRIN4
X1	0.75289	-0.02766	0.29923	-0.44446
X2	0.72656	-0.30268	0.43969	-0.04949
X3	0.60720	0.39440	-0.55393	-0.33955
X4	0.78335	-0.02175	-0.10986	0.53676
X5	0.88222	-0.07646	-0.17931	0.16171
X6	0.09457	0.90675	0.37909	0.14412

8.14 S is given in Example 5.2.

$$\hat{\lambda}_1 = 200.5, \quad \hat{\lambda}_2 = 4.5, \quad \hat{\lambda}_3 = 1.3$$

The first sample principal component explains a proportion $200.5/(200.5 + 4.5 + 1.3) = .97$ of the total sample variance. Also,

$$\hat{\mathbf{e}}_1' = [-.051, -.998, .029]$$

Hence $\hat{y}_1 = -.051x_1 - .998x_2 + .029x_3$

The first principal component is essentially X_2 = sodium content. (Note the (relatively) large sample variance for sodium in S). A Q-Q plot of the $\hat{y}_1$ values is shown below. These data appear to be approximately normal with no suspect observations.

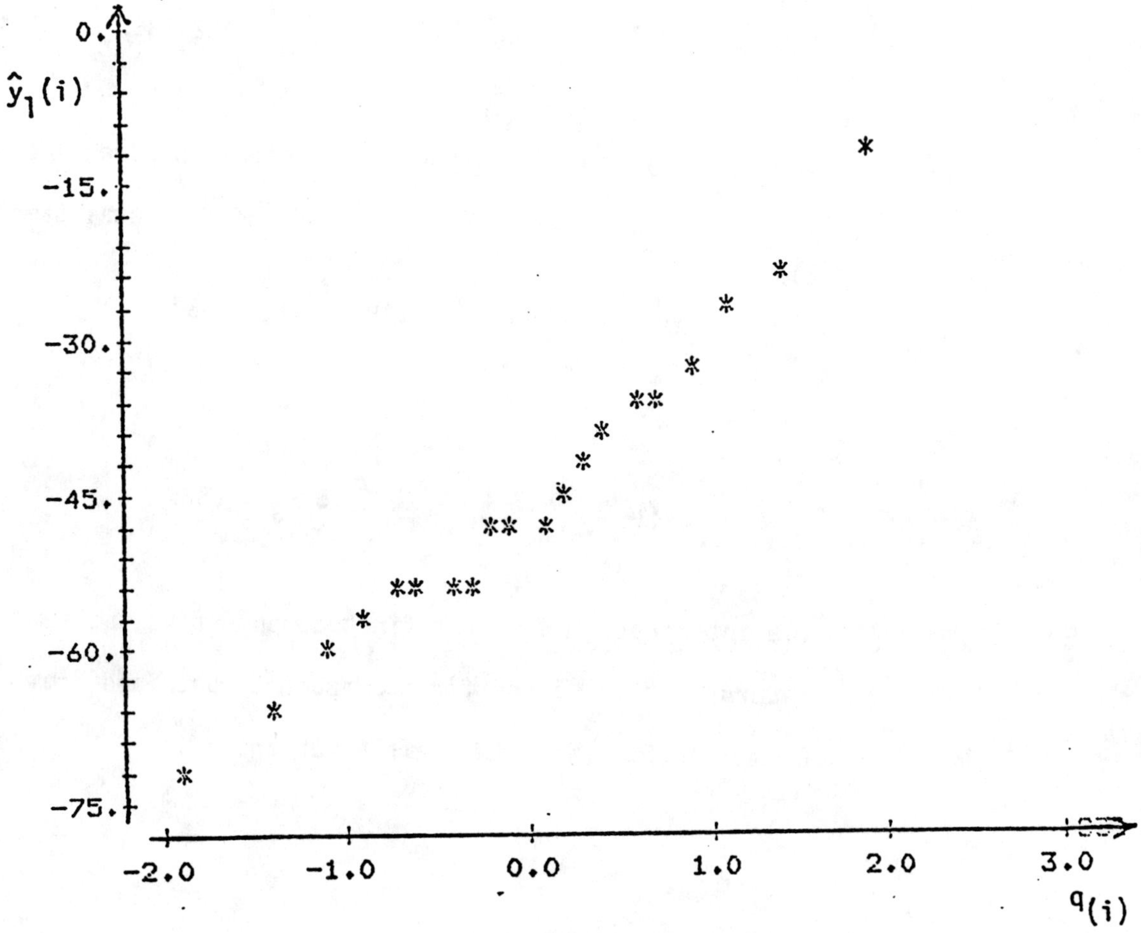

Q-Q plot for $\hat{y}_1$.

8.15

$$S = \begin{bmatrix} 1088.40 & 831.28 & 763.23 & 784.09 \\ & 1128.41 & 850.32 & 926.73 \\ & & 1336.15 & 904.53 \\ \text{(Symmetric)} & & & 1395.15 \end{bmatrix}$$

$$\hat{\lambda}_1 = 3779.01; \quad \hat{\lambda}_2 = 468.25; \quad \hat{\lambda}_3 = 452.13; \quad \hat{\lambda}_4 = 248.72$$

Consequently, the first sample principal component accounts for a proportion $3779.01/4948.11 \doteq .76$ of the total sample variance. Also,

$$\hat{\underset{\sim}{e}}_1' = [.45, \; .49, \; .51, \; .53]$$

Consequently,

$$\hat{y}_1 = .45x_1 + .49x_2 + .51x_3 + .53x_4$$

The interpretation of the first component is the same as the interpretation of the first component, obtained from R, in Example 8.6. (Note the sample variances in S are nearly equal).

8.16. Principal component analysis of Wisconsin fish data

(a) All are positively correlated.

(b) Principal component analysis using $x1 - x4$

```
Eigenvalues of R
2.1539 0.7875 0.6157 0.4429

Eigenvectors of R
0.7032  0.4295  0.1886 -0.7071
0.6722  0.3871 -0.4652  0.4702
0.5914 -0.7126 -0.2787 -0.3216
0.6983 -0.2016  0.4938  0.5318

                     pc1    pc2    pc3    pc4
         St. Dev. 1.4676 0.8874 0.7846 0.6655
     Prop. of Var. 0.5385 0.1969 0.1539 0.1107
Cumulative Prop. 0.5385 0.7354 0.8893 1.0000
```

The first principal component is essentially a total of all four. The second contrasts the Bluegill and Crappie with the two bass.

(c) Principal component analysis using $x1 - x6$

```
Eigenvalues of R
2.3549 1.0719 0.9842 0.6644 0.5004 0.4242

Eigenvectors of R
-0.6716  0.0114  0.5284 -0.0471  0.3765 -0.7293
-0.6668 -0.0100  0.2302 -0.7249 -0.1863  0.5172
-0.5555 -0.2927 -0.2911  0.1810 -0.6284 -0.3081
-0.7013 -0.0403  0.0355  0.6231  0.3407  0.5972
 0.3621 -0.4203  0.0143 -0.2250  0.5074  0.0872
-0.4111  0.0917 -0.8911 -0.2530  0.4021 -0.1731

                     pc1    pc2    pc3    pc4    pc5    pc6
         St. Dev. 1.5346 1.0353 0.9921 0.8151 0.7074 0.6513
     Prop. of Var. 0.3925 0.1786 0.1640 0.1107 0.0834 0.0707
Cumulative Prop. 0.3925 0.5711 0.7352 0.8459 0.9293 1.0000
```

The Walleye is contrasted with all the others in the first principal component (look at the covariance pattern). The second principal component is essentially the Walleye and somewhat the largemouth bass. The third principal component is nearly a contrast between Northern pike and Bluegill.

8.17

COVARIANCE MATRIX
----

x1	.0130016					
x2	.0103784	.0114179				
x3	.0223500	.0185352	.0803572			
x4	.0200857	.0210995	.0667762	.0694845		
x5	.0912071	.0085298	.0168369	.0177355	.0115684	
x6	.0079578	.0089085	.0128470	.0167936	.0080712	.0105991

The eigenvalues are

0.164 0.018 0.008 0.003 0.002 0.001

and the first two principal components are

$$[\ .218\ ,\ .204\ ,\ .673\ ,\ .633\ ,\ .181\ ,\ .159\ ]\ \underset{\sim}{X}$$

$$[\ .337\ ,\ .432\ ,\ -.500\ ,\ .024\ ,\ .430\ ,\ .514\ ]\ \underset{\sim}{X}$$

8.18

(a) Correlation Matrix

	X1	X2	X3	X4	X5	X6	X7
X1	1.0000	0.9528	0.8347	0.7277	0.7284	0.7417	0.6863
X2	0.9528	1.0000	0.8570	0.7241	0.6984	0.7099	0.6856
X3	0.8347	0.8570	1.0000	0.8984	0.7878	0.7776	0.7054
X4	0.7277	0.7241	0.8984	1.0000	0.9016	0.8636	0.7793
X5	0.7284	0.6984	0.7878	0.9016	1.0000	0.9692	0.8779
X6	0.7417	0.7099	0.7776	0.8636	0.9692	1.0000	0.8998
X7	0.6863	0.6856	0.7054	0.7793	0.8779	0.8998	1.0000

Eigenvalues of the Correlation Matrix

	Eigenvalue	Difference	Proportion	Cumulative
PRIN1	5.80569	5.15204	0.829384	0.82938
PRIN2	0.65365	0.35376	0.093378	0.92276
PRIN3	0.29988	0.17440	0.042840	0.96560
PRIN4	0.12548	0.07166	0.017925	0.98353
PRIN5	0.05382	0.01477	0.007688	0.99122
PRIN6	0.03905	0.01661	0.005578	0.99679
PRIN7	0.02244	.	0.003206	1.00000

Eigenvectors

	PRIN1	PRIN2	PRIN3	PRIN4	PRIN5	PRIN6	PRIN7
X1	0.368356	0.490060	0.286012	-.319386	0.231169	0.619825	0.052177
X2	0.365364	0.536580	0.229819	0.083302	0.041455	-.710765	-.109225
X3	0.381610	0.246538	-.515367	0.347377	-.572178	0.190946	0.208497
X4	0.384559	-.155402	-.584526	0.042076	0.620324	-.019089	-.315210
X5	0.389104	-.360409	-.012912	-.429539	0.030261	-.231248	0.692562
X6	0.388866	-.347539	0.152728	-.363120	-.463355	0.009277	-.598359
X7	0.367004	-.369208	0.484370	0.672497	0.130536	0.142281	0.069598

(b) $\hat{y}_1 = 0.3684z_1 + 0.3654z_2 + 0.3816z_3 + 0.3846z_4 + 0.3891z_5 + 0.3889z_6 + 0.3670z_7$.
$\hat{y}_2 = 0.4901z_1 + 0.5366z_2 + 0.2465z_3 - 0.1554z_4 - 0.3604z_5 - 0.3475z_6 + 0.3692z_7$.
The cumulative percentage of total sample variance explained by the two components is given in (a). The correlations of the components with standardized variables are

	PRIN1	PRIN2
PRIN1	1.00000	0.00000
PRIN2	0.00000	1.00000
Z1	0.88755	0.39621
Z2	0.88034	0.43382

Z3	0.91949	0.19932
Z4	0.92660	-0.12564
Z5	0.93755	-0.29138
Z6	0.93697	-0.28098
Z7	0.88430	-0.29850

(c) The first principal component has a strong positive correlation with all standardized variables. That is, small values of the first component are related to small values of standardized variables (athletic excellence). On the other hand, the second principal component has a weak positive correlation with $z_1 - z_3$ (100m, 200m, 400m) and has a weak negative correlation with $z_4 - z_7$ (800m, 1500m, 3000m, Marathon). The small values of the second component are weakly related to athletic excellence in running short distances, and its large values are related to athletic excellence in running long distances.

(d)

RANK	NATION	PRIN1
1	gdr	-3.50602
2	ussr	-3.46469
3	usa	-3.33581
4	czech	-3.05380
5	frg	-2.92578
6	gbni	-2.78316
7	poland	-2.67210
8	canada	-2.60813
9	finland	-2.18184
10	italy	-2.13954
11	australi	-2.09355
12	rumania	-2.02983
13	france	-1.89217
14	sweden	-1.82775
15	netherla	-1.79443
16	nz	-1.51126
17	belgium	-1.50999
18	norway	-1.48301
19	hungary	-1.47721
20	austria	-1.38044
21	switzerl	-1.34665
22	ireland	-1.11735
23	denmark	-1.11638
24	taipei	-0.50012
25	kenya	-0.43089
26	spain	-0.35565
27	portugal	-0.22428
28	israel	-0.14297
29	brazil	-0.11840
30	mexico	-0.06348
31	japan	-0.05923
32	columbia	0.14157
33	bermuda	0.38782
34	dprkorea	0.46230
35	argentin	0.52726
36	chile	0.54783
37	china	0.64127
38	greece	0.81425
39	india	1.01454
40	korea	1.23386
41	luxembou	1.30174
42	turkey	1.60820
43	philippi	1.64019
44	burma	1.68204
45	thailand	1.95318
46	singapor	1.97013
47	indonesi	2.11236
48	domrep	2.29544
49	malaysia	2.34054
50	costa	2.61923
51	guatemal	3.22730
52	png	3.98086
53	mauritiu	4.23385
54	cookis	6.07728
55	wsamoa	8.33288

8.19. Principal component analysis of national track records for women

S

100m	200m	400m	800m	1500m	3000m	Marathon
0.1096	0.1238	0.1039	0.0795	0.0991	0.1032	0.1348
0.1238	0.1533	0.1265	0.0940	0.1137	0.1174	0.1583
0.1039	0.1265	0.1408	0.1112	0.1217	0.1222	0.1518
0.0795	0.0940	0.1112	0.1085	0.1220	0.1199	0.1468
0.0991	0.1137	0.1217	0.1220	0.1625	0.1618	0.1964
0.1032	0.1174	0.1222	0.1199	0.1618	0.1734	0.2097
0.1348	0.1583	0.1518	0.1468	0.1964	0.2097	0.3216

Eigenvalues of S

0.9791 0.0986 0.0531 0.0232 0.0072 0.0052 0.0034

Eigenvectors of S

0.2908	-0.4270	0.2504	0.3291	-0.1558	-0.7279	-0.0895
0.3419	-0.5582	0.3202	0.1320	-0.0997	0.6288	0.2152
0.3386	-0.3818	-0.3208	-0.5370	0.4768	-0.0538	-0.3435
0.3054	-0.0079	-0.4753	-0.3093	-0.5046	-0.1372	0.5583
0.3859	0.1971	-0.3725	0.3622	-0.3645	0.2259	-0.5987
0.3996	0.2540	-0.2146	0.4748	0.5907	-0.0233	0.3935
0.5310	0.5069	0.5668	-0.3655	-0.0440	-0.0396	-0.0527

	pc1	pc2	pc3	pc4	pc5	pc6	pc7
St. Dev.	0.9895	0.3140	0.2303	0.1525	0.0849	0.0722	0.0584
Prop. of Var.	0.8370	0.0843	0.0454	0.0199	0.0062	0.0045	0.0029
Cumulative Prop.	0.8370	0.9212	0.9666	0.9865	0.9926	0.9971	1.0000

The marathon is weighted somewhat higher in the first principal component, and the second principal component is a 'length' variable. This interpretation uses **S** which is a point in its favor.

Score on the first principal component

1	usa	1.4996	21	ireland	0.5871	41	india	-0.5537
2	ussr	1.4550	22	denmark	0.5587	42	thailand	-0.7376
3	gdr	1.3593	23	austria	0.5054	43	burma	-0.7465
4	frg	1.2494	24	portugal	0.2867	44	turkey	-0.7773
5	gbni	1.1897	25	spain	0.1810	45	singapor	-0.8238
6	czech	1.1419	26	japan	0.1776	46	philippi	-0.8455
7	canada	1.1107	27	israel	0.0857	47	malaysia	-0.9535
8	poland	1.0044	28	mexico	0.0746	48	costa	-0.9611
9	italy	0.9387	29	taipei	0.0199	49	indonesi	-0.9920
10	finland	0.8976	30	kenya	0.0170	50	domrep	-1.0594
11	australi	0.8607	31	brazil	-0.0574	51	guatemal	-1.4305
12	norway	0.8248	32	columbia	-0.0743	52	png	-1.7285
13	nz	0.7799	33	bermuda	-0.2201	53	mauritiu	-1.8892
14	rumania	0.7679	34	chile	-0.2331	54	cookis	-2.3059
15	netherla	0.7669	35	china	-0.2436	55	wsamoa	-3.0960
16	sweden	0.7618	36	dprkorea	-0.2609			
17	france	0.7590	37	argentin	-0.3155			
18	switzerl	0.6313	38	korea	-0.4138			
19	hungary	0.6136	39	greece	-0.4599			
20	belgium	0.6126	40	luxembou	-0.5393			

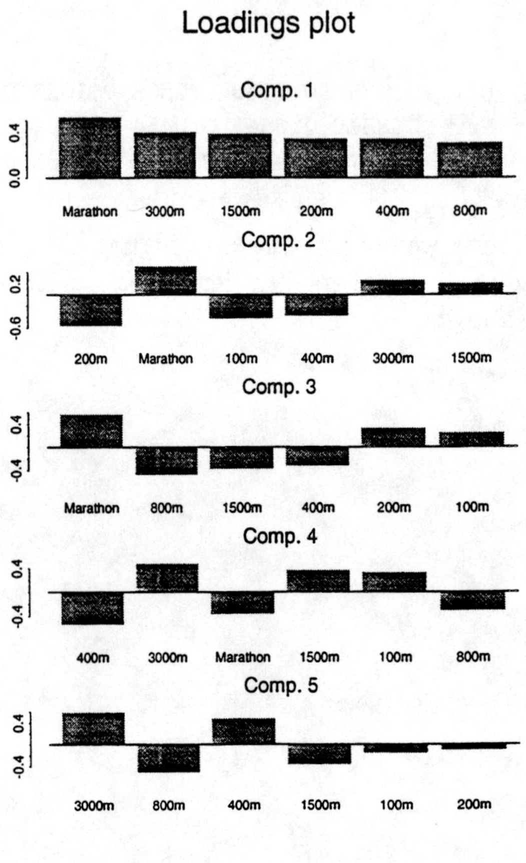

8.20

(a) Correlation Matrix

	X1	X2	X3	X4	X5	X6	X7	X8
X1	1.0000	0.9226	0.8411	0.7560	0.7002	0.6195	0.6325	0.5199
X2	0.9226	1.0000	0.8507	0.8066	0.7750	0.6954	0.6965	0.5962
X3	0.8411	0.8507	1.0000	0.8702	0.8353	0.7786	0.7872	0.7050
X4	0.7560	0.8066	0.8702	1.0000	0.9180	0.8636	0.8690	0.8065
X5	0.7002	0.7750	0.8353	0.9180	1.0000	0.9281	0.9347	0.8655
X6	0.6195	0.6954	0.7786	0.8636	0.9281	1.0000	0.9746	0.9322
X7	0.6325	0.6965	0.7872	0.8690	0.9347	0.9746	1.0000	0.9432
X8	0.5199	0.5962	0.7050	0.8065	0.8655	0.9322	0.9432	1.0000

Eigenvalues of the Correlation Matrix

	Eigenvalue	Difference	Proportion	Cumulative
PRIN1	6.62215	5.74453	0.827768	0.82777
PRIN2	0.87762	0.71830	0.109702	0.93747
PRIN3	0.15932	0.03527	0.019915	0.95739
PRIN4	0.12405	0.04417	0.015506	0.97289
PRIN5	0.07988	0.01192	0.009985	0.98288
PRIN6	0.06797	0.02155	0.008496	0.99137
PRIN7	0.04642	0.02382	0.005802	0.99717
PRIN8	0.02260	.	0.002825	1.00000

Eigenvectors

	PRIN1	PRIN2	PRIN3	PRIN4
X1	0.317556	0.566878	0.332262	0.127628
X2	0.336979	0.461626	0.360657	-.259116
X3	0.355645	0.248273	-.560467	0.652341
X4	0.368684	0.012430	-.532482	-.479999
X5	0.372810	-.139797	-.153443	-.404510
X6	0.364374	-.312030	0.189764	0.029588
X7	0.366773	-.306860	0.181752	0.080069
X8	0.341926	-.438963	0.263209	0.299512

	PRIN5	PRIN6	PRIN7	PRIN8
X1	0.262555	-.593704	0.136241	0.105542
X2	-.153957	0.656137	-.112640	-.096054
X3	-.218323	0.156625	-.002854	-.000127
X4	0.540053	-.014692	-.238016	-.038165
X5	-.487715	-.157843	0.610011	0.139291
X6	-.253979	-.141299	-.591299	0.546697
X7	-.133176	-.219017	-.176871	-.796795
X8	0.497928	0.315285	0.398822	0.158164

(b) $\hat{y}_1 = 0.3176z_1 + 0.3370z_2 + 0.3556z_3 + 0.3687z_4 + 0.3728z_5 + 0.3644z_6 + 0.3668z_7 + 0.3419z_8$.
$\hat{y}_2 = 0.5669z_1 + 0.4616z_2 + 0.2483z_3 + 0.0124z_4 - 0.1398z_5 - 0.3120z_6 + 0.3069z_7 - 0.4390z_8$.
The cumulative percentage of total sample variance explained by the two components is given in (a). This is consistent with the results obtained from the women's data. The correlations of the components with standardized variables are

	PRIN1	PRIN2	Z1	Z2	Z3
PRIN1	1.00000	0.00000	0.81718	0.86717	0.91520
PRIN2	0.00000	1.00000	0.53106	0.43246	0.23259
Z1	0.81718	0.53106	1.00000	0.92264	0.84115
Z2	0.86717	0.43246	0.92264	1.00000	0.85073
Z3	0.91520	0.23259	0.84115	0.85073	1.00000
Z4	0.94875	0.01164	0.75603	0.80663	0.87017
Z5	0.95937	-0.13096	0.70024	0.77495	0.83527
Z6	0.93766	-0.29231	0.61946	0.69538	0.77861
Z7	0.94384	-0.28747	0.63254	0.69654	0.78720
Z8	0.87990	-0.41123	0.51995	0.59618	0.70499

(c) The first principal component has a strong positive correlation with all standardized variables. That is, small values of the first component are related to small values of standardized variables (athletic excellence). On the other hand, the second principal component has a weak positive correlation with $z_1 - z_4$ (100m, 200m, 400m, 800m) and has a weak negative correlation with $z_5 - z_8$ (1500m, 5000m, 10000m, Marathon). The small values of the second component are weakly related to athletic excellence in running short distances, and its large values are related to athletic excellence in running long distances. The results are consistent with those of the women's data except 800m is considered to be a long distance for women and a short distance for men.

(d)

RANK	NATION	PRIN1
1	usa	-3.4306
2	gbni	-3.0242
3	italy	-2.7269
4	ussr	-2.6269
5	gdr	-2.5901
6	frg	-2.5527
7	australi	-2.4464
8	france	-2.1719
9	kenya	-2.1683
10	belgium	-2.0413
11	poland	-2.0006
12	canada	-1.7464
13	finland	-1.6920

15	sweden	-1.6032
16	nz	-1.5997
17	brazil	-1.5583
18	netherla	-1.5554
19	spain	-1.4806
20	czech	-1.3726
21	japan	-1.2379
22	hungary	-1.2052
23	rumania	-1.1965
24	denmark	-1.1132
25	portugal	-0.9164
26	ireland	-0.8842
27	norway	-0.8115
28	austria	-0.8076
29	mexico	-0.6785
30	columbia	-0.3901
31	chile	-0.3811
32	greece	-0.3796
33	india	-0.1652
34	korea	0.2075
35	luxembou	0.2205
36	argentin	0.2619
37	turkey	0.2661
38	china	0.4090
39	israel	0.4346
40	bermuda	0.7393
41	taipei	0.9505
42	dprkorea	1.6837
43	malaysia	1.7083
44	domrep	1.7149
45	burma	1.9719
46	philippi	2.0704
47	costa	2.2966
48	guatemal	2.6724
49	indonesi	2.7478
50	thailand	2.7618
51	singapor	3.1221
52	png	3.9092
53	mauritiu	4.2587
54	wsamoa	7.2312
55	cookis	10.5556

8.21. Principal component analysis of national track records for men

S

100m	200m	400m	800m	1500m	5000m	10,000m	Marathon
0.0904	0.0799	0.0644	0.0565	0.0558	0.0579	0.0605	0.0482
0.0799	0.0827	0.0631	0.0579	0.0597	0.0636	0.0651	0.0540
0.0644	0.0631	0.0670	0.0565	0.0579	0.0643	0.0665	0.0582
0.0565	0.0579	0.0565	0.0627	0.0612	0.0690	0.0708	0.0638
0.0558	0.0597	0.0579	0.0612	0.0722	0.0802	0.0824	0.0739
0.0579	0.0636	0.0643	0.0690	0.0802	0.1042	0.1034	0.0957
0.0605	0.0651	0.0665	0.0708	0.0824	0.1034	0.1086	0.0992
0.0482	0.0540	0.0582	0.0638	0.0739	0.0957	0.0992	0.1021

Eigenvalues of S

0.5655 0.0834 0.0124 0.0093 0.0068 0.0059 0.0042 0.0026

Eigenvectors of S

0.3152	-0.6027	0.3456	-0.1821	-0.0448	0.5963	-0.0957	-0.1343
0.3248	-0.4700	0.2923	0.2411	-0.0888	-0.7004	0.1177	0.1327
0.3095	-0.2308	-0.5885	-0.5254	0.4379	-0.1901	-0.0236	-0.0137
0.3121	-0.0560	-0.5376	0.2076	-0.6179	0.1897	0.3733	0.1015
0.3428	0.0790	-0.2204	0.4849	0.0428	0.0101	-0.7367	-0.2190
0.4063	0.2955	0.1398	0.2356	0.3809	0.0697	0.5095	-0.5128
0.4178	0.2965	0.1606	0.0527	0.2493	0.1959	-0.0018	0.7800
0.3805	0.4218	0.2566	-0.5443	-0.4556	-0.1957	-0.1866	-0.1859

	pc1	pc2	pc3	pc4	pc5	pc6	pc7	pc8
St. Dev.	0.7520	0.2888	0.1112	0.0967	0.0824	0.0765	0.0645	0.0509
Prop. of Var.	0.8195	0.1209	0.0179	0.0135	0.0098	0.0085	0.0060	0.0038
Cumulative Prop.	0.8195	0.9404	0.9583	0.9719	0.9817	0.9902	0.9962	1.0000

Again, the first principal component is an overall total and the second is a 'length' variable. This analysis is based on **S** which is a point in its favor.

Score on the first principal component

1	usa	1.0488	21	czech	0.3863	41	taipei	-0.3466
2	gbni	0.9218	22	rumania	0.3724	42	dprkorea	-0.4833
3	italy	0.8342	23	hungary	0.3468	43	domrep	-0.6103
4	ussr	0.8022	24	portugal	0.3460	44	malaysia	-0.6147
5	gdr	0.7908	25	denmark	0.3305	45	costa	-0.6228
6	frg	0.7544	26	ireland	0.2827	46	burma	-0.6241
7	australi	0.7404	27	norway	0.2670	47	philippi	-0.7095
8	kenya	0.6624	28	austria	0.2638	48	guatemal	-0.7740
9	france	0.6346	29	mexico	0.2370	49	indonesi	-0.8658
10	belgium	0.6203	30	columbia	0.1552	50	thailand	-0.9105
11	poland	0.6032	31	chile	0.0857	51	singapor	-0.9831
12	canada	0.5202	32	greece	0.0673	52	png	-1.1646
13	finland	0.5143	33	india	0.0178	53	mauritiu	-1.3063
14	switzerl	0.5054	34	turkey	-0.0576	54	wsamoa	-2.0289
15	nz	0.4988	35	luxembou	-0.0835	55	cookis	-2.8330
16	sweden	0.4926	36	korea	-0.0937			
17	netherla	0.4703	37	argentin	-0.1211			
18	spain	0.4513	38	china	-0.1302			
19	brazil	0.4171	39	israel	-0.1488			
20	japan	0.4049	40	bermuda	-0.3337			

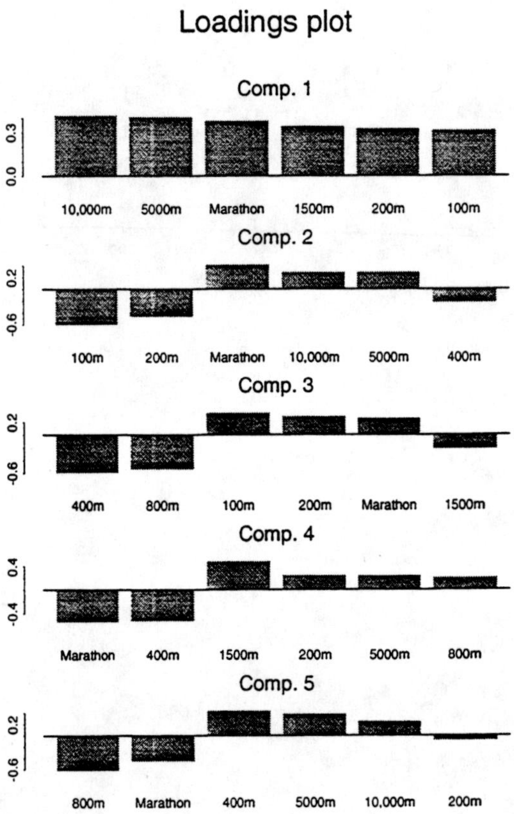

8.22 Using S

Eigenvalues of the Covariance Matrix

	Eigenvalue	Difference	Proportion	Cumulative
PRIN1	20579.6	15704.9	0.808198	0.80820
PRIN2	4874.7	4869.2	0.191437	0.99964
PRIN3	5.4	2.1	0.000213	0.99985
PRIN4	3.3	2.8	0.000130	0.99998
PRIN5	0.5	0.4	0.000018	1.00000
PRIN6	0.1	0.1	0.000003	1.00000
PRIN7	0.0	.	0.000000	1.00000

Eigenvectors

	PRIN1	PRIN2	PRIN3	PRIN4	PRIN5	PRIN6	PRIN7	
X3	0.005887	0.009680	0.286337	0.608787	0.535569	-.509727	0.024592	yrhgt
X4	0.487047	0.872697	-.034277	-.003227	0.000444	-.000457	-.000253	ftfrbody
X5	0.008526	0.029196	0.904389	-.425175	0.008388	0.010389	0.014293	prctffb
X6	0.003112	0.004886	0.133267	0.311194	0.390573	0.855204	-.037984	frame
X7	0.000069	-.000493	-.018864	-.005278	0.011906	0.043786	0.998778	bkfat
X8	0.009330	0.008577	0.284215	0.593037	-.748598	0.082331	0.013820	saleht
X9	0.873259	-.487193	0.004847	-.005597	0.002665	-.000341	-.000256	salewt

Plot of Y1*Y2. Symbol is value of X1.
(NOTE: 10 obs hidden.)

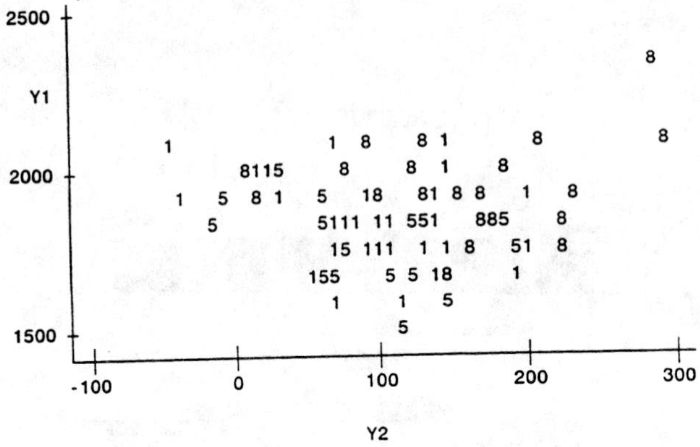

8.22 Using R

Eigenvalues of the Correlation Matrix

	Eigenvalue	Difference	Proportion	Cumulative
PRIN1	4.12070	2.78357	0.588671	0.58867
PRIN2	1.33713	0.59575	0.191018	0.77969
PRIN3	0.74138	0.31996	0.105912	0.88560
PRIN4	0.42143	0.23562	0.060204	0.94580
PRIN5	0.18581	0.03930	0.026544	0.97235
PRIN6	0.14650	0.09945	0.020929	0.99328
PRIN7	0.04706	.	0.006722	1.00000

Eigenvectors

	PRIN1	PRIN2	PRIN3	PRIN4	PRIN5	PRIN6	PRIN7	
X3	0.449931	-.042790	-.415709	0.113356	0.065871	-.072234	0.774926	yrhgt
X4	0.412326	0.129837	0.450292	0.247479	-.719343	-.177061	0.017768	ftfrbody
X5	0.355562	-.315508	0.568273	0.314787	0.579367	0.127800	-.002397	prctffb
X6	0.433957	0.007728	-.452345	0.242818	0.142995	-.434144	-.582337	frame
X7	-.186705	0.714719	-.038732	0.618117	0.160238	0.208017	0.042442	bkfat
X8	0.452854	0.101315	-.176650	-.215769	-.109535	0.799288	-.236723	saleht
X9	0.269947	0.600515	0.253312	-.582433	0.290547	-.276561	0.047036	salewt

Plot of Y1*Y2. Symbol is value of X1.
(NOTE: 27 obs hidden.)

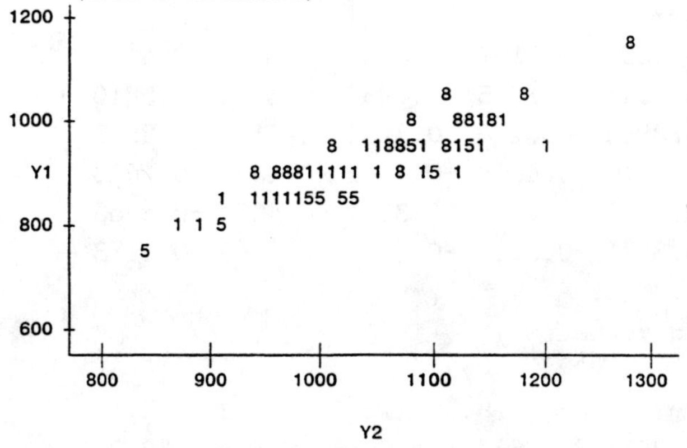

Plot of Y1*Q2. Symbol used is '*'.
(NOTE: 36 obs hidden.) FOR S

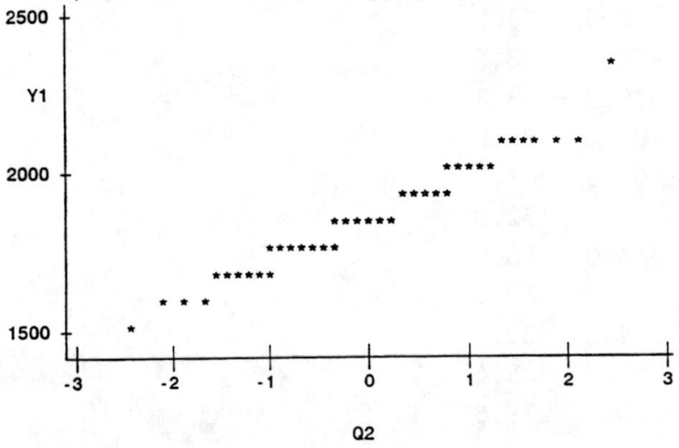

Plot of Y1*Q2. Symbol used is '*'.
(NOTE: 38 obs hidden.) FOR R

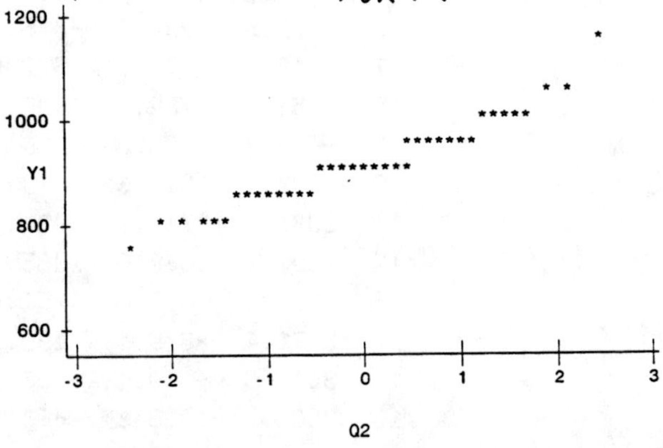

8.23. An ellipse format chart based on the first two principal components of the Madison, Wisconsin, Police Department data

XBAR

3557.8 1478.4 2676.9 13563.6 800 7141

S

367884.7	-72093.8	85714.8	222491.4	-44908.3	101312.9
-72093.8	1399053.1	43399.9	139692.2	110517.1	1161018.3
85714.8	43399.9	1458543.0	-1113809.8	330923.8	1079573.3
222491.4	139692.2	-1113809.8	1698324.4	-244785.9	-462615.6
-44908.3	110517.1	330923.8	-244785.9	224718.0	427767.5
101312.9	1161018.3	1079573.3	-462615.6	427767.5	2488728.4

Eigenvalues of S

4045921.9 2265078.9 761592.1 288919.3 181437.0 94302.6

Eigenvectors of S

-0.0008	-0.0567	-0.5157	0.6122	0.4311	-0.4126
-0.3092	-0.5541	0.5615	0.4932	-0.1796	-0.0810
-0.4821	0.3862	-0.3270	0.3404	-0.5696	0.2667
0.3675	-0.6415	-0.4898	-0.0642	-0.4308	0.1543
-0.1544	0.0359	-0.0316	-0.3071	-0.4062	-0.8453
-0.7163	-0.3575	-0.2662	-0.4094	0.3269	0.1173

Principal components

	y1	y2	y3	y4	y5	y6
1	1745.4	-1479.3	618.7	222.6	7.2	178.1
2	-1096.6	2011.8	652.5	-69.5	636.9	560.2
3	210.6	490.6	365.8	-899.8	-293.5	-15.2
4	-1360.1	1448.1	420.1	523.5	-972.2	88.5
5	-1255.9	502.1	-422.4	-893.8	359.9	-273.7
6	971.6	284.7	-316.9	-942.8	-83.5	-70.1
7	1118.5	123.7	572.9	319.9	-60.8	-598.5
8	-1151.6	1752.0	-1322.1	700.2	-242.2	-158.8
9	-497.3	-593.0	209.5	-149.2	101.6	-586.2
10	-2397.1	1819.6	-9.5	-147.6	-109.9	207.8
11	-3931.9	-3715.7	924.1	35.1	-274.2	152.9
12	-1392.4	-1688.0	-2285.1	372.1	444.0	85.2
13	326.8	650.8	1251.6	728.8	809.5	-140.0
14	3371.4	-379.1	-499.9	-114.6	-324.3	286.9
15	3076.6	-199.1	-105.7	419.8	-122.3	3.4
16	2261.9	-1029.3	-53.7	-104.5	123.8	279.6

$$2.5\times 10^{-7}y_1^2 + 4.4\times 10^{-7}y_2^2 = 5.99$$

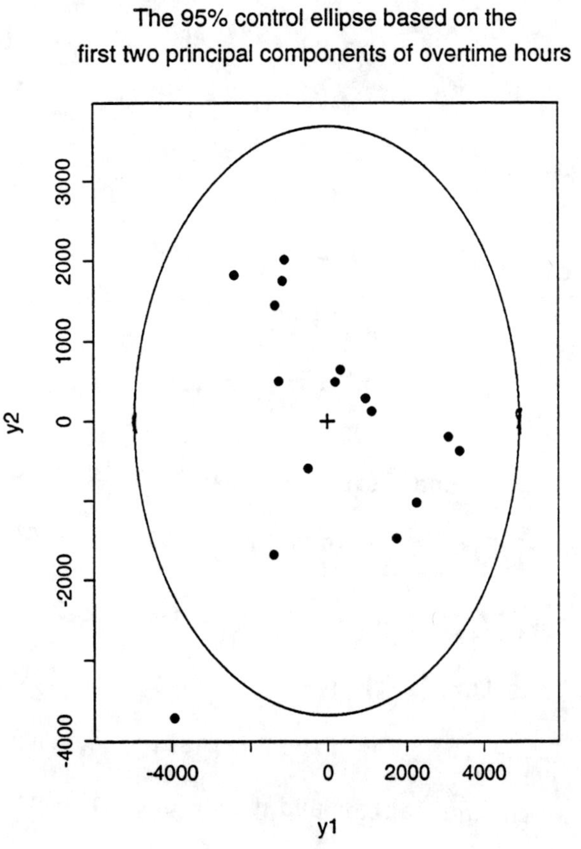

8.24. A control chart based on the sum of squares d_{Uj}^2. Period 12 looks unusual.

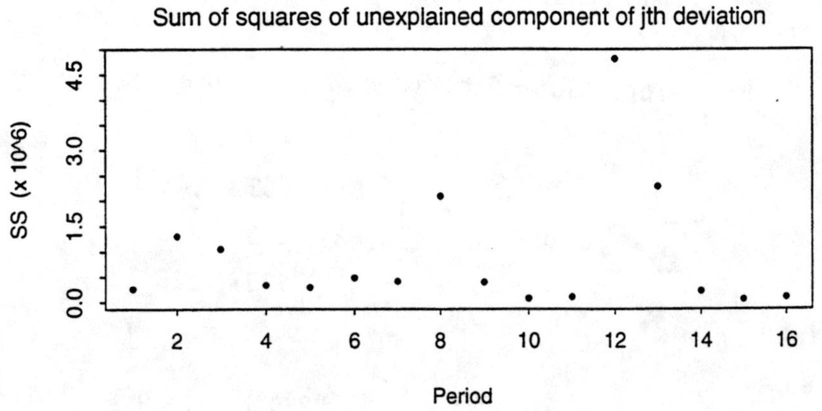

Chapter 9

9.1

$$L' = [.9 \quad .7 \quad .5]; \qquad LL' = \begin{bmatrix} .81 & .63 & .45 \\ .63 & .49 & .35 \\ .45 & .35 & .25 \end{bmatrix}$$

so $\underset{\sim}{\rho} = LL' + \Psi$

9.2 a) For $m = 1$

$$h_1^2 = \ell_{11}^2 = .81$$
$$h_2^2 = \ell_{21}^2 = .49$$
$$h_3^2 = \ell_{31}^2 = .25$$

The communalities are those parts of the variances of the variables explained by the single factor.

b) $\text{Corr}(Z_i, F_1) = \text{Cov}(Z_i, F_1)$, $i = 1,2,3$. By (9-5) $\text{Cov}(Z_i, F_1) = \ell_{i1}$. Thus $\text{Corr}(Z_1, F_1) = \ell_{11} = .9$; $\text{Corr}(Z_2, F_1) = \ell_{21} = .7$; $\text{Corr}(Z_3, F_1) = \ell_{31} = .5$. The first variable, Z_1, has the largest correlation with the factor and therefore will probably carry the most weight in naming the factor.

9.3 a) $L = \sqrt{\lambda_1}\, \underset{\sim}{e}_1 = \sqrt{1.96} \begin{bmatrix} .625 \\ .593 \\ .507 \end{bmatrix} = \begin{bmatrix} .876 \\ .831 \\ .711 \end{bmatrix}$. Slightly different from result in Exercise 9.1.

b) Proportion of total variance explained $= \dfrac{\lambda_1}{p} = \dfrac{1.96}{3} = .65$

9.4

$$\tilde{\underset{\sim}{\rho}} = \underset{\sim}{\rho} - \Psi = LL' = \begin{bmatrix} .81 & .63 & .45 \\ .63 & .49 & .35 \\ .45 & .35 & .25 \end{bmatrix}$$

$$L = \sqrt{\lambda_1}\, \underset{\sim}{e}_1 = \sqrt{1.55} \begin{bmatrix} .7229 \\ .5623 \\ .4016 \end{bmatrix} = \begin{bmatrix} .9 \\ .7 \\ .5 \end{bmatrix}$$

Result is consistent with results in Exercise 9.1. It should be since $m = 1$ common factor completely determines $\tilde{\boldsymbol{\rho}} = \boldsymbol{\rho} - \Psi$.

9.5

Since $\tilde{\Psi}$ is diagonal and $S - \tilde{L}\tilde{L}' - \tilde{\Psi}$ has zeros on the diagonal, (sum of squared entries of $S - \tilde{L}\tilde{L}' - \tilde{\Psi}$) $\le$ (sum of squared entries of $S - \tilde{L}\tilde{L}$). By the hint, $S - \tilde{L}\tilde{L} = \hat{P}_{(2)}\hat{\Lambda}_{(2)}\hat{P}'_{(3)}$ which has sum of squared entries

$$\operatorname{tr}[\hat{P}_{(2)}\hat{\Lambda}_{(2)}\hat{P}'_{(2)}(\hat{P}_{(2)}\hat{\Lambda}_{(2)}\hat{P}'_{(2)})'] = \operatorname{tr}[\hat{P}_{(2)}\hat{\Lambda}_{(2)}\hat{\Lambda}'_{(2)}\hat{P}'_{(2)}]$$

$$= \operatorname{tr}[\hat{\Lambda}_{(2)}\hat{\Lambda}'_{(2)}\hat{P}'_{(2)}\hat{P}_{(2)}] = \operatorname{tr}[\hat{\Lambda}_{(2)}\hat{\Lambda}'_{(2)}]$$

$$= \hat{\lambda}^2_{m+1} + \hat{\lambda}^2_{m+2} + \cdots + \hat{\lambda}^2_p$$

Therefore,

$$(\text{sum of squared entries of } S - \tilde{L}\tilde{L}' - \tilde{\Psi}) \le \hat{\lambda}^2_{m+1} + \hat{\lambda}^2_{m+2} + \cdots + \hat{\lambda}^2_p$$

9.6

a) Follows directly from hint.

b) Using the hint, we post multiply by $(LL' + \Psi)$ to get

$$I = (\Psi^{-1} - \Psi^{-1}L(I + L'\Psi^{-1}L)^{-1}L'\Psi^{-1})(LL' + \Psi)$$

$$= \Psi^{-1}(LL' + \Psi) - \Psi^{-1}\underbrace{L(I + L'\Psi^{-1}L)^{-1}L'\Psi^{-1}(LL'}_{\text{(use part (a))}} + \Psi)$$

$$= \Psi^{-1}(LL' + \Psi) - \Psi^{-1}L(I - (I + L'\Psi^{-1}L)^{-1})L'$$

$$- \Psi^{-1}L(I + L'\Psi^{-1}L)^{-1}L'$$

$$= \Psi^{-1}LL' + I - \Psi^{-1}LL' + \Psi^{-1}L(I + L'\Psi^{-1}L)^{-1}L'$$

$$- \Psi^{-1}L(I + L'\Psi^{-1}L)^{-1}L' = I$$

Note all these multiplication steps are reversible.

c) Multiplying the result in (b) by L we get

$$(LL' + \Psi)^{-1}L = \Psi^{-1}L - \Psi^{-1}\underbrace{L(I + L'\Psi^{-1}L)^{-1}L'\Psi^{-1}L}_{\text{(use part (a))}}$$

$$= \Psi^{-1}L - \Psi^{-1}L(I - (I + L'\Psi^{-1}L)^{-1}) = \Psi^{-1}L(I + L'\Psi^{-1}L)^{-1}$$

Result follows by taking the transpose of both sides of the final equality.

9.7 From the equation $\Sigma = LL' + \Psi$, $m = 1$, we have

$$\begin{bmatrix} \sigma_{11} & \sigma_{12} \\ \sigma_{12} & \sigma_{22} \end{bmatrix} = \begin{bmatrix} \ell_{11}^2 + \psi_1 & \ell_{11}\ell_{21} \\ \ell_{11}\ell_{21} & \ell_{21}^2 + \psi_2 \end{bmatrix}$$

so $\sigma_{11} = \ell_{11}^2 + \psi_1$, $\sigma_{22} = \ell_{21}^2 + \psi_2$ and $\sigma_{12} = \ell_{11}\ell_{21}$. Let $\rho = \sigma_{12}/\sqrt{\sigma_{11}}\sqrt{\sigma_{22}}$. Then, for any choice $|\rho|\sqrt{\sigma_{22}} \le \ell_{21} \le \sqrt{\sigma_{22}}$, set $\ell_{11} = \sigma_{12}/\ell_{21}$ and check $\sigma_{12} = \ell_{11}\ell_{21}$. We obtain $\psi_1 = \sigma_{11} - \ell_{11}^2 = \sigma_{11} - \frac{\sigma_{12}^2}{\ell_{21}^2} \ge \sigma_{11} - \frac{\sigma_{12}^2}{\rho^2\sigma_{22}} = \sigma_{11} - \sigma_{11} = 0$ and $\psi_2 = \sigma_{22} - \ell_{21}^2 \ge \sigma_{22} - \sigma_{22} = 0$. Since ℓ_{21} was arbitrary within a suitable interval, there are an infinite number of solutions to the factorization.

9.8 $\Sigma = LL' + \Psi$ for $m = 1$ implies

$$\begin{pmatrix} 1 = \ell_{11}^2 + \psi_1 & .4 = \ell_{11}\ell_{21} & .9 = \ell_{11}\ell_{31} \\ & 1 = \ell_{21}^2 + \psi_2 & .7 = \ell_{21}\ell_{31} \\ & & 1 = \ell_{31}^2 + \psi_3 \end{pmatrix}$$

Now $\frac{\ell_{11}}{\ell_{21}} = \frac{.9}{.7}$ and $\ell_{11}\ell_{21} = .4$, so $\ell_{11}^2 = (\frac{.9}{.7})(.4)$ and $\ell_{11} = \pm .717$. Thus $\ell_{21} = \pm .558$. Finally, from $.9 = \ell_{11}\ell_{31}$, we have $\ell_{31} = \pm .9/.717 = \pm 1.255$.

Note all the loadings must be of the same sign because all the covariances are positive. We have

$$LL' = \begin{bmatrix} .717 \\ .558 \\ 1.255 \end{bmatrix} [.717 \ .558 \ 1.255] = \begin{bmatrix} .514 & .4 & .9 \\ .4 & .3111 & .7 \\ .9 & .7 & 1.575 \end{bmatrix}$$

so $\psi_3 = 1 - 1.575 = -.575$, which is inadmissible as a variance.

9.9

(a) Stoetzel's interpretation seems reasonable. The first factor seems to contrast sweet with strong liquors.

(b)

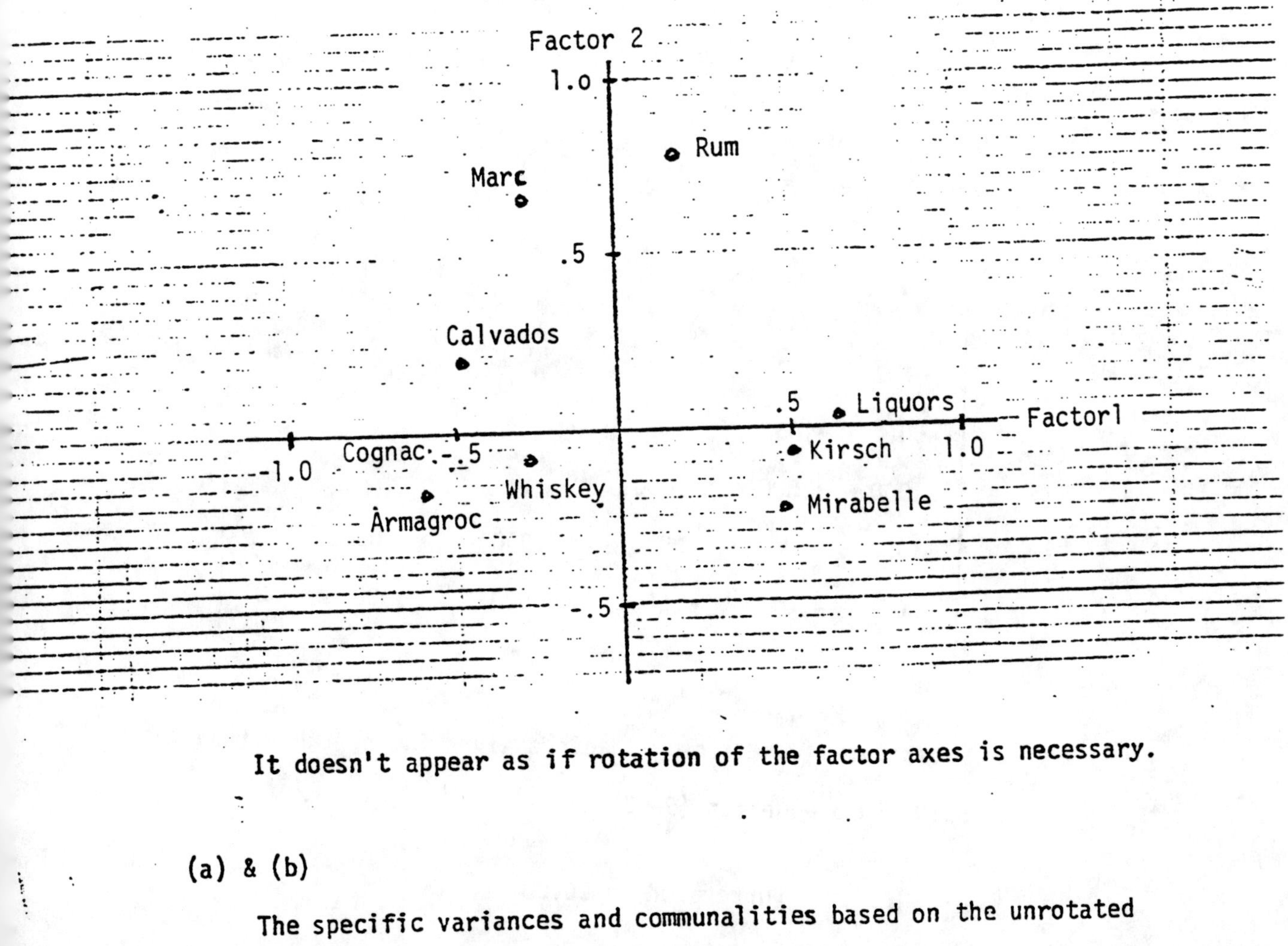

It doesn't appear as if rotation of the factor axes is necessary.

(a) & (b)

The specific variances and communalities based on the unrotated factors, are given in the following table:

Variable	Specific Variance	Communality
Skull length	.5976	.4024
Skull breadth	.7582	.2418
Femur length	.1221	.8779
Tibia length	.0000	1.0000
Humerus length	.0095	.9905
Ulna length	.0938	.9062

(c) The proportion of variance explained by each factor is:

Factor 1 : $\frac{1}{6} \sum_{i=1}^{p} \ell_{1i}^2 = \frac{4.0001}{6}$ or 66.7%

Factor 2 : $\frac{1}{6} \sum_{i=1}^{p} \ell_{2i}^2 = \frac{.4177}{6}$ or 6.7%

(d) $R - \hat{L}_z \hat{L}_z' - \hat{\Psi} =$

$$\begin{bmatrix} 0 & & & & & \\ .193 & 0 & & & & \\ -.017 & -.032 & 0 & & & \\ .000 & .000 & .000 & 0 & & \\ -.000 & .001 & .000 & .000 & 0 & \\ -.001 & -.018 & .003 & .000 & .000 & 0 \end{bmatrix}$$

9.11 Substituting the factor loadings given in the table (Exercise 9.10) into equation (9-45) gives.

$$V \text{ (unrotated)} = .01087$$
$$V \text{ (rotated)} = .04692$$

Although the rotated loadings are to be preferred by the varimax ("simple structure") criterion, interpretation of the factors seems clearer with the unrotated loadings.

9.12

The covariance matrix for the logarithms of turtle measurements is:

$$\mathbf{S} = 10^{-3} \times \begin{bmatrix} 11.0720040 & 8.0191419 & 8.1596480 \\ 8.0191419 & 6.4167255 & 6.0052707 \\ 8.1596480 & 6.0052707 & 6.7727585 \end{bmatrix}$$

The maximum likelihood estimates of the factor loadings for an m=1 model are

Variable	Estimated factor loadings F_1
1. ln(length)	0.1021632
2. ln(width)	0.0752017
3. ln(height)	0.0765267

Therefore,

$$\hat{L} = \begin{bmatrix} 0.1021632 \\ 0.0752017 \\ 0.0765267 \end{bmatrix}, \quad \hat{L}\hat{L}' = 10^{-3} \times \begin{bmatrix} 10.4373 & 7.6828 & 7.8182 \\ 7.6828 & 5.6553 & 5.7549 \\ 7.8182 & 5.7549 & 5.8563 \end{bmatrix}$$

(b) Since $\hat{h}_i^2 = \hat{l}_{i1}^2$ for an m=1 model, the communalities are

$$\hat{h}_1^2 = 0.0104373, \quad \hat{h}_2^2 = 0.0056553, \quad \hat{h}_3^2 = 0.0058563$$

(a) To find specific variances ψ_i's, we use the equation

$$\hat{\psi}_i = s_{ii} - \hat{h}_i^2$$

Note that in this case, we should use S_n to get s_{ii}, not S because the maximum likelihood estimation method is used.

$$S_n = \frac{n-1}{n} S = \frac{23}{24} S = 10^{-3} \times \begin{bmatrix} 10.6107 & 7.685 & 7.8197 \\ 7.685 & 6.1494 & 5.7551 \\ 7.8197 & 5.7551 & 6.4906 \end{bmatrix}$$

Thus we get

$$\hat{\psi}_1 = 0.0001734, \quad \hat{\psi}_2 = 0.0004941, \quad \hat{\psi}_3 = 0.0006342$$

(c) The proportion explained by the factor is

$$\frac{\hat{h}_1^2 + \hat{h}_2^2 + \hat{h}_3^2}{s_{11} + s_{22} + s_{33}} = \frac{0.0219489}{0.0232507} = .9440$$

(d) From (a)-(c), the residual matrix is:

$$S_n - \hat{L}\hat{L}' - \hat{\Psi} = 10^{-6} \times \begin{bmatrix} 0 & 2.1673 & 1.4474 \\ 2.1673 & 0 & 0.112497 \\ 1.4474 & 0.112497 & 0 \end{bmatrix}.$$

9.13

Equation (9-40) requires $m < \frac{1}{2}(2p+1 - \sqrt{8p+1})$. Here we have $m = 1$, $p = 3$ and the strict inequality does not hold.

9.14 Since

$$\hat{\Psi}^{\frac{1}{2}}\hat{\Psi}^{-1}\hat{\Psi}^{\frac{1}{2}} = I, \quad \hat{\Delta}^{\frac{1}{2}}\hat{\Delta}^{\frac{1}{2}} = \hat{\Delta} \text{ and } \hat{E}'\hat{E} = I,$$

$$\hat{L}'\hat{\Psi}^{-1}\hat{L} = \hat{\Delta}^{\frac{1}{2}}\hat{E}'\hat{\Psi}^{\frac{1}{2}}\hat{\Psi}^{-1}\hat{\Psi}^{\frac{1}{2}}\hat{E}\hat{\Delta}^{\frac{1}{2}} = \hat{\Delta}^{\frac{1}{2}}\hat{E}'\hat{E}\hat{\Delta}^{\frac{1}{2}} = \hat{\Delta}^{\frac{1}{2}}\hat{\Delta}^{\frac{1}{2}} = \hat{\Delta}.$$

9.15

(a)

variable	variance	communality
HRA	0.188966	0.811034
HRE	0.133955	0.866045
HRS	0.068971	0.931029
RRA	0.100611	0.899389
RRE	0.079682	0.920318
RRS	0.096522	0.903478
Q	0.02678	0.97322
REV	0.039634	0.960366

(b) **Residual Matrix**

0	0.021205	0.014563	-0.022111	-0.093691	-0.078402	-0.02145	-0.015523
0.021205	0	0.063146	-0.107308	-0.058312	-0.052289	-0.005516	0.035712
0.014563	0.063146	0	-0.065101	-0.009639	-0.070351	0.005454	0.013953
-0.022111	-0.107308	-0.065101	0	0.036263	0.058415	0.00695	-0.033857
-0.093691	-0.058312	-0.009639	0.036263	0	0.032645	0.008854	0.00065
-0.078402	-0.052289	-0.070351	0.058415	0.032645	0	0.002626	-0.004011
-0.02145	-0.005516	0.005454	0.00695	0.008854	0.002626	0	-0.02449
-0.015523	0.035712	0.013953	-0.033857	0.00065	-0.004011	-0.02449	0

The m=3 factor model appears appropriate.

(c) The first factor is related to market-value measures (Q, REV). The second factor is related to accounting historical measures on equity (HRE, RRE). The third factor is related to accounting historical measures on sales (HRS, RRS). Accounting historical measures on assets (HRA,RRA) are weakly related to all factors. Therefore, market-value measures provide evidence of profitability distinct from that provided by the accounting measures. However, we cannot separate accounting historical measures of profitability from accounting replacement measures.

PROBLEM 9.15

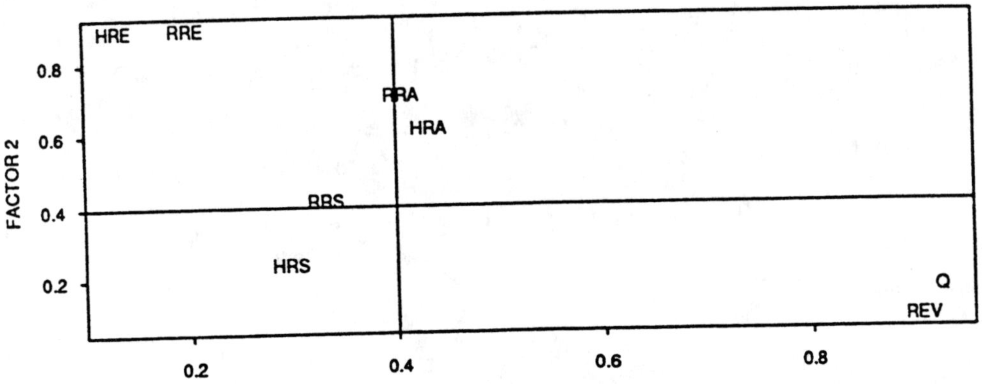

FACTOR 1
Rotated Factor Pattern

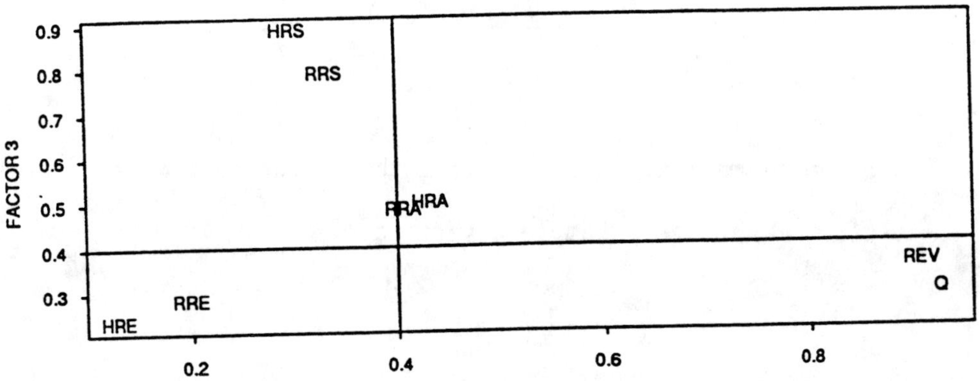

FACTOR 1
Rotated Factor Pattern

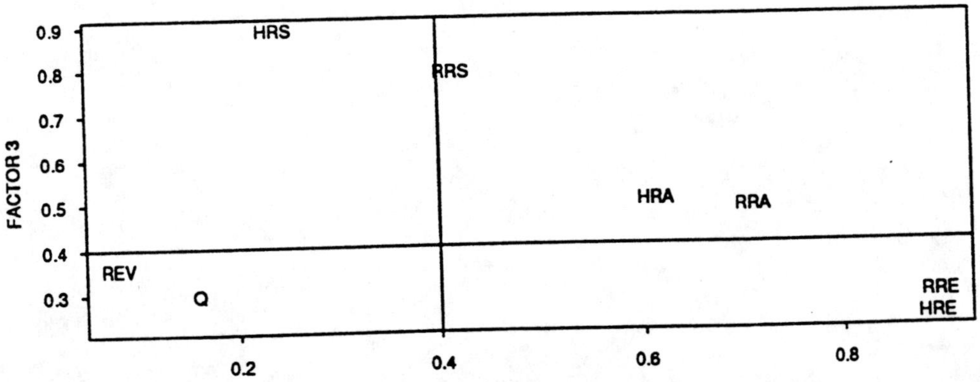

FACTOR 2
Rotated Factor Pattern

9.16 From (9-50) $\hat{\underset{\sim}{f}}_j = \hat{\Delta}^{-1}\hat{L}'\hat{\Psi}^{-1}(\underset{\sim}{x}_j - \bar{\underset{\sim}{x}})$ and

$$\sum_{j=1}^{n} \underset{\sim}{f}_j = \hat{\Delta}^{-1}\hat{L}'\hat{\Psi}^{-1} \sum_{j=1}^{n} (\underset{\sim}{x}_j - \bar{\underset{\sim}{x}}) = \underset{\sim}{0} .$$

Since $\hat{\underset{\sim}{f}}_j\hat{\underset{\sim}{f}}'_j = \hat{\Delta}^{-1}\hat{L}'\hat{\Psi}^{-1}(\underset{\sim}{x}_j - \bar{\underset{\sim}{x}})(\underset{\sim}{x}_j - \bar{\underset{\sim}{x}})'\hat{\Psi}^{-1}\hat{L}\hat{\Delta}^{-1}$,

$$\sum_{j=1}^{n} \hat{\underset{\sim}{f}}_j\hat{\underset{\sim}{f}}'_j = \hat{\Delta}^{-1}\hat{L}'\hat{\Psi}^{-1} \sum_{j=1}^{n} (\underset{\sim}{x}_j - \bar{\underset{\sim}{x}})(\underset{\sim}{x}_j - \bar{\underset{\sim}{x}})'\hat{\Psi}^{-1}\hat{L}\hat{\Delta}^{-1}$$

$$= n\, \hat{\Delta}^{-1}\hat{L}'\hat{\Psi}^{-1}S_n\hat{\Psi}^{-1}\hat{L}\hat{\Delta}^{-1} .$$

Using (9A-1),

$$\sum_{j=1}^{n} \underset{\sim}{f}_j\underset{\sim}{f}'_j = n\, \hat{\Delta}^{-1}\hat{L}'\hat{\Psi}^{-\frac{1}{2}}\hat{\Psi}^{-\frac{1}{2}}\hat{L}(I + \hat{\Delta})\hat{\Delta}^{-1}$$

$$= n\, \hat{\Delta}^{-1}\hat{\Delta}(I + \hat{\Delta})\hat{\Delta}^{-1} = n(I + \hat{\Delta}^{-1}),$$

a diagonal matrix. Consequently, the factor scores have sample mean vector $\underset{\sim}{0}$ and zero sample covariances.

9.17

$$\begin{aligned}\eta &= \gamma\xi+\zeta\\ \begin{bmatrix}Y_1\\Y_2\end{bmatrix} &= \begin{bmatrix}\lambda_1\\1\end{bmatrix}\eta+\begin{bmatrix}\epsilon_1\\\epsilon_2\end{bmatrix},\quad \begin{bmatrix}X_1\\X_2\end{bmatrix}=\begin{bmatrix}1\\\lambda_2\end{bmatrix}\eta+\begin{bmatrix}\delta_1\\\delta_2\end{bmatrix}\end{aligned}$$

Additionally, let $Var(\xi)=\phi$, $Var(\zeta)=\psi$;

$$Cov(\epsilon)=\begin{bmatrix}\theta_1 & 0\\ 0 & \theta_2\end{bmatrix} \text{ and } Cov(\delta)=\begin{bmatrix}\theta_3 & 0\\ 0 & \theta_4\end{bmatrix}$$

Using

$$Cov\left(\begin{bmatrix}\tilde{Y}\\\tilde{X}\end{bmatrix}\right)=\mathbf{S}=\left[\begin{array}{cc|cc}14.3 & -27.6 & 6.4 & 3.2\\ -27.6 & 55.4 & -12.8 & -6.4\\ \hline 6.4 & -12.8 & 3.7 & 1.6\\ 3.2 & -6.4 & 1.6 & 1.1\end{array}\right]$$

We obtain

$$\begin{aligned}
\text{(i) } \hat{Var}(Y_1) &= \hat{\lambda}_1^2(\hat{\gamma}^2\hat{\phi}+\hat{\psi})+\hat{\theta}_1=14.3\\
\text{(ii) } \hat{Var}(Y_2) &= (\hat{\gamma}^2\hat{\phi}+\hat{\psi})+\hat{\theta}_2=55.4\\
\text{(iii) } \hat{Cov}(Y_1,Y_2) &= \hat{\lambda}_1(\hat{\gamma}^2\hat{\phi}+\hat{\psi})=-27.6\\
\text{(iv) } \hat{Cov}(Y_1,X_1) &= \hat{\lambda}_1\hat{\gamma}\hat{\phi}=6.4\\
\text{(v) } \hat{Cov}(Y_1,X_2) &= \hat{\lambda}_1\hat{\lambda}_2\hat{\gamma}\hat{\phi}=3.2\\
\text{(vi) } \hat{Cov}(Y_2,X_1) &= \hat{\gamma}\hat{\phi}=-12.8\\
\text{(vii) } \hat{Cov}(Y_2,X_2) &= \hat{\lambda}_2\hat{\gamma}\hat{\phi}=-6.4\\
\text{(viii) } \hat{Var}(X_1) &= \hat{\phi}+\hat{\theta}_3=3.7\\
\text{(ix) } \hat{Var}(X_2) &= \hat{\lambda}_2^2\hat{\phi}+\hat{\theta}_4=1.1\\
\text{(x) } \hat{Cov}(X_1,X_2) &= \hat{\lambda}_2\hat{\phi}=1.6
\end{aligned}$$

By solving (i)-(x), we obtain $\hat{\phi}=3.2$, $\hat{\gamma}=-4$, $\hat{\psi}=4$ and

$$\hat{\Lambda}_y=\begin{bmatrix}-0.5\\1.0\end{bmatrix},\quad \hat{\Lambda}_x=\begin{bmatrix}1\\0.5\end{bmatrix},\quad Cov(\epsilon)=\begin{bmatrix}0.5 & 0\\0 & 0.2\end{bmatrix},\quad Cov(\delta)=\begin{bmatrix}0.5 & 0\\0 & 0.3\end{bmatrix}$$

Compare the fitted equations from Example 9.16 and Exercise 9.17;

$$\text{Example 9.16: }\left\{\begin{aligned}\eta &= 4\xi+\zeta\\ Y_1 &= \eta+\epsilon_1\\ Y_2 &= -2\eta+\epsilon_2\\ X_1 &= 2\eta+\delta_1\\ X_2 &= \eta+\delta_2\end{aligned}\right.\qquad \text{Exercise 9.17: }\left\{\begin{aligned}\eta &= -4\xi+\zeta\\ Y_1 &= -0.5\eta+\epsilon_1\\ Y_2 &= \eta+\epsilon_2\\ X_1 &= \eta+\delta_1\\ X_2 &= 0.5\eta+\delta_2\end{aligned}\right.$$

If η and ξ are not scaled, λ's can't be determined uniquely without additional constraints. Note that a change of scaling in this case changes the "interpretation" of results. Of course, the data were artificial and the story in Example 9.16 was constructed to "fit" the model.

9.18. Factor analysis of Wisconsin fish data

(a) Principal component solution using $x_1 - x_4$

Initial Factor Method: Principal Components

	1	2	3	4
Eigenvalue	2.1539	0.7876	0.6157	0.4429
Difference	1.3663	0.1719	0.1728	
Proportion	0.5385	0.1969	0.1539	0.1107
Cumulative	0.5385	0.7354	0.8893	1.0000

Factor Pattern (m = 1)

	FACTOR1
BLUEGILL	0.77273
BCRAPPIE	0.73867
SBASS	0.64983
LBASS	0.76738

Factor Pattern (m = 2)

	FACTOR1	FACTOR2
BLUEGILL	0.77273	-0.40581
BCRAPPIE	0.73867	-0.36549
SBASS	0.64983	0.67309
LBASS	0.76738	0.19047

(b) Maximum likelihood solution using $x_1 - x_4$

Initial Factor Method: Maximum Likelihood

Factor Pattern (m = 1)

	FACTOR1
BLUEGILL	0.70812
BCRAPPIE	0.63002
SBASS	0.48544
LBASS	0.65312

Factor Pattern (m = 2)

	FACTOR1	FACTOR2
BLUEGILL	0.98748	-0.02251
BCRAPPIE	0.50404	0.25907
SBASS	0.28186	0.65863
LBASS	0.48073	0.41799

(c) Varimax rotation. Note that rotation is not possible with 1 factor.

Principal Components
Varimax Rotated Factor Pattern

	FACTOR1	FACTOR2
BLUEGILL	0.85703	0.16518
BCRAPPIE	0.80526	0.17543
SBASS	0.08767	0.93147
LBASS	0.48072	0.62774

Maximum Likelihood
Varimax Rotated Factor Pattern

	FACTOR1	FACTOR2
BLUEGILL	0.96841	0.19445
BCRAPPIE	0.43501	0.36324
SBASS	0.13066	0.70439
LBASS	0.37743	0.51319

For both solutions, Bluegill and Crappie load heavily on the first factor, while large-mouth and smallmouth bass load heavily on the second factor.

(d) Factor analysis using $x_1 - x_6$

Initial Factor Method: Principal Components

	1	2	3	4	5	6
Eigenvalue	2.3549	1.0719	0.9843	0.6644	0.5004	0.4242
Difference	1.2830	0.0876	0.3199	0.1640	0.0762	
Proportion	0.3925	0.1786	0.1640	0.1107	0.0834	0.0707
Cumulative	0.3925	0.5711	0.7352	0.8459	0.9293	1.0000

Factor Pattern (m = 3)

	FACTOR1	FACTOR2	FACTOR3
BLUEGILL	0.72944	-0.02285	-0.47611
BCRAPPIE	0.72422	0.01989	-0.20739
SBASS	0.60333	0.58051	0.26232
LBASS	0.76170	0.07998	-0.03199
WALLEYE	-0.39334	0.83342	-0.01286
NPIKE	0.44657	-0.18156	0.80285

Varimax Rotated Factor Pattern

	FACTOR1	FACTOR2	FACTOR3
BLUEGILL	0.85090	-0.12720	-0.13806
BCRAPPIE	0.74189	0.11256	-0.06957
SBASS	0.51192	0.46222	0.54231
LBASS	0.71176	0.28458	0.00311
WALLEYE	-0.24459	-0.21480	0.86227
NPIKE	0.05282	0.92348	-0.14613

Initial Factor Method: Maximum Likelihood
Factor Pattern

	FACTOR1	FACTOR2	FACTOR3
BLUEGILL	0.00000	1.00000	0.00000
BCRAPPIE	0.18979	0.49190	0.23481
SBASS	0.96466	0.26350	0.00000
LBASS	0.29875	0.46530	0.29435
WALLEYE	0.12927	-0.22770	-0.49746
NPIKE	0.24062	0.06520	0.46665

Varimax Rotated Factor Pattern

	FACTOR1	FACTOR2	FACTOR3
BLUEGILL	0.99637	0.06257	0.05767
BCRAPPIE	0.46485	0.21097	0.26931
SBASS	0.20017	0.97853	0.04905
LBASS	0.42801	0.31567	0.33099
WALLEYE	-0.20771	0.13392	-0.50492
NPIKE	0.02359	0.22600	0.47779

The first principal component factor influences the Bluegill, Crappie and the Bass. The Northern Pike alone loads heavily on the second factor, and the Walleye and smallmouth bass on the third factor. The MLE solution is different.

9.19 (a), (b) and (c) Maximum Likelihood (m = 3)

UNROTATED FACTOR LOADINGS (PATTERN)
FOR MAXIMUM LIKELIHOOD CANONICAL FACTORS

		Factor 1	Factor 2	Factor 3
Growth	1	0.772	0.295	0.527
Profits	2	0.570	0.347	0.721
Newaccts	3	0.774	0.433	0.355
Creative	4	0.389	0.921	0.000
Mechanic	5	0.509	0.426	0.334
Abstract	6	0.968	-0.250	0.000
Math	7	0.632	0.181	0.729
	VP	3.267	1.520	1.566

ROTATED FACTOR LOADINGS (PATTERN)

		Factor 1	Factor 2	Factor 3
Growth	1	0.794	0.374	0.437
Profits	2	0.912	0.316	0.184
Newaccts	3	0.653	0.544	0.437
Creative	4	0.255	0.967	0.019
Mechanic	5	0.541	0.464	0.208
Abstract	6	0.390	0.054	0.953
Math	7	0.919	0.179	0.295
	VP	3.180	1.720	1.454

		Communalities	Specific Variances
1	Growth	0.9615	.0385
2	Profits	0.9648	.0352
3	Newaccts	0.9124	.0876
4	Creative	1.0000	.0000
5	Mechanic	0.5519	.4481
6	Abstract	1.0000	.0000
7	Math	0.9631	.0369

$$R = \begin{bmatrix} 1.0 & .926 & .884 & .572 & .708 & .674 & .927 \\ & 1.0 & .843 & .542 & .746 & .465 & .944 \\ & & 1.0 & .700 & .637 & .641 & .853 \\ & & & 1.0 & .591 & .147 & .413 \\ & & & & 1.0 & .386 & .575 \\ \text{(Symmetric)} & & & & & 1.0 & .566 \\ & & & & & & 1.0 \end{bmatrix}$$

$$\hat{L}\hat{L}' + \hat{\Psi} = \begin{bmatrix} 1.0 & .923 & .912 & .572 & .694 & .674 & .925 \\ & 1.0 & .848 & .542 & .679 & .465 & .948 \\ & & 1.0 & .700 & .696 & .641 & .826 \\ & & & 1.0 & .591 & .147 & .413 \\ & & & & 1.0 & .386 & .646 \\ & & & & & 1.0 & .566 \\ \text{(Symmetric)} & & & & & & 1.0 \end{bmatrix}$$

It is clear from an examination of the residual matrix $R - (\hat{L}\hat{L}' + \hat{\Psi})$ that an $m = 3$ factor solution represents the observed correlations quite well. However, it is difficult to provide interpretations for the factors. If we consider the rotated loadings, we see that the last two factors are dominated by the single variables "creative" and "abstract" respectively. The first factor links the salespeople performance variables with math ability.

(d) Using (9-39) with $n = 50$, $p = 7$, $m = 3$ we have

$$43.833 \ln\left(\frac{.000075933}{.000018427}\right) = 62.1 > \chi_3^2(.01) = 11.3$$

so we reject $H_0: \Sigma = LL' + \Psi$ for $m = 3$. Neither of the $m = 2$, $m = 3$ factor models appear to fit by the χ^2 criterion. We note that the matrices R, $\hat{L}\hat{L}' + \hat{\Psi}$ have small determinants and rounding error could affect the calculation of the test statistic. Again, the residual matrix above indicates a good fit for $m = 3$.

(e) $\underset{\sim}{z}' = [1.522, -.852, .465, .957, 1.129, .673, .497]$

Using the regression method for computing factor scores, we have; with $\underset{\sim}{f} = \hat{L}_z R_z^{-1}$:

Principal components (m = 3)	Maximum likelihood (m = 3)
$\underset{\sim}{f}' = [.686, .271, 1.395]$	$\underset{\sim}{f}' = [-.702, .679, -.751]$

Factor scores using weighted least squares can only be computed for the principal component solutions since $\hat{\Psi}^{-1}$ cannot be computed for the maximum likelihood solutions. ($\hat{\Psi}$ has zeros on the main diagonal for the maximum likelihood solutions). Using (9-50),

Principal components (m = 3)

$\underset{\sim}{f}' = [.344, .233, 1.805]$

9.20

$$S = \begin{bmatrix} & X_1 & X_2 & X_5 & X_6 \\ & 2.50 & -2.77 & -.59 & -2.23 \\ & & 300.52 & 6.78 & 30.78 \\ & & & 11.36 & 3.13 \\ & \text{(symmetric)} & & & 31.98 \end{bmatrix}$$

(a) Principal components (m = 2)

	Factor 1 loadings	Factor 2 loadings
X_1 (wind)	-.17	-.37
X_2 (solar rad.)	17.32	-.61
X_5 (NO_2)	.42	.74
X_6 (O_3)	1.96	5.19

(b) Maximum likelihood estimates of the loadings are obtained from $\hat{L} = \hat{V}^{1/2}\hat{L}_z$ where $\hat{L}_z$ are the loadings obtained from the sample correlation matrix R. (For $\hat{L}_z$ see problem 9.23). Note: Maximum likelihood estimates of the loadings for m = 2 may be difficult to obtain for some computer packages without good estimates of the communalities. One choice for initial estimates of the communalities are the communalities from the m = 2 principal components solution.

(c) Maximum likelihood estimation (with m = 2) does a better job of accounting for the covariances in S than the m = 2 principal component solution. On the other hand, the principal component solution generally produces uniformly smaller estimates of the specific variances. For the unrotated m = 2 solution, the first factor is dominated by X_2 = solar radiation and X_6 = O_3. The second factor seems to be a contrast between the pair X_1 = wind; X_2 = solar radiation and the pair X_5 = NO_2 and X_6 = O_3.

9.21 Principal components (m = 2)

	Rotated loadings	
	Factor 1	Factor 2
X_1 (wind)	.10	-.46
X_2 (solar rad.)	2.00	.05
X_5 (NO_2)	.05	.87
X_6 (O_3)	.71	5.49

Again the first factor is dominated by solar radiation and, to some extent, ozone. The second factor might be interpreted as a contrast between wind and the pair of pollutants NO_2 and O_3. Recall solar radiation and ozone have the largest sample variances. This will affect the estimated loadings obtained by the principal component method.

9.22 (a) Since, for maximum likelihood estimates, $\hat{L} = D^{\frac{1}{2}}\hat{L}_z$ and $S = D^{\frac{1}{2}}RD^{\frac{1}{2}}$, the factor scores generated by the equations for $\underset{\sim}{f}_j$ in (9-58) will be identical. Similarly, the factor scores generated by the weighted least squares formulas in (9-50) will be identical.

The factor scores generated by the regression method with maximum likelihood estimates (m = 2; see problem 9.23) are given below for the first 10 cases.

Case	$\hat{f}_1$	$\hat{f}_2$
1	0.316	-0.544
2	0.252	-0.546
3	0.129	-0.509
4	0.332	-0.790
5	0.492	-0.012
6	0.515	-0.370
7	0.530	-0.456
8	1.070	0.724
9	0.384	-0.023
10	-0.179	0.105

(b) Factor scores using principal component estimates (m = 2) and (9-51) for the first 10 cases are given below:

Case	$\hat{f}_1$	$\hat{f}_2$
1	1.203	-0.368
2	1.646	-1.029
3	1.447	-0.937
4	0.717	0.795
5	0.856	-0.049
6	0.811	0.394
7	0.518	0.950
8	-0.083	1.168
9	0.410	0.259
10	-0.492	0.072

(c) The sets of factor scores are quite different. Factor scores depend heavily on the method used to estimate loadings and specific variances as well as the method used to generate them.

9.23

Principal components (m = 2)

	Factor 1 loadings	Factor 2 loadings	Rotated loadings Factor 1	Rotated loadings Factor 2
X_1 (wind)	-.56	-.24	-.31	-.53
X_2 (solar rad.)	.65	-.52	.83	-.04
X_5 (NO_2)	.48	.74	-.05	.88
X_6 (O_3)	.77	-.20	.74	.30

Maximum likelihood (m = 2)

	Factor 1 loadings	Factor 2 loadings	Rotated loadings Factor 1	Factor 2
X_1 (wind)	-.38	.32	-.09	.49
X_2 (solar rad.)	.50	.27	.56	-.10
X_5 (NO_2)	.25	-.04	.17	-.19
X_6 (O_3)	.65	-.03	.49	-.43

Examining the rotated loadings, we see that both solution methods yield similar estimated loadings for the first factor. It might be called a "ozone pollution factor". There are some differences for the second factor. However, the second factor appears to compare one of the pollutants with wind. It might be called a "pollutant transport" factor. We note that the interpretations of the factors might differ depending upon the choice of R or S (see problems 9.20 and 9.21) for analysis. Also the two solution methods give somewhat different results indicating the solution is not very stable. Some of the observed correlations between the variables are very small implying that a $m = 1$ or $m = 2$ factor model for these four variables will not be a completely satisfactory description of the underlying structure. We may need about as many factors as variables. If this is the case, there is nothing to be gained by proposing a factor model.

9.24

$$R = \begin{bmatrix} 1.0 & .610 & .971 & .740 & -.172 \\ & 1.0 & .494 & .095 & .186 \\ & & 1.0 & .848 & -.249 \\ & & & 1.0 & -.358 \\ \text{(Symmetric)} & & & & 1.0 \end{bmatrix}$$

Principal Components (m = 2)

	Factor 1 loadings	Factor 2 loadings	Rotated loadings Factor 1	Rotated loadings Factor 2	Communalities
Population	.97	.15	.98	.09	.97
Schooling	.55	.72	.59	.68	.81
Employment	.99	.01	.99	-.06	.98
Health emp.	.85	-.35	.82	-.41	.84
Home value	-.30	.80	-.25	.82	.73
Percentage Variance Explained	60.6%	25.8%	60.4%	26.0%	

Maximum Likelihood (m = 2)

	Factor 1 loadings	Factor 2 loadings	Rotated loadings Factor 1	Rotated loadings Factor 2	Communalities
Population	.97	.18	.84	.52	.98
Schooling	.49	.72	.85	-.20	.77
Employment	1.00	.00	.74	.68	1.00
Health emp.	.85	-.45	.32	.90	.92
Home value	-.25	.38	.07	-.45	.20
Percentage Variance Explained	59.4%	18.0%	41.6%	35.7%	

Residual matrices; $R - \hat{L}\hat{L}' - \hat{\Psi}$:

Principal Components (m = 2)

$$\begin{bmatrix} 0 & -.026 & .009 & -.031 & .004 \\ & 0 & -.049 & -.115 & -.219 \\ & & 0 & .012 & .047 \\ & & & 0 & .179 \\ \text{(Symmetric)} & & & & 0 \end{bmatrix}$$

Maximum Likelihood (m = 2)

$$\begin{bmatrix} 0 & -.002 & .000 & -.002 & .001 \\ & 0 & .000 & .000 & .036 \\ & & 0 & .000 & .000 \\ & & & 0 & .022 \\ & & & & 0 \end{bmatrix}$$

A m = 3 factor solution does not substantially improve the factor model fit.

Factor scores (m = 2) using the regression method:

Principal Components Factor 1	Principal Components Factor 2	Maximum Likelihood Factor 1	Maximum Likelihood Factor 2
0.379	0.522	0.704	-0.248
-1.363	-0.033	-1.155	-0.982
-0.882	-1.023	-0.813	-0.297
-0.269	1.251	0.565	-1.112
0.381	0.015	0.280	0.289
1.903	0.215	1.416	1.248
-0.792	-0.536	-0.578	-0.541
1.168	0.727	1.551	-0.584
1.251	-1.461	-0.299	2.294
-0.525	-1.334	-0.918	0.425
-0.013	-1.150	-0.623	0.951
-1.295	1.657	-0.853	-0.978
-0.631	0.023	-0.696	-0.261
0.688	1.128	1.420	-0.203

9.25

$$S = \begin{bmatrix} 105{,}625 & 94{,}734 & 87{,}242 & 94{,}280 \\ & 101{,}761 & 76{,}186 & 81{,}204 \\ & & 91{,}809 & 90{,}343 \\ (\text{Symmetric}) & & & 104{,}329 \end{bmatrix}$$

A $m = 1$ factor model appears to represent these data quite well.

	Principal Components Factor 1 loadings	Maximum Likelihood Factor 1 loadings
Shock wave	317.	320.
Vibration	293.	291.
Static test 1	287.	275.
Static test 2	307.	297.
Proportion Variance Explained	90.1%	86.9%

Factor scores ($m = 1$) using the regression method for the first few cases are:

Principal Components	Maximum Likelihood
-.009	-.033
1.530	1.524
.808	.719
-.804	-.802

The factor scores produced from the two solution methods are very similar. The correlation between the two sets of scores is .992.

The outliers, specimens 9 and 16, were identified in Example 4.14.

9.26

a) Principal Components

	m = 1		m = 2		
	Factor 1 loadings	$\tilde{\Psi}_i$	Factor 1 loadings	Factor 2 loadings	$\tilde{\Psi}_i$
Litter 1	27.9	309.0	27.9	-6.2	271.2
Litter 2	30.4	205.7	30.4	-4.9	182.2
Litter 3	31.5	344.3	31.5	18.5	1.7
Litter 4	32.9	310.0	32.9	-8.0	245.8
Percentage Variance Explained	76.4%		76.4%	9.4%	

b) Maximum Likelihood

	m = 1	
	Factor loadings	$\hat{\Psi}_i$
Litter 1	26.8	370.2
Litter 2	30.5	198.2
Litter 3	28.4	529.6
Litter 4	30.4	471.0
Percentage Variance Explained	68.8%	

The maximum likelihood estimates of the factor loadings for $m = 2$ were not obtained due to convergence difficulties in the computer program.

c) It is only necessary to rotate the $m = 2$ solution.

Principal Components (m = 2)

	Rotated loadings Factor 1	Factor 2
Litter 1	26.2	11.4
Litter 2	27.5	13.8
Litter 3	14.7	33.4
Litter 4	31.4	12.8
Percentage Variance Explained	53.5%	32.4%

9.27 Principal Components (m = 2)

	Factor 1 loadings	Factor 2 loadings	$\tilde{\Psi}_i$	Rotated loadings Factor 1	Factor 2
Litter 1	.86	.44	.06	.33	.91
Litter 2	.91	.12	.15	.59	.71
Litter 3	.85	-.36	.14	.87	.32
Litter 4	.87	-.21	.20	.78	.44
Percentage Variance Explained	76.5%	9.5%		45.4%	40.6%

Maximum Likelihood (m = 1)

	Factor 1 loadings	$\hat{\Psi}_i$
Litter 1	.81	.34
Litter 2	.91	.17
Litter 3	.78	.39
Litter 4	.81	.34
Percentage Variance Explained	68.8%	

$$\hat{f} = \hat{L}_z R^{-1} \tilde{z} = .297$$

9.28

(a) Eigenvalues of the Covariance Matrix: Total = 935.371389

	1	2	3	4
Eigenvalue	930.8658	4.0507	0.3187	0.1152
Difference	926.8151	3.7320	0.2035	0.1009
Proportion	0.9952	0.0043	0.0003	0.0001
Cumulative	0.9952	0.9995	0.9999	1.0000

	5	6	7
Eigenvalue	0.0143	0.0059	0.0009
Difference	0.0084	0.0050	
Proportion	0.0000	0.0000	0.0000
Cumulative	1.0000	1.0000	1.0000

Initial Factor Method: Principal Components

$m = 3$ factors

Rotated Factor Pattern

	FACTOR1	FACTOR2	FACTOR3
X1	0.36663	0.83326	0.31421
X2	0.34746	0.88221	0.31545
X3	0.38535	0.55660	0.73584
X4	0.55613	0.36477	0.65724
X5	0.72499	0.34778	0.43539
X6	0.75284	0.36437	0.39485
X7	0.91435	0.33642	0.22535

SAS scales the loadings obtained from a covariance matrix and then rotates scaled loadings. Scaling is: $\hat{l}_{ij}/\sqrt{s_{ii}}$

Communality:

X1	X2	X3	X4	X5	X6	X7
0.927475	0.998530	0.999761	0.874301	0.836124	0.855444	1.0000

The first factor is related to athletic excellence in running long distances (X4,X5,X6,X7). The second factor is related to athletic excellence in running short distances (X1,X2,X3). The 55th observation (wsamoa) is an outlier. (Three factors probably too many.)

ROTATED FACTOR PATTERN

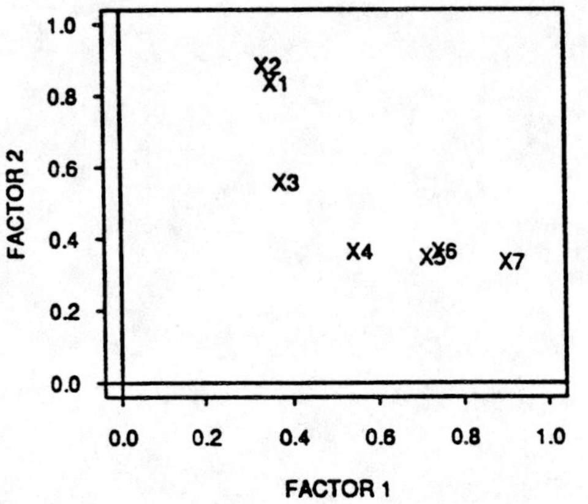

FACTOR SCORES

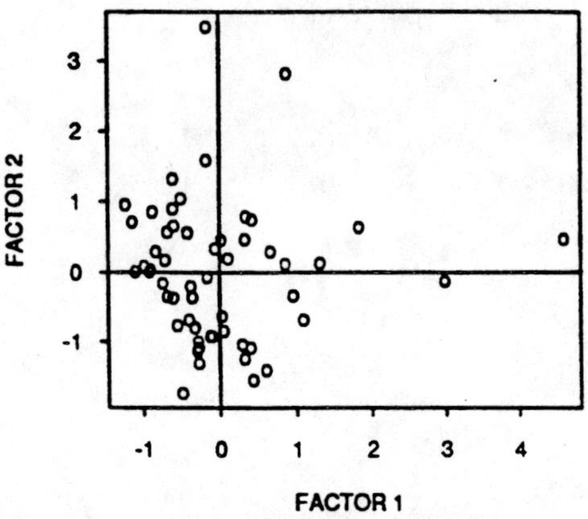

(b) Eigenvalues of the Correlation Matrix: Total = 7 Average = 1

	1	2	3	4
Eigenvalue	5.8057	0.6536	0.2999	0.1255
Difference	5.1520	0.3538	0.1744	0.0717
Proportion	0.8294	0.0934	0.0428	0.0179
Cumulative	0.8294	0.9228	0.9656	0.9835

	5	6	7
Eigenvalue	0.0538	0.0390	0.0224
Difference	0.0148	0.0166	
Proportion	0.0077	0.0056	0.0032
Cumulative	0.9912	0.9968	1.0000

Initial Factor Method: Principal Components $m = 3$ factors

Rotated Factor Pattern

	FACTOR1	FACTOR2	FACTOR3
X7	0.87271	0.36846	0.20986
X6	0.82430	0.36014	0.39331
X5	0.79589	0.32641	0.47325
X2	0.33828	0.88051	0.29885
X1	0.37953	0.86513	0.27705
X4	0.55805	0.35513	0.73437
X3	0.35652	0.60801	0.68415

Communality : Total = 6.759214

X1	X2	X3	X4	X5	X6	X7
0.969261	0.979042	0.964840	0.976825	0.963947	0.963862	0.941436

The first factor is related to athletic excellence in running long distances (X4,X5,X6,X7). The second factor is related to athletic excellence in running short distances(X1,X2,X3). Again, $m = 3$ factors is probably too many. The 55th observation (wsamoa) is an outlier.

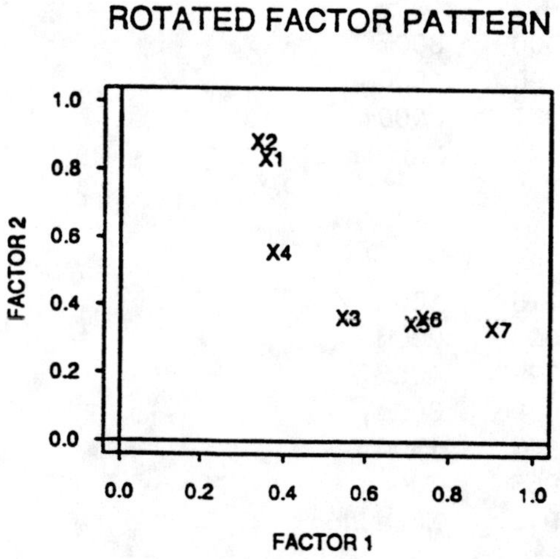

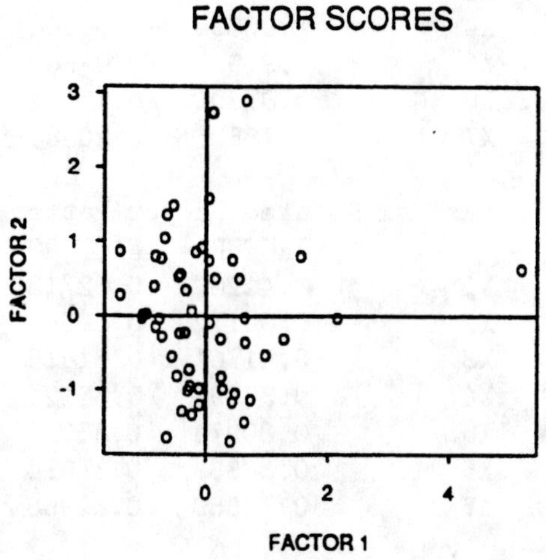

9.29. Factor analysis of national track records for women (Speeds measured in m/sec.)

Factor analysis using sample covariance matrix S

Initial Factor Method: Principal Components

	1	2	3	4	5	6	7
Eigenvalue	0.9791	0.0986	0.0531	0.0232	0.0072	0.0052	0.0034
Difference	0.8805	0.0455	0.0298	0.0160	0.0020	0.0018	
Proportion	0.8370	0.0843	0.0454	0.0199	0.0062	0.0045	0.0029
Cumulative	0.8370	0.9212	0.9666	0.9865	0.9926	0.9971	1.0000

Factor Pattern

	FACTOR1	FACTOR2	FACTOR3	
X1	0.86930	0.40491	-0.17419	100m
X2	0.86408	0.44756	-0.18835	200m
X3	0.89276	0.31939	0.19692	400m
X4	0.91731	0.00755	0.33227	800m
X5	0.94726	-0.15355	0.21285	1500m
X6	0.94947	-0.19146	0.11869	3000m
X7	0.92656	-0.28062	-0.23021	Marathon

Varimax Rotated Factor Pattern

	FACTOR1	FACTOR2	FACTOR3	
X1	0.85466	0.32551	0.33702	100m
X2	0.88870	0.30681	0.31383	200m
X3	0.68624	0.64708	0.21947	400m
X4	0.41962	0.80159	0.36509	800m
X5	0.35101	0.73871	0.54524	1500m
X6	0.35284	0.66781	0.61788	3000m
X7	0.38127	0.38185	0.83611	Marathon

Initial Factor Method: Maximum Likelihood

Factor Pattern

	FACTOR1	FACTOR2	FACTOR3	
X1	0.72932	-0.61880	0.07578	100m
X2	0.72857	-0.68474	-0.00221	200m
X3	0.89901	-0.30027	-0.09978	400m
X4	1.00000	-0.00000	0.00000	800m
X5	0.91907	-0.07509	0.33390	1500m
X6	0.87401	-0.12302	0.45258	3000m
X7	0.78586	-0.20584	0.38423	Marathon

Varimax Rotated Factor Pattern

	FACTOR1	FACTOR2	FACTOR3	
X1	0.41267	0.82713	0.25715	100m
X2	0.34172	0.89694	0.28003	200m
X3	0.41761	0.61415	0.59729	400m
X4	0.59693	0.36121	0.71639	800m
X5	0.80274	0.36259	0.43122	1500m
X6	0.86333	0.37684	0.31065	3000m
X7	0.74559	0.42958	0.25913	Marathon

Factor analysis using sample correlation matrix R

Initial Factor Method: Principal Components

	1	2	3	4	5	6	7
Eigenvalue	5.8491	0.6047	0.3026	0.1291	0.0518	0.0395	0.0231
Difference	5.2444	0.3021	0.1734	0.0773	0.0123	0.0164	
Proportion	0.8356	0.0864	0.0432	0.0184	0.0074	0.0056	0.0033
Cumulative	0.8356	0.9220	0.9652	0.9836	0.9911	0.9967	1.0000

Factor Pattern

	FACTOR1	FACTOR2	FACTOR3	
X1	0.89366	0.38938	0.13802	100m
X2	0.88834	0.42515	0.09791	200m
X3	0.92166	0.17838	-0.28914	400m
X4	0.92951	-0.16459	-0.29235	800m
X5	0.94108	-0.27525	-0.02782	1500m
X6	0.93550	-0.27822	0.08275	3000m
X7	0.88718	-0.24545	0.31185	Marathon

Varimax Rotated Factor Pattern

	FACTOR1	FACTOR2	FACTOR3	
X1	0.38716	0.85873	0.28632	100m
X2	0.34264	0.87427	0.31264	200m
X3	0.32973	0.61414	0.69209	400m
X4	0.54384	0.35164	0.74642	800m
X5	0.74579	0.33264	0.54342	1500m
X6	0.79709	0.35221	0.44724	3000m
X7	0.85640	0.40138	0.22373	Marathon

Initial Factor Method: Maximum Likelihood
Factor Pattern

	FACTOR1	FACTOR2	FACTOR3	
X1	0.72932	-0.61880	0.07578	100m
X2	0.72857	-0.68474	-0.00221	200m
X3	0.89901	-0.30027	-0.09978	400m
X4	1.00000	-0.00000	0.00000	800m
X5	0.91907	-0.07509	0.33390	1500m
X6	0.87401	-0.12302	0.45258	3000m
X7	0.78586	-0.20584	0.38423	Marathon

Varimax Rotated Factor Pattern

	FACTOR1	FACTOR2	FACTOR3	
X1	0.41267	0.82713	0.25715	100m
X2	0.34172	0.89694	0.28003	200m
X3	0.41761	0.61415	0.59729	400m
X4	0.59693	0.36121	0.71639	800m
X5	0.80274	0.36259	0.43122	1500m
X6	0.86333	0.37684	0.31065	3000m
X7	0.74559	0.42958	0.25913	Marathon

Factor scores for the first two factors using S
and varimax rotated PC estimates of factor loadings

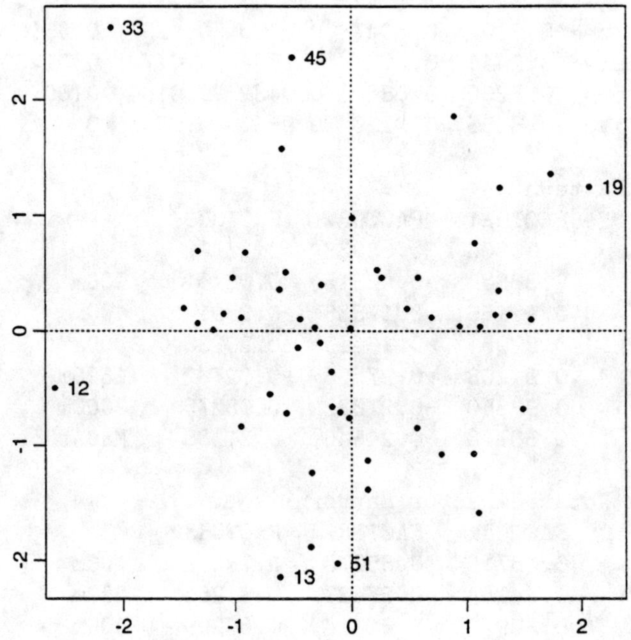

Factor scores for the first two factors using R
and varimax rotated PC estimates of factor loadings

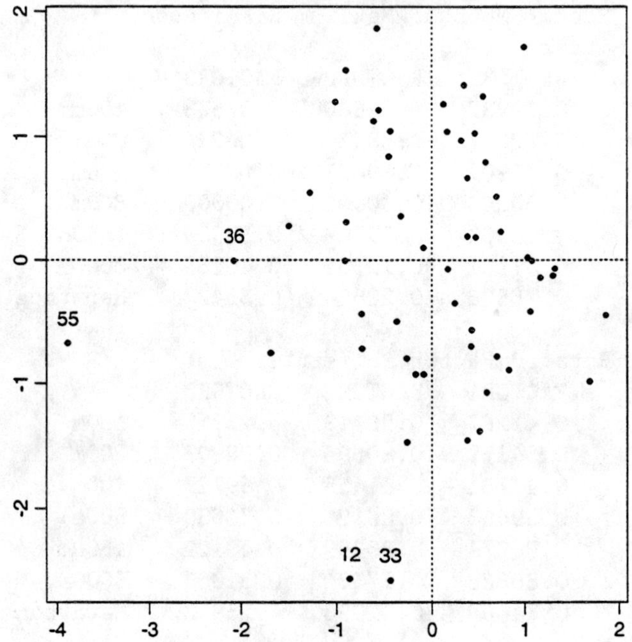

9.30

(a)

```
Eigenvalues of the Covariance Matrix:   Total = 91.7382348
                       1           2           3           4
Eigenvalue       89.9136      1.4126      0.2598      0.1094
Difference       88.5010      1.1528      0.1504      0.0821
Proportion        0.9801      0.0154      0.0028      0.0012
Cumulative        0.9801      0.9955      0.9983      0.9995
                       5           6           7           8
Eigenvalue        0.0273      0.0127      0.0022      0.0004
Difference        0.0146      0.0105      0.0018
Proportion        0.0003      0.0001      0.0000      0.0000
Cumulative        0.9998      1.0000      1.0000      1.0000
```

Initial Factor Method: Principal Components   m=2 factors

```
Rotated Factor Pattern
          FACTOR1    FACTOR2
X8        0.95970    0.28092
X7        0.84564    0.47527
X6        0.83602    0.46833
X5        0.73059    0.58968
X3        0.48134    0.86183
X1        0.29473    0.84153
X2        0.38182    0.81593
X4        0.64748    0.65989
Communality:
      X1        X2        X3        X4        X5        X6        X7        X8
0.795042  0.811526  0.974441  0.854692  0.881486  0.918272  0.940988  0.999938
```

SAS scales the loadings obtained from a covariance matrix and then rotates scaled loadings. Scaling is: $\hat{l}_{ij}/\sqrt{s_{ii}}$

The first factor is related to athletic excellence in running long distances (X4,X5,X6,X7,X8). The second factor is related to athletic excellence in running short distances (X1,X2,X3,X4). The 12th observation (wsamoa) is an outlier.

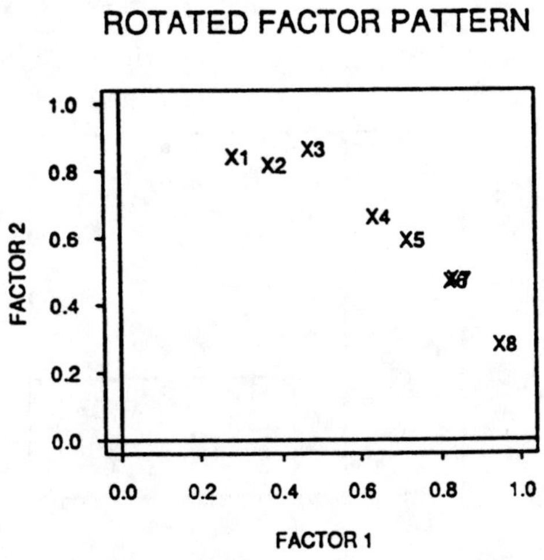

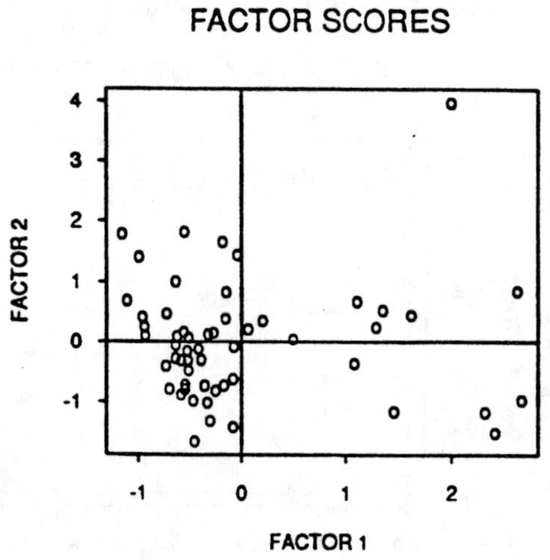

(b) **Eigenvalues of the Correlation Matrix: Total = 8 Average = 1**

	1	2	3	4
Eigenvalue	6.6221	0.8776	0.1593	0.1240
Difference	5.7445	0.7183	0.0353	0.0442
Proportion	0.8278	0.1097	0.0199	0.0155
Cumulative	0.8278	0.9375	0.9574	0.9729

	5	6	7	8
Eigenvalue	0.0799	0.0680	0.0464	0.0226
Difference	0.0119	0.0215	0.0238	
Proportion	0.0100	0.0085	0.0058	0.0028
Cumulative	0.9829	0.9914	0.9972	1.0000

Initial Factor Method: Principal Components $m = 2$ factors

Rotated Factor Pattern

	FACTOR1	FACTOR2
X8	0.93554	0.26095
X7	0.90346	0.39651
X6	0.90194	0.38881
X5	0.81333	0.52539
X4	0.71240	0.62670
X1	0.27430	0.93519
X2	0.37644	0.89291
X3	0.54306	0.77252

Final Communality Estimates: Total = 7.499764

X1	X2	X3	X4	X5	X6	X7	X8
0.949814	0.938997	0.891689	0.900271	0.937545	0.964660	0.973464	0.943324

The first factor is related to athletic excellence in running long distances (X4,X5,X6,X7,X8). The second factor is related to athletic excellence in running short distances(X1,X2,X3, X4). When **R** is factored, the position of variables is the same as before. The 12th observation (wsamoa) is an outlier.

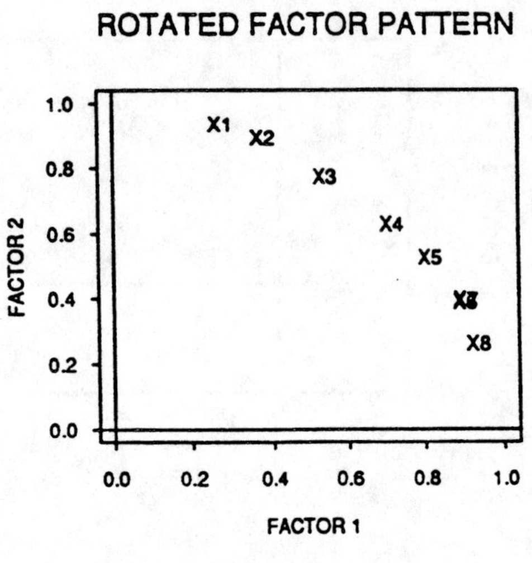

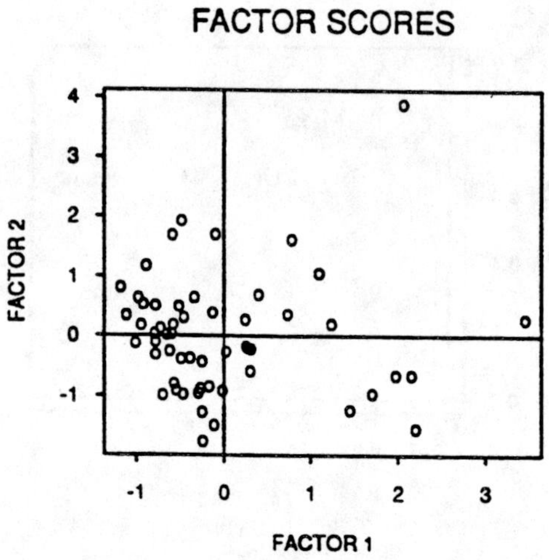

9.31. Factor analysis of national track records for men (Speeds measured in m/sec)

Factor analysis using sample covariance matrix S

Initial Factor Method: Principal Components

	1	2	3	4	5	6	7	8
Eigenvalue	0.5655	0.0834	0.0124	0.0093	0.0068	0.0059	0.0042	0.0026
Difference	0.4821	0.0710	0.0030	0.0026	0.0009	0.0017	0.0016	
Proportion	0.8195	0.1209	0.0179	0.0135	0.0098	0.0085	0.0060	0.0038
Cumulative	0.8195	0.9404	0.9583	0.9719	0.9817	0.9902	0.9962	1.0000

Factor Pattern

	FACTOR1	FACTOR2	FACTOR3	
X1	0.78816	0.57873	-0.12781	100m
X2	0.84930	0.47193	-0.11305	200m
X3	0.89921	0.25748	0.25290	400m
X4	0.93704	0.06455	0.23875	800m
X5	0.95959	-0.08496	0.09125	1500m
X6	0.94664	-0.26443	-0.04818	5000m
X7	0.95334	-0.25979	-0.05420	10000m
X8	0.89517	-0.38117	-0.08932	Marathon

Varimax Rotated Factor Pattern

	FACTOR1	FACTOR2	FACTOR3	
X1	0.26461	0.93751	0.15337	100m
X2	0.37496	0.88603	0.17652	200m
X3	0.48071	0.65926	0.52262	400m
X4	0.63196	0.53416	0.50449	800m
X5	0.76847	0.46633	0.35826	1500m
X6	0.89506	0.35268	0.20701	5000m
X7	0.89837	0.36162	0.20359	10000m
X8	0.93569	0.24188	0.14342	Marathon

Initial Factor Method: Maximum Likelihood
Factor Pattern

	FACTOR1	FACTOR2	FACTOR3	
X1	1.00000	0.00000	0.00000	100m
X2	0.92318	0.16634	-0.11639	200m
X3	0.82684	0.36137	-0.19600	400m
X4	0.74954	0.52704	-0.27373	800m
X5	0.69011	0.65885	-0.16183	1500m
X6	0.59617	0.77593	-0.00099	5000m
X7	0.61046	0.78403	0.04811	10000m
X8	0.50178	0.80728	0.04617	Marathon

Varimax Rotated Factor Pattern

	FACTOR1	FACTOR2	FACTOR3	
X1	0.26628	0.96368	-0.02055	100m
X2	0.39081	0.85251	0.11820	200m
X3	0.54203	0.71303	0.22466	400m
X4	0.67046	0.59946	0.32501	800m
X5	0.79423	0.50163	0.23261	1500m
X6	0.90062	0.37171	0.09046	5000m
X7	0.91812	0.38069	0.04257	10000m
X8	0.91118	0.26998	0.04977	Marathon

Factor analysis using sample correlation matrix R

Initial Factor Method: Principal Components

	1	2	3	4	5	6	7	8
Eigenvalue	6.5701	0.9094	0.1686	0.1239	0.0863	0.0668	0.0502	0.0247
Difference	5.6607	0.7409	0.0447	0.0376	0.0195	0.0167	0.0255	
Proportion	0.8213	0.1137	0.0211	0.0155	0.0108	0.0084	0.0063	0.0031
Cumulative	0.8213	0.9349	0.9560	0.9715	0.9823	0.9906	0.9969	1.0000

Factor Pattern

	FACTOR1	FACTOR2	FACTOR3	
X1	0.80689	0.54561	0.13750	100m
X2	0.86520	0.43765	0.14727	200m
X3	0.91373	0.22687	-0.24263	400m
X4	0.94574	0.02367	-0.21734	800m
X5	0.95839	-0.12904	-0.03843	1500m
X6	0.93312	-0.30233	0.07848	5000m
X7	0.93976	-0.29754	0.07865	10000m
X8	0.87679	-0.41423	0.08960	Marathon

Varimax Rotated Factor Pattern

	FACTOR1	FACTOR2	FACTOR3	
X1	0.26502	0.93349	0.16134	100m
X2	0.37998	0.88949	0.16195	200m
X3	0.47924	0.65893	0.53045	400m
X4	0.63969	0.53110	0.50094	800m
X5	0.78202	0.46937	0.32369	1500m
X6	0.89814	0.35328	0.19187	5000m
X7	0.89995	0.36094	0.19414	10000m
X8	0.93154	0.23765	0.15535	Marathon

Initial Factor Method: Maximum Likelihood

Factor Pattern

	FACTOR1	FACTOR2	FACTOR3	
X1	1.00000	0.00000	0.00000	100m
X2	0.92318	0.16634	-0.11639	200m
X3	0.82684	0.36137	-0.19600	400m
X4	0.74954	0.52704	-0.27373	800m
X5	0.69011	0.65885	-0.16183	1500m
X6	0.59617	0.77593	-0.00099	5000m
X7	0.61046	0.78403	0.04811	10000m
X8	0.50178	0.80728	0.04617	Marathon

Varimax Rotated Factor Pattern

	FACTOR1	FACTOR2	FACTOR3	
X1	0.26628	0.96368	-0.02055	100m
X2	0.39081	0.85251	0.11820	200m
X3	0.54203	0.71303	0.22466	400m
X4	0.67046	0.59946	0.32501	800m
X5	0.79423	0.50163	0.23261	1500m
X6	0.90062	0.37171	0.09046	5000m
X7	0.91812	0.38069	0.04257	10000m
X8	0.91118	0.26998	0.04977	Marathon

Both **S** and **R** lead to similar conclusions. So, we would choose the analysis based on the **S** for the speed data.

Factor scores for the first two factors using S
and varimax rotated PC estimates of factor loadings

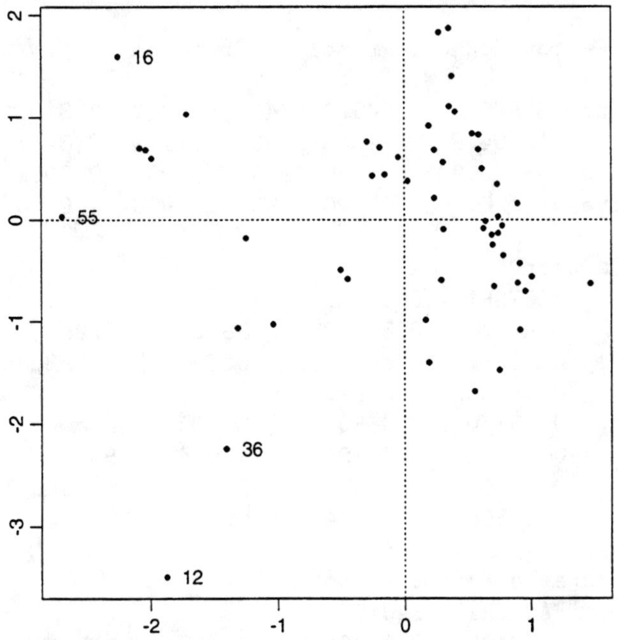

Factor scores for the first two factors using R
and varimax rotated PC estimates of factor loadings

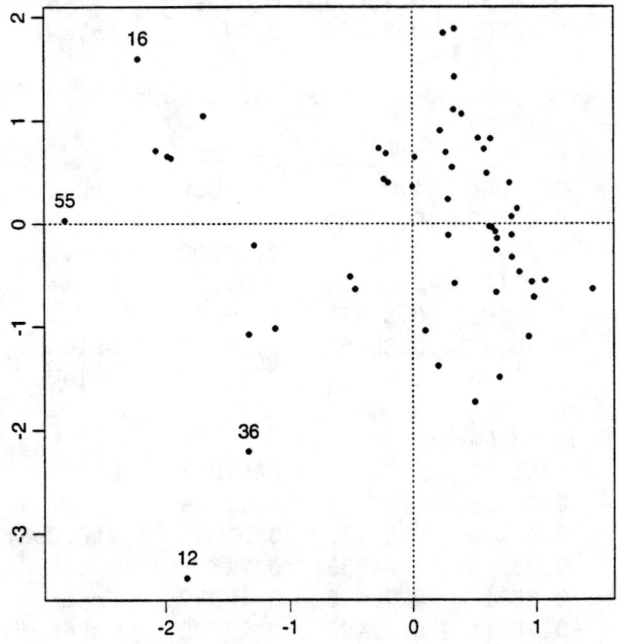

9.32. Factor analysis of data on bulls

Factor analysis using sample covariance matrix S

Initial Factor Method: Principal Components

	1	2	3	4	5	6	7
Eigenvalue	20579.6126	4874.6748	5.4292	3.3163	0.4688	0.0741	0.0045
Difference	15704.9378	4869.2456	2.1129	2.8475	0.3948	0.0695	
Proportion	0.8082	0.1914	0.0002	0.0001	0.0000	0.0000	0.0000
Cumulative	0.8082	0.9996	0.9998	1.0000	1.0000	1.0000	1.0000

Factor Pattern

	FACTOR1	FACTOR2	FACTOR3	
X3	0.48777	0.39033	0.38532	YrHgt
X4	0.75367	0.65725	-0.00086	FtFrBody
X5	0.37408	0.62342	0.64446	PrctFFB
X6	0.48170	0.36809	0.33505	Frame
X7	0.11083	-0.38394	-0.49074	BkFat
X8	0.66769	0.29875	0.33038	SaleHt
X9	0.96506	-0.26204	0.00009	SaleWt

Varimax Rotated Factor Pattern

	FACTOR1	FACTOR2	FACTOR3	
X3	0.50195	0.42460	0.32637	YrHgt
X4	0.25853	0.90600	0.33514	FtFrBody
X5	0.83816	0.45576	0.18354	PrctFFB
X6	0.44716	0.42166	0.31943	Frame
X7	-0.60974	-0.06913	0.15478	BkFat
X8	0.40890	0.46689	0.50894	SaleHt
X9	-0.13508	0.30219	0.94363	SaleWt

Initial Factor Method: Maximum Likelihood
Factor Pattern

	FACTOR1	FACTOR2	FACTOR3	
X3	0.00000	1.00000	0.00000	YrHgt
X4	0.42819	0.62380	0.39838	FtFrBody
X5	0.85244	0.52282	0.00000	PrctFFB
X6	-0.01180	0.94025	0.03120	Frame
X7	-0.36162	-0.34428	0.39308	BkFat
X8	0.08393	0.85951	0.28992	SaleHt
X9	0.00598	0.36843	0.83599	SaleWt

Varimax Rotated Factor Pattern

	FACTOR1	FACTOR2	FACTOR3	
X3	0.94438	0.28442	0.16509	YrHgt
X4	0.41219	0.50159	0.55648	FtFrBody
X5	0.23003	0.94883	0.21635	PrctFFB
X6	0.88812	0.25026	0.18382	Frame
X7	-0.25711	-0.51405	0.27102	BkFat
X8	0.75340	0.26667	0.43720	SaleHt
X9	0.25282	-0.05273	0.87634	SaleWt

Factor analysis using sample correlation matrix R

Initial Factor Method: Principal Components

	1	2	3	4	5	6	7
Eigenvalue	4.1207	1.3371	0.7414	0.4214	0.1858	0.1465	0.0471
Difference	2.7836	0.5957	0.3200	0.2356	0.0393	0.0994	
Proportion	0.5887	0.1910	0.1059	0.0602	0.0265	0.0209	0.0067
Cumulative	0.5887	0.7797	0.8856	0.9458	0.9723	0.9933	1.0000

Factor Pattern

	FACTOR1	FACTOR2	FACTOR3	
X3	0.91334	-0.04948	-0.35794	YrHgt
X4	0.83700	0.15014	0.38772	FtFrBody
X5	0.72177	-0.36484	0.48930	PrctFFB
X6	0.88091	0.00894	-0.38949	Frame
X7	-0.37900	0.82646	-0.03335	BkFat
X8	0.91927	0.11715	-0.15210	SaleHt
X9	0.54798	0.69440	0.21811	SaleWt

Varimax Rotated Factor Pattern

	FACTOR1	FACTOR2	FACTOR3	
X3	0.94188	0.27085	-0.06532	YrHgt
X4	0.44792	0.78354	0.24262	FtFrBody
X5	0.26505	0.87071	-0.25513	PrctFFB
X6	0.93812	0.21799	-0.01382	Frame
X7	-0.23541	-0.37460	0.79502	BkFat
X8	0.83365	0.41206	0.13094	SaleHt
X9	0.34932	0.39692	0.74194	SaleWt

Initial Factor Method: Maximum Likelihood

Factor Pattern

	FACTOR1	FACTOR2	FACTOR3	
X3	0.00000	1.00000	0.00000	YrHgt
X4	0.42819	0.62380	0.39838	FtFrBody
X5	0.85244	0.52282	0.00000	PrctFFB
X6	-0.01180	0.94025	0.03120	Frame
X7	-0.36162	-0.34428	0.39308	BkFat
X8	0.08393	0.85951	0.28992	SaleHt
X9	0.00598	0.36843	0.83599	SaleWt

Varimax Rotated Factor Pattern

	FACTOR1	FACTOR2	FACTOR3	
X3	0.94438	0.28442	0.16509	YrHgt
X4	0.41219	0.50159	0.55648	FtFrBody
X5	0.23003	0.94883	0.21635	PrctFFB
X6	0.88812	0.25026	0.18382	Frame
X7	-0.25711	-0.51405	0.27102	BkFat
X8	0.75340	0.26667	0.43720	SaleHt
X9	0.25282	-0.05273	0.87634	SaleWt

The interpretation of factors from **R** is different of the interpretation of factors from S.

Factor scores for the first two factors using S
and varimax rotated PC estimates of factor loadings

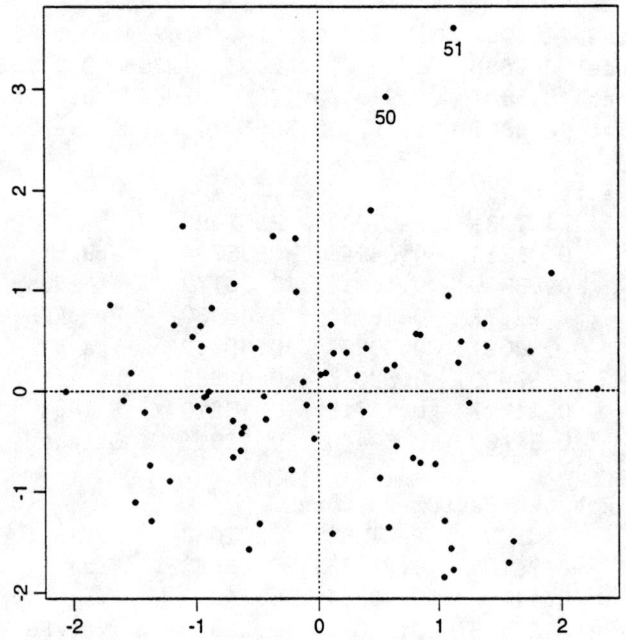

Factor scores for the first two factors using R
and varimax rotated PC estimates of factor loadings

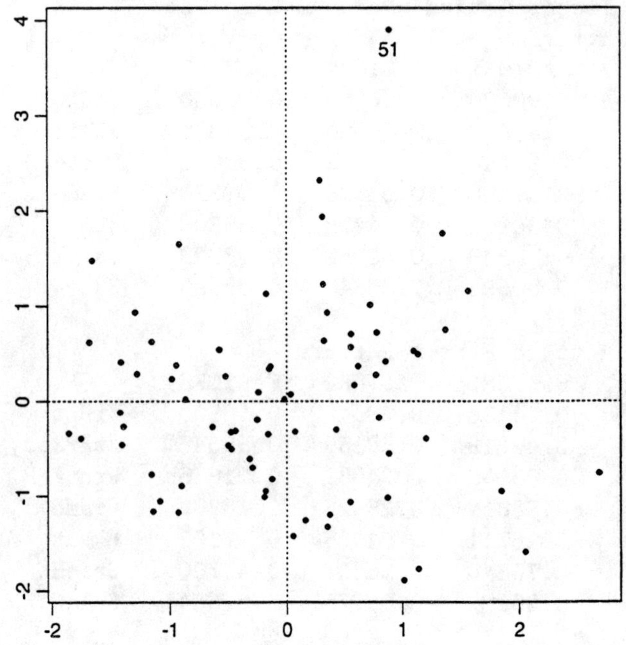

Chapter 10

10.1. $\Sigma_{11}^{-1/2}\Sigma_{12}\Sigma_{22}^{-1}\Sigma_{21}\Sigma_{11}^{-1/2} = \begin{bmatrix} 0 & 0 \\ 0 & (.95)^2 \end{bmatrix}$

which has eigenvalues $\rho_1^{*2} = (.95)^2$ and $\rho_2^{*2} = 0.$

The normalized eigenvectors are $\mathbf{e}_1 = \begin{bmatrix} 0 \\ 1 \end{bmatrix}$ and $\mathbf{e}_2 = \begin{bmatrix} 1 \\ 0 \end{bmatrix}$.

Thus

$$U_1 = \mathbf{e}_1' \Sigma_{11}^{1/2}\mathbf{X}^{(1)} = [0\ 1]\begin{bmatrix} .1 & 0 \\ 0 & 1 \end{bmatrix}\begin{bmatrix} X_1^{(1)} \\ X_2^{(1)} \end{bmatrix} = X_2^{(1)}$$

Since $\mathbf{f}_1' \Sigma_{22}^{-1/2} = [1\ \ 0]$, $V_1 = X_1^{(2)}$.

Thus $U_1 = X_2^{(1)}$, $V_1 = X_1^{(2)}$ and $\rho_1^* = .95$.

10.2 (i) $\rho_1^* = .55$, $\rho_2^* = .49$

(ii) $U_1 = .32X_1^{(1)} - .36X_2^{(1)}$

$V_1 = .36X_1^{(2)} - .10X_2^{(2)}$

$U_2 = .20X_1^{(1)} + .30X_2^{(1)}$

$V_2 = .23X_1^{(2)} + .30X_2^{(1)}$

(iii) $E\begin{bmatrix} U_1 \\ U_2 \\ V_1 \\ V_2 \end{bmatrix} = \begin{bmatrix} -1.675 \\ .015 \\ -.095 \\ .386 \end{bmatrix}$

$$\text{Cov}\begin{bmatrix} \underset{\sim}{U} \\ \underset{\sim}{V} \end{bmatrix} = \begin{bmatrix} 1 & 0 & .55 & 0 \\ 0 & 1 & 0 & .49 \\ .55 & 0 & 1 & 0 \\ 0 & .49 & 0 & 1 \end{bmatrix} = \begin{bmatrix} 1 & 0 & \rho_1^* & 0 \\ 0 & 1 & 0 & \rho_2^* \\ \rho_1^* & 0 & 1 & 0 \\ 0 & \rho_2^* & 0 & 1 \end{bmatrix}$$

10.5 (i) $\Sigma_{11}^{-1}\Sigma_{12}\Sigma_{22}^{-1}\Sigma_{21} = \boldsymbol{\rho}_{11}^{-1}\boldsymbol{\rho}_{12}\boldsymbol{\rho}_{22}^{-1}\boldsymbol{\rho}_{21} = \begin{bmatrix} .45189 & .28919 \\ .14633 & .17361 \end{bmatrix}$

$$\begin{vmatrix} .45189-\lambda & .28919 \\ .14633 & .17361-\lambda \end{vmatrix} = \lambda^2 - .5467\lambda + .0005$$
$$= (\lambda - .5457)(\lambda - .0009)$$

The characteristic equation is the same as that of

$^{-1/2}_{11}\ _{12}\ ^{-1}_{22}\ _{21}\ ^{-1/2}_{11}$ (see Example 10.1) and consequently the

eigenvalues are the same.

(ii) $U_2 = -.677Z_1^{(1)} + 1.055Z_2^{(1)}$

$V_2 = -.863Z_1^{(2)} + .706Z_2^{(2)}$

$\text{Var}(U_2) = (-0.677)^2 + (1.055)^2 - 2(.677)(1.055)(.4) = 1.0$

$\text{Var}(V_2) = 1.0$

$\text{Corr}(U_2, V_2) = (-.677)(-.863)(.5) + (-.863)(1.055)(.3)$

$+ (.706)(-.677)(.6) + (.706)(1.055)(.4) = .03 = \rho_2^*$

10.7 (i) $\rho_1^* = \frac{2\rho}{1+\rho} \qquad 0 < \rho < 1$

$$U_1 = \frac{1}{\sqrt{2(1+\rho)}}(X_1^{(1)}+X_2^{(1)})$$

$$V_1 = \frac{1}{\sqrt{2(1+\rho)}}(X_1^{(2)}+X_2^{(2)})$$

10.8 (iii) $\hat{\rho}_1^* = .72$

$\hat{V}_1 = .20X_1^{(2)}+.70X_2^{(2)}$

$\hat{\beta} = 45° \equiv \frac{\pi}{4}$ radians

(v) $\hat{\rho}_1^* = .57$

$\hat{U}_1 = 1.03 \cos \theta_1 + .46 \text{ Sin } \theta_1$

$V_1 = .49 \cos \theta_2 + .78 \text{ Sin } \theta_2$

10.9 (i) $\hat{\rho}_1^* = .39 \quad ; \quad \hat{\rho}_2^* = .07$

$\hat{U}_1 = 1.26Z_1^{(1)}-1.03Z_2^{(1)}; \quad \hat{U}_2 = .30Z_1^{(1)}+.79Z_2^{(1)}$

$\hat{V}_1 = 1.10Z_1^{(2)}-.45Z_2^{(2)} \quad ; \quad V_2 = -.02Z_1^{(2)}+1.01Z_2^{(2)}$

(ii) $n = 140,\ p=2,\ q=2,\ n-1-\frac{1}{2}(p+q+1) = 136.5$

Null hypothesis	Value of test statistic	Degrees of Freedom	Upper 5% point of χ^2 distribution
H_o: $\Sigma_{12} = P_{12} = 0$	$-136.5\ \ln(.8444)(.9953) = 23.74$	4	9.48
$H_o^{(1)}$: $\rho_1^* \neq 0,\ \rho_2^* = 0$	$-136.5\ \ln(.9953) = .65$	1	3.84

Therefore, reject H_o but do not reject $H_o^{(1)}$. Reading ability (summarized by $\hat{U}_1$) does correlate with arithmetic ability (summarized by $\hat{V}_1$) but the correlation (represented by $\rho_1 = .39$) is not particularly strong.

10.10 (i) $\hat{\rho}_1^* = .33,\ \hat{\rho}_2^* = .17$

(ii) $\hat{U}_1 = 1.002Z_1^{(1)} - .003Z_2^{(1)}$

$\hat{V}_1 = -.602Z_1^{(2)} - .977Z_2^{(2)}$

$\hat{U}_1 \doteq Z_1^{(1)}$ = 1973 nonprimary homicides (standardized)

$\hat{V}_1 \doteq \frac{3}{5} Z_1^{(2)} + Z_2^{(2)}$ = a "punishment index"

Punishment appears to be correlated with nonprimary homicides but not primary homicides.

10.11 $\hat{\rho}_1^* = .566;\quad \hat{\rho}_2^* = .155$

$\hat{U}_1 = .546Z_1^{(1)} + .065Z_2^{(1)} + .553Z_3^{(1)}$

$\hat{V}_1 = .515Z_1^{(2)} + .630Z_2^{(2)}$

$\hat{U}_2 = -1.008Z_1^{(1)} + 1.117Z_2^{(1)} + .088Z_3^{(1)}$

$\hat{V}_2 = 1.054Z_1^{(2)} - .990Z_2^{(2)}$

$\hat{U}_1$ is essentially an average of two chemical companies (Allied Chemical and Union Carbide). $\hat{V}_1$ is essentially an average of the oil companies.

Here H_0: $\Sigma_{12} = \rho_{12} = 0$ is rejected at the 5% level

but $H_0^{(1)}$: $\rho_1^* \neq 0,\ \rho_2^* = 0$ is not rejected (5% level).

10.12 (i) $\hat{\rho}_1^* = .69, \quad \hat{\rho}_2^* = .19$

Reject $H_0: \rho_{12} = 0$ at the 5% level but do not reject

$H_0^{(1)} = \rho_1^* \neq 0, \ \rho_2^* = 0$ at the 5% level.

(ii) $\hat{U}_1 = .77Z_1^{(1)} + .27Z_2^{(1)}$

$\hat{V}_1 = .05Z_1^{(2)} + .90Z_2^{(2)} + .19Z_3^{(2)}$

(iii) Sample Correlations Between Original Variables and Canonical Variables

$\underset{\sim}{X}^{(1)}$ Variables	$\hat{U}_1$	$\hat{V}_1$	$\underset{\sim}{X}^{(2)}$ Variables	$\hat{U}_1$	$\hat{V}_1$
1. annual frequency of restaurant dining	.99	.68	1. age of head of household	.29	.42
2. annual frequency of attending movies	.89	.61	2. annual family income	.68	.98
			3. educational level of head of household	.35	.51

(iv) $\hat{U}_1$ is a measure of family entertainment outside the home. $\hat{V}_1$ may be considered a measure of family "status" which is dominated by family income. Essentially, family entertainment outside the home is positively associated with family income.

10.13 (i) $\hat{\rho}_1^* = .909, \quad \hat{\rho}_2^* = .636, \quad \hat{\rho}_3^* = .256, \quad \hat{\rho}_4^* = .094$

	Null hypothesis	Value of test statistic	Degrees of freedom	Conclusion at 1% level
1.	$H_0: \Sigma_{12} = \rho_{12} = 0$	309.98	20	Reject H_0
2	$H_0: \rho_1 \neq 0, \ \rho_2 = \ldots = \rho_4 = 0$	78.63	12	Reject H_0
3.	$H_0: \rho_1 \neq 0, \ \rho_2 \neq 0, \ \rho_3 = 0, \ \rho_4 = 0$	16.81	6	Do not reject H_0.

$$\begin{bmatrix} \hat{U}_1 \\ \hat{U}_2 \end{bmatrix} = \begin{bmatrix} .21 & .17 & -.33 & -.26 & .30 \\ .92 & -.58 & .65 & .34 & .55 \end{bmatrix} \begin{bmatrix} Z_1^{(1)} \\ Z_2^{(1)} \\ Z_3^{(1)} \\ Z_4^{(1)} \\ Z_5^{(1)} \end{bmatrix}$$

$$\begin{bmatrix} \hat{V}_1 \\ \hat{V}_2 \end{bmatrix} = \begin{bmatrix} -.54 & -.29 & .46 & .03 \\ 1.01 & .03 & .98 & -.18 \end{bmatrix} \begin{bmatrix} Z_1^{(2)} \\ Z_2^{(2)} \\ Z_3^{(2)} \\ Z_4^{(2)} \end{bmatrix}$$

(ii) $\hat{U}_1$ appears to measure quality of wheat as a "contrast" between "good" aspects ($Z_1^{(1)}$, $Z_2^{(1)}$ and $Z_5^{(1)}$) and "bad" aspects ($Z_3^{(1)}$, $Z_4^{(1)}$).

$\hat{V}_1$ is harder to interpret. It appears to measure the quality of the flour as represented by $Z_1^{(2)}$, $Z_2^{(2)}$ and $Z_3^{(2)}$.

10.14

(a) $\hat{\rho}_1^* = 0.7520$, $\hat{\rho}_2^* = 0.5395$. And the sample canonical variates are

Raw Canonical Coefficients for the Accounting measures of profitability

	U1	U2
HRA	-0.494697741	1.9655018549
HRE	0.2133051339	-0.794353012
HRS	0.7228315515	-0.538822808
RRA	2.7749354333	-4.38345956
RRE	-1.383659039	1.6471230054
RRS	-1.032933813	2.5190103052

Raw Canonical Coefficients for the Market measures of profitability

	V1	V2
Q	1.3930601511	-2.500804367
REV	-0.431692979	2.8298904995

U_1 is most highly correlated with RRA and HRA and also HRS and RRS. V_1 is highly correlated with both of its components. The second pair does not correlate well with their respective components.

(b) Standardized Variance of the Accounting measures of profitability Explained by

	Their Own Canonical Variables			The Opposite Canonical Variables	
	Proportion	Cumulative Proportion	Canonical R-Squared	Proportion	Cumulative Proportion
1	0.5041	0.5041	0.5655	0.2851	0.2851
2	0.0905	0.5946	0.2910	0.0263	0.3114

Standardized Variance of the Market measures of profitability Explained by

	Their Own Canonical Variables			The Opposite Canonical Variables	
	Proportion	Cumulative Proportion	Canonical R-Squared	Proportion	Cumulative Proportion
1	0.8702	0.8702	0.5655	0.4921	0.4921
2	0.1298	1.0000	0.2910	0.0378	0.5299

Market measures can be well explained by its canonical variate $\hat{V}_1$. However, accounting measures cannot be well explained. In fact, from the correlation between measures and canonical variates, accounting measures on equity have weak correlation with $\hat{U}_1$.

Correlations Between the Accounting measures of profitability and Their Canonical Variables

	U1	U2
HRA	0.8110	0.2711
HRE	0.4225	0.0968
HRS	0.7184	0.5526
RRA	0.8605	-0.0089

RRE	0.5741	-0.0959
RRS	0.7761	0.3814

Correlations Between the Market measures of profitability and Their Canonical Variables

	V1	V2
Q	0.9886	0.1508
REV	0.8736	0.4866

10.15

$\hat{\rho}_1^* = 0.9129$, $\hat{\rho}_2^* = 0.0681$. And the sample canonical variates are

Raw Canonical Coefficients for the dynamic measure

	U1	U2
X1	0.0036015621	-0.006663216
X2	-0.000595735	0.0077029513

Raw Canonical Coefficients for the static measures

	V1	V2
X3	0.0013448038	0.008471035
X4	0.0018933921	-0.007828962

Standardized Variance of the dynamic measure Explained by

	Their Own Canonical Variables			The Opposite Canonical Variables	
	Proportion	Cumulative Proportion	Canonical R-Squared	Proportion	Cumulative Proportion
1	0.8840	0.8840	0.8334	0.7367	0.7367
2	0.1160	1.0000	0.0046	0.0005	0.7373

Standardized Variance of the static measures Explained by

	Their Own Canonical Variables			The Opposite Canonical Variables	
	Proportion	Cumulative Proportion	Canonical R-Squared	Proportion	Cumulative Proportion
1	0.9601	0.9601	0.8334	0.8002	0.8002
2	0.0399	1.0000	0.0046	0.0002	0.8003

Static measures can be well explained by its canonical variate $\hat{U}_1$. Also, dynamic measures can be well explained by its canonical variate $\hat{V}_1$.

10.16 From the computer output below, the first two canonical correlations are $\hat{\rho}_1^* = 0.517345$ and $\hat{\rho}_2^* = 0.125508$. The large sample tests

$$-(n-1-\frac{1}{2}(p+q-1))\ln[(1-\hat{\rho*}_1^2)(1-\hat{\rho*}_1^2)] \geq \chi^2_{pq}(.05)$$

or

$$-(46-1-\frac{1}{2}(3+2-1))\ln[(1-(.517345)^2)(1-(.125508)^2)] = 13.50 \geq \chi^2_6(.05) = 12.59$$

and

$$-(n-1-\frac{1}{2}(p+q-1))\ln[(1-\hat{\rho*}_1^2)] \geq \chi^2_{(p-1)(q-1)}(.05)$$

or

$$-(46-1-\frac{1}{2}(3+2-1))\ln[(1-(.125508)^2)] = 0.667 \geq \chi^2_2(.05) = 5.99$$

suggest that only the first pair of canonical variables are important. Even if the variables means were given, we prefer to interpret the canonical variables obtained from **S** in terms of coefficients of standardized variables.

$$\begin{aligned}\hat{U}_1 &= .4357z_1^{(1)} - .7047z_2^{(1)} + 1.0815z_3^{(1)} \\ \hat{V}_1 &= 1.020z_1^{(2)} - .1609z_2^{(2)}\end{aligned}$$

The two insulin responses dominate $\hat{U}_1$ while $\hat{V}_1$ consists primarily of the relative weight variable.

Canonical Correlation Analysis

	Canonical Correlation	Adjusted Canonical Correlation	Approx Standard Error	Squared Canonical Correlation
1	0.517345	0.517145	0.007324	0.267646
2	0.125508	0.125158	0.009843	0.015752

Canonical Correlation Analysis

Raw Canonical Coefficients for the Glucose and Insulin

GLUCOSE	0.0131006541	0.0247524811
INSULIN	-0.014438254	-0.009317525
INSULRES	0.023399723	-0.008667216

Raw Canonical Coefficients for the Weight and Fasting

WEIGHT	8.0655750801	-0.375167814
FASTING	-0.019159052	0.1200675138

Standardized Canonical Coefficients for the Glucose and Insulin

GLUCOSE	0.4357	0.8232
INSULIN	-0.7047	-0.4547
INSULRES	1.0815	-0.4006

Standardized Canonical Coefficients for the Weight and Fasting

	SECONDA1	SECONDA2
WEIGHT	1.0202	-0.0475
FASTING	-0.1609	1.0086

Correlations Between the Glucose and Insulin and Their Canonical Variables

	PRIMARY1	PRIMARY2
GLUCOSE	0.3397	0.6838
INSULIN	-0.0502	-0.4565
INSULRES	0.7551	-0.5729

Correlations Between the Weight and Fasting and Their Canonical Variables

	SECONDA1	SECONDA2
WEIGHT	0.9875	0.1576
FASTING	0.0465	0.9989

10.17 The computer output below suggests maybe two canonical pairs of variables. the canonical correlations are 0.521594, 0.375256, 0.242181 and 0.136568. $\hat{U}_1$ ignores the first smoking question and $\hat{U}_2$ ignores the third. $\hat{V}_1$ depends heavily on the difference of annoyance and tenseness.

Even the second pairs do not explain their own variances very well. $R^2_{\boldsymbol{z}^{(1)}|\hat{U}_2} = .1249$ and $R^2_{\boldsymbol{z}^{(1)}|\hat{V}_2} = 0.0879$

Canonical Correlation Analysis

	Canonical Correlation	Adjusted Canonical Correlation	Approx Standard Error	Squared Canonical Correlation
1	0.521594	0.520771	0.007280	0.272060
2	0.375256	0.374364	0.008592	0.140817
3	0.242181	0.241172	0.009414	0.058652
4	0.136568	0.135586	0.009814	0.018651

Standardized Canonical Coefficients for the Smoking

	SMOKING1	SMOKING2	SMOKING3	SMOKING4
SMOK1	-0.0430	1.0898	1.1161	-1.0092
SMOK2	1.1622	0.6988	-1.4170	0.1732
SMOK3	-1.3753	0.2081	0.0156	1.6899
SMOK4	0.8909	-1.6506	0.8325	-0.2630

Standardized Canonical Coefficients for the Psych and Physical State

	STATE1	STATE2	STATE3	STATE4
CONCEN	0.4733	-0.8141	0.4946	-0.1604
ANNOY	-0.7806	-0.4510	0.5909	-0.7193
SLEEP	0.2567	-0.6052	0.6981	0.6246
TENSE	0.6919	0.3800	-0.4190	0.4376
ALERT	-0.1451	-0.1840	-1.5191	-0.7253
IRRITAB	-0.0704	0.6255	-0.3343	0.8760
TIRED	0.3127	0.5898	0.2276	0.1861
CONTENT	0.3364	0.4869	0.8334	-0.6557

Canonical Structure

Correlations Between the Smoking and Their Canonical Variables

	SMOKING1	SMOKING2	SMOKING3	SMOKING4
SMOK1	0.4458	0.5278	0.6615	0.2917
SMOK2	0.7305	0.3822	0.1487	0.5461
SMOK3	0.2910	0.2664	0.4668	0.7915
SMOK4	0.6403	-0.0620	0.5586	0.5236

Correlations Between the Psychological and Physical
State and Their Canonical Variables

	STATE1	STATE2	STATE3	STATE4
CONCEN	0.7199	-0.3579	0.0125	-0.3137
ANNOY	0.3035	0.1365	0.3906	-0.4058
SLEEP	0.5995	-0.3490	0.3709	0.2586
TENSE	0.7015	0.3305	0.0053	-0.1861
ALERT	0.7290	-0.1539	-0.1459	-0.3681
IRRITAB	0.4585	0.3342	0.1211	-0.0805
TIRED	0.6905	-0.0267	0.2544	0.0749
CONTENT	0.5323	0.4350	0.3207	-0.5601

Canonical Redundancy Analysis

Raw Variance of the Smoking
Explained by

	Their Own Canonical Variables			The Opposite Canonical Variables	
	Proportion	Cumulative Proportion	Canonical R-Squared	Proportion	Cumulative Proportion
1	0.3068	0.3068	0.2721	0.0835	0.0835
2	0.1249	0.4316	0.1408	0.0176	0.1010
3	0.2474	0.6790	0.0587	0.0145	0.1155
4	0.3210	1.0000	0.0187	0.0060	0.1215

Raw Variance of the Psychological and Physical State
Explained by

	Their Own Canonical Variables			The Opposite Canonical Variables	
	Proportion	Cumulative Proportion	Canonical R-Squared	Proportion	Cumulative Proportion
1	0.3705	0.3705	0.2721	0.1008	0.1008
2	0.0879	0.4583	0.1408	0.0124	0.1132
3	0.0617	0.5201	0.0587	0.0036	0.1168
4	0.1032	0.6233	0.0187	0.0019	0.1187

Chapter 11

11.1 (a) The linear discriminant function given in (11-19) is

$$\hat{y} = (\bar{\boldsymbol{x}}_1 - \bar{\boldsymbol{x}}_2)' \boldsymbol{S}_{\text{pooled}}^{-1} \boldsymbol{x} = \hat{\boldsymbol{a}}' \boldsymbol{x}$$

where

$$\boldsymbol{S}_{\text{pooled}}^{-1} = \begin{bmatrix} 2 & -1 \\ -1 & 1 \end{bmatrix}$$

so the the linear discriminant function is

$$\left(\begin{bmatrix} 3 \\ 6 \end{bmatrix} - \begin{bmatrix} 5 \\ 8 \end{bmatrix} \right)' \begin{bmatrix} 2 & -1 \\ -1 & 1 \end{bmatrix} \boldsymbol{x} = [-2 \quad 0] = -2x_1$$

(b)

$$\hat{m} = \frac{1}{2}(\hat{y}_1 + \hat{y}_2) = \frac{1}{2}(\hat{\boldsymbol{a}}'\bar{\boldsymbol{x}}_1 + \hat{\boldsymbol{a}}'\bar{\boldsymbol{x}}_2) = -8$$

Assign $\boldsymbol{x}_0'$ to π_1 if

$$\hat{y}_0 = [2 \quad 7]\boldsymbol{x}_0 \geq \hat{m} = -8$$

and assign $\boldsymbol{x}_0$ to π_2 otherwise.

Since $[-2 \quad 0]\boldsymbol{x}_0 = -4$ is greater than $\hat{m} = -8$, assign $\boldsymbol{x}_0'$ to population π_1.

11.2 (a) $\pi_1 \equiv$ Riding-mower owners; $\pi_2 \equiv$ Nonowners

Here are some summary statistics for the data in Example 11.1:

$$\overline{\boldsymbol{x}}_1 = \begin{bmatrix} 79.475 \\ 20.267 \end{bmatrix}, \qquad \overline{\boldsymbol{x}}_2 = \begin{bmatrix} 57.400 \\ 17.633 \end{bmatrix}$$

$$\boldsymbol{S}_1 = \begin{bmatrix} 352.644 & -11.818 \\ -11.818 & 4.082 \end{bmatrix}, \qquad \boldsymbol{S}_2 = \begin{bmatrix} 200.705 & -2.589 \\ -2.589 & 4.464 \end{bmatrix}$$

$$\boldsymbol{S}_{\text{pooled}} = \begin{bmatrix} 276.675 & -7.204 \\ -7.204 & 4.273 \end{bmatrix}, \qquad \boldsymbol{S}_{\text{pooled}}^{-1} = \begin{bmatrix} .00378 & .00637 \\ .00637 & .24475 \end{bmatrix}$$

The linear classification function for the data in Example 11.1 using (11-19) is

$$\left(\begin{bmatrix} 79.475 \\ 20.267 \end{bmatrix} - \begin{bmatrix} 57.400 \\ 17.633 \end{bmatrix} \right)' \begin{bmatrix} .00378 & .00637 \\ .00637 & .24475 \end{bmatrix} \boldsymbol{x} = \begin{bmatrix} .100 & .785 \end{bmatrix} \boldsymbol{x}$$

where

$$\hat{m} = \frac{1}{2}(\overline{y}_1 + \overline{y}_2) = \frac{1}{2}(\hat{\boldsymbol{a}}'\overline{\boldsymbol{x}}_1 + \hat{\boldsymbol{a}}'\overline{\boldsymbol{x}}_2) = 21.739$$

(b) Assign an observation $\boldsymbol{x}$ to π_1 if

$$0.100x_1 + 0.79x_2 \geq 21.74$$

Otherwise, assign $\boldsymbol{x}$ to π_2

Here are the observations and their classifications:

Owners			Nonowners		
Observation	$\boldsymbol{a}'\boldsymbol{x}_0$	Classification	Observation	$\boldsymbol{a}'\boldsymbol{x}_0$	Classification
1	20.461	nonowner	1	22.907	owner
2	21.761	owner	2	21.624	owner
3	23.455	owner	3	20.000	nonowner
4	22.496	owner	4	20.348	nonowner
5	27.250	owner	5	22.239	nonowner
6	26.111	owner	6	18.751	owner
7	24.644	owner	7	18.517	nonowner
8	25.887	owner	8	21.063	nonowner
9	22.620	owner	9	17.628	nonowner
10	25.653	owner	10	18.069	nonowner
11	22.386	owner	11	16.104	nonowner
12	23.822	owner	12	17.935	nonowner

From this, we can construct the confusion matrix:

		Predicted Membership		
		π_1	π_2	Total
Actual membership	π_1	11	1	12
	π_2	2	10	12

(c) The apparent error rate is $\frac{1+2}{12+12} = 0.125$

(d) The assumptions are that the observations from π_1 and π_2 are from multivariate normal distributions with equal covariance matrices, $\Sigma_1 = \Sigma_2 = \Sigma$.

11.3 We need to show that the regions R_1 and R_2 that minimize the ECM are defined by the values $\boldsymbol{x}$ for which the following inequalities hold:

$$R_1 : \frac{f_1(\boldsymbol{x})}{f_2(\boldsymbol{x})} \geq \left(\frac{c(1|2)}{c(2|1)}\right)\left(\frac{p_2}{p_1}\right)$$

$$R_2 : \frac{f_1(\boldsymbol{x})}{f_2(\boldsymbol{x})} < \left(\frac{c(1|2)}{c(2|1)}\right)\left(\frac{p_2}{p_1}\right)$$

Substituting the expressions for $P(2|1)$ and $P(1|2)$ into (11-5) gives

$$\text{ECM} = c(2|1)p_1 \int_{R_2} f_1(\boldsymbol{x})d\boldsymbol{x} + c(1|2)p_2 \int_{R_1} f_2(\boldsymbol{x})d\boldsymbol{x}$$

And since $\Omega = R_1 \cup R_2$,

$$1 = \int_{R_1} f_1(\boldsymbol{x})d\boldsymbol{x} + \int_{R_2} f_1(\boldsymbol{x})d\boldsymbol{x}$$

and thus,

$$\text{ECM} = c(2|1)p_1 \left[1 - \int_{R_1} f_1(\boldsymbol{x})d\boldsymbol{x}\right] + c(1|2)p_2 \int_{R_1} f_2(\boldsymbol{x})d\boldsymbol{x}$$

Since both of the integrals above are over the same region, we have

$$\text{ECM} = \int_{R_1} [c(1|2)p_2 f_2(\boldsymbol{x})d\boldsymbol{x} - c(2|1)p_1 f_1(\boldsymbol{x})]d\boldsymbol{x} + c(2|1)p_1$$

The minimum is obtained when R_1 is chosen to be the region where the term in brackets is less than or equal to 0. So choose R_1 so that

$$c(2|1)p_1 f_1(\boldsymbol{x}) \geq c(1|2)p_2 f_2(\boldsymbol{x}) \quad \text{or}$$

$$\frac{f_1(\boldsymbol{x})}{f_2(\boldsymbol{x})} \geq \left(\frac{c(1|2)}{c(2|1)}\right)\left(\frac{p_2}{p_1}\right)$$

11.4 (a) The minimum ECM rule is given by assigning an observation $\boldsymbol{x}$ to π_1 if

$$\frac{f_1(\boldsymbol{x})}{f_2(\boldsymbol{x})} \geq \left(\frac{c(1|2)}{c(2|1)}\right)\left(\frac{p_1}{p_2}\right) = \left(\frac{100}{50}\right)\left(\frac{.2}{.8}\right) = .5$$

and assigning $\boldsymbol{x}$ to π_2 if

$$\frac{f_1(\boldsymbol{x})}{f_2(\boldsymbol{x})} < \left(\frac{c(1|2)}{c(2|1)}\right)\left(\frac{p_1}{p_2}\right) = \left(\frac{100}{50}\right)\left(\frac{.2}{.8}\right) = .5$$

(b) Since $f_1(\boldsymbol{x}) = .3$ and $f_2(\boldsymbol{x}) = .5$,

$$\frac{f_1(\boldsymbol{x})}{f_2(\boldsymbol{x})} = .6 \geq .5$$

and assign $\boldsymbol{x}$ to π_1.

11.5 $-\frac{1}{2}(\underset{\sim}{x}-\underset{\sim}{\mu}_1)'\Sigma^{-1}(\underset{\sim}{x}-\underset{\sim}{\mu}_1) + \frac{1}{2}(\underset{\sim}{x}-\underset{\sim}{\mu}_2)'\Sigma^{-1}(\underset{\sim}{x}-\underset{\sim}{\mu}_2) =$

$$-\frac{1}{2}[\underset{\sim}{x}'\Sigma^{-1}\underset{\sim}{x}-2\underset{\sim}{\mu}_1'\Sigma^{-1}\underset{\sim}{x}+\underset{\sim}{\mu}_1'\Sigma^{-1}\underset{\sim}{\mu}_1-\underset{\sim}{x}'\Sigma^{-1}\underset{\sim}{x}+2\underset{\sim}{\mu}_2'\Sigma^{-1}\underset{\sim}{x}-\underset{\sim}{\mu}_2'\Sigma^{-1}\underset{\sim}{\mu}_2]$$

$$= -\frac{1}{2}[-2(\underset{\sim}{\mu}_1-\underset{\sim}{\mu}_2)'\Sigma^{-1}\underset{\sim}{x}+\underset{\sim}{\mu}_1'\Sigma^{-1}\underset{\sim}{\mu}_1-\underset{\sim}{\mu}_2'\Sigma^{-1}\underset{\sim}{\mu}_2]$$

$$= (\underset{\sim}{\mu}_1-\underset{\sim}{\mu}_2)'\Sigma^{-1}\underset{\sim}{x} - \frac{1}{2}(\underset{\sim}{\mu}_1-\underset{\sim}{\mu}_2)'\Sigma^{-1}(\underset{\sim}{\mu}_1+\underset{\sim}{\mu}_2)\ .$$

11.6 a) $E(\underset{\sim}{\ell}'\underset{\sim}{X}|\pi_1) - m = \underset{\sim}{\ell}'\underset{\sim}{\mu}_1 - m = \underset{\sim}{\ell}'\underset{\sim}{\mu}_1 - \frac{1}{2}\underset{\sim}{\ell}'(\underset{\sim}{\mu}_1 + \underset{\sim}{\mu}_2)$

$$= \frac{1}{2}\underset{\sim}{\ell}'(\underset{\sim}{\mu}_1 - \underset{\sim}{\mu}_2) = \frac{1}{2}(\underset{\sim}{\mu}_1 - \underset{\sim}{\mu}_2)'\Sigma^{-1}(\underset{\sim}{\mu}_1 - \underset{\sim}{\mu}_2) > 0 \text{ since}$$

Σ^{-1} is positive definite.

b) $E(\underset{\sim}{\ell}'\underset{\sim}{X}|\pi_2) - m = \underset{\sim}{\ell}'\underset{\sim}{\mu}_2 - m = \frac{1}{2}\underset{\sim}{\ell}'(\underset{\sim}{\mu}_2 - \underset{\sim}{\mu}_1)$

$$= -\frac{1}{2}(\underset{\sim}{\mu}_1 - \underset{\sim}{\mu}_2)'\Sigma^{-1}(\underset{\sim}{\mu}_1 - \underset{\sim}{\mu}_2) < 0 .$$

11.7 (a) Here are the densities:

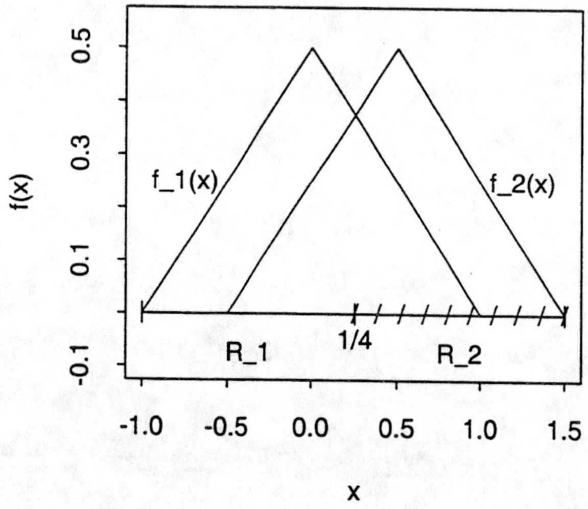

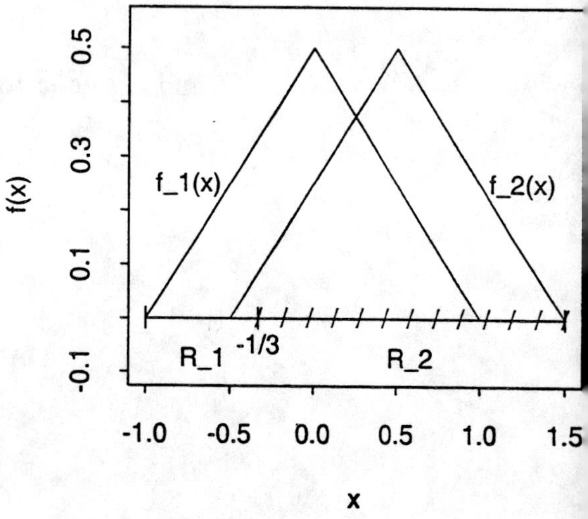

(b) When $p_1 = p_2$ and $c(1|2) = c(2|1)$, the classification regions are

$$R_1 : \frac{f_1(x)}{f_2(x)} \geq 1 \qquad R_2 : \frac{f_1(x)}{f_2(x)} < 1$$

These regions are given by $R_1 : -1 \leq x \leq .25$ and $R_2 : .25 < x \leq 1.5$.

(c) When $p_1 = .2$, $p_2 = .8$, and $c(1|2) = c(2|1)$, the classification regions are

$$R_1 : \frac{f_1(x)}{f_2(x)} \geq \frac{p_2}{p_1} = .4 \qquad R_2 : \frac{f_1(x)}{f_2(x)} < .4$$

These regions are given by $R_1 : -1 \leq x \leq -1/3$ and $R_2 : -1/3 < x \leq 1.5$.

11.8 (a) Here are the densities:

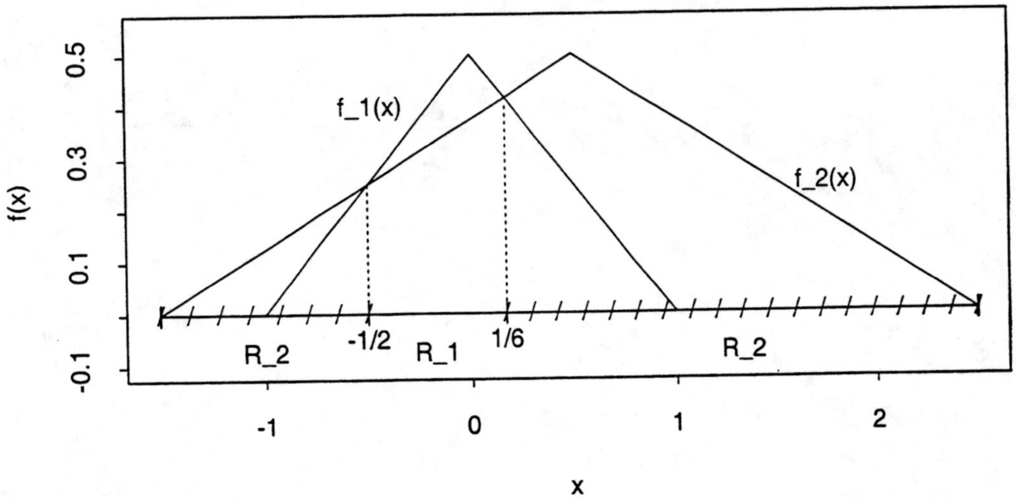

(b) When $p_1 = p_2$ and $c(1|2) = c(2|1)$, the classification regions are

$$R_1 : \frac{f_1(x)}{f_2(x)} \geq 1 \qquad R_2 : \frac{f_1(x)}{f_2(x)} < 1$$

These regions are given by

$$R_1 : -1/2 \le x < 1/6 \quad \text{and} \quad R_2 = -1.5 \le x < -1/2, \quad 1/6 \le x \le 2.5$$

11.9

$$\frac{\boldsymbol{\ell}'B_0\boldsymbol{\ell}}{\boldsymbol{\ell}'\boldsymbol{\Sigma}\boldsymbol{\ell}} = \frac{\boldsymbol{\ell}'[(\boldsymbol{\mu}_1-\bar{\boldsymbol{\mu}})(\boldsymbol{\mu}_1-\bar{\boldsymbol{\mu}})' + (\boldsymbol{\mu}_2-\bar{\boldsymbol{\mu}})(\boldsymbol{\mu}_2-\bar{\boldsymbol{\mu}})']\boldsymbol{\ell}}{\boldsymbol{\ell}'\boldsymbol{\Sigma}\boldsymbol{\ell}}$$

where $\bar{\boldsymbol{\mu}} = \frac{1}{2}(\boldsymbol{\mu}_1 + \boldsymbol{\mu}_2)$. Thus $\boldsymbol{\mu}_1 - \bar{\boldsymbol{\mu}} = \frac{1}{2}(\boldsymbol{\mu}_1 - \boldsymbol{\mu}_2)$ and $\boldsymbol{\mu}_2 - \bar{\boldsymbol{\mu}} = \frac{1}{2}(\boldsymbol{\mu}_2 - \boldsymbol{\mu}_1)$ so

$$\frac{\boldsymbol{\ell}'B_0\boldsymbol{\ell}}{\boldsymbol{\ell}'\boldsymbol{\Sigma}\boldsymbol{\ell}} = \frac{\frac{1}{2}\,\boldsymbol{\ell}'(\boldsymbol{\mu}_1-\boldsymbol{\mu}_2)(\boldsymbol{\mu}_1-\boldsymbol{\mu}_2)'\boldsymbol{\ell}}{\boldsymbol{\ell}'\boldsymbol{\Sigma}\boldsymbol{\ell}} .$$

11.10 **(a)** Hotelling's two-sample T^2-statistic is

$$T^2 = (\bar{\boldsymbol{x}}_1 - \bar{\boldsymbol{x}}_2)' \left[\left(\frac{1}{n_1} + \frac{1}{n_2}\right) \boldsymbol{S}_{\text{pooled}}\right]^{-1} (\bar{\boldsymbol{x}}_1 - \bar{\boldsymbol{x}}_2)$$

$$= [-3 \quad -2] \left[\left(\frac{1}{11} + \frac{1}{12}\right) \begin{bmatrix} 7.3 & -1.1 \\ -1.1 & 4.8 \end{bmatrix}\right]^{-1} \begin{bmatrix} -3 \\ -2 \end{bmatrix} = 14.52$$

Under $H_0 : \mu_1 = \mu_2$,

$$T^2 \sim \frac{(n_1 + n_2 - 2)p}{n_1 + n_2 - p - 1} F_{p, n_1 + n_2 - p - 1}$$

Since $T^2 = 14.52 \geq \frac{(11+12-2)2}{11+12-2-1} F_{2,20}(.1) = 5.44$, we reject the null hypothesis $H_0 : \boldsymbol{\mu}_1 = \boldsymbol{\mu}_2$ at the $\alpha = 0.1$ level of significance.

(b) Fisher's linear discriminant function is

$$\hat{y}_0 = \hat{\boldsymbol{a}}' \boldsymbol{x}_0 = -.49x_1 - .53x_2$$

(c) Here, $\hat{m} = -.25$. Assign $\boldsymbol{x}_0'$ to π_1 if $-.49x_1 - .53x_2 + .25 \geq 0$. Otherwise assign $\boldsymbol{x}_0'$ to π_2.

For $\boldsymbol{x}_0' = [0 \quad 1], \hat{y}_0 = -.53(1) = -.53$ and $\hat{y}_0 - \hat{m} = -.28 < 0$. Thus, assign $\boldsymbol{x}_0$ to π_2.

11.11 Assuming equal prior probabilities $p_1 = p_2 = \frac{1}{2}$, and equal misclassification costs $c(2|1) = c(1|2) = \$10$:

c	$P(B1\|A2)$	$P(B2\|A1)$	$P(A2 \text{ and } B1)$	$P(A1 \text{ and } B2)$	$P(\text{error})$	Expected cost
9	.006	.691	.346	.003	.349	3.49
10	.023	.500	.250	.011	.261	2.61
11	.067	.309	.154	.033	.188	1.88
12	.159	.159	.079	.079	.159	1.59
13	.309	.067	.033	.154	.188	1.88
14	.500	.023	.011	.250	.261	2.61

Using (11-28), the expected cost is minimized for $c = 12$ and the minimum expected cost is \$1.59.

11.12 Assuming equal prior probabilities $p_1 = p_2 = \frac{1}{2}$, and misclassification costs $c(2|1) =$ \$5 and $c(1|2) = \$10$,

expected cost $= \$5P(A1 \text{ and } B2) + \$15P(A2 \text{ and } B1)$.

c	$P(B1\|A2)$	$P(B2\|A1)$	$P(A2 \text{ and } B1)$	$P(A1 \text{ and } B2)$	$P(\text{error})$	Expected cost
9	0.006	0.691	0.346	0.003	0.349	1.78
10	0.023	0.500	0.250	0.011	0.261	1.42
11	0.067	0.309	0.154	0.033	0.188	1.27
12	0.159	0.159	0.079	0.079	0.159	1.59
13	0.309	0.067	0.033	0.154	0.188	2.48
14	0.500	0.023	0.011	0.250	0.261	3.81

Using (11-28), the expected cost is minimized for $c = 10.90$ and the minimum expected cost is \$1.27.

11.13 Assuming prior probabilities $P(A1) = 0.25$ and $P(A2) = 0.75$, and misclassification costs $c(2|1) = \$5$ and $c(1|2) = \$10$,

expected cost $= \$5P(B2|A1)(.25) + \$15P(B1|A2)(.75)$.

c	$P(B1\|A2)$	$P(B2\|A1)$	$P(A2 \text{ and } B1)$	$P(A1 \text{ and } B2)$	$P(\text{error})$	Expected cost
9	0.006	0.691	0.173	0.005	0.178	0.93
10	0.023	0.500	0.125	0.017	0.142	0.88
11	0.067	0.309	0.077	0.050	0.127	1.14
12	0.159	0.159	0.040	0.119	0.159	1.98
13	0.309	0.067	0.017	0.231	0.248	3.56
14	0.500	0.023	0.006	0.375	0.381	5.65

Using (11-28), the expected cost is minimized for $c = 9.80$ and the minimum expected cost is \$0.88.

11.14 Using (11-21),

$$\hat{\boldsymbol{a}}_1^* = \frac{\hat{\boldsymbol{a}}}{\sqrt{\hat{\boldsymbol{a}}'\hat{\boldsymbol{a}}}} = \begin{bmatrix} .79 \\ -.61 \end{bmatrix} \quad \text{and} \quad \hat{m}_1^* = -0.10$$

Since $\hat{\boldsymbol{a}}_1^* \boldsymbol{x}_0 = -0.14 < \hat{m}_1^* = -0.1$, classify $\boldsymbol{x}_0$ as π_2.

Using (11-22),

$$\hat{\boldsymbol{a}}_2^* = \frac{\hat{\boldsymbol{a}}}{\hat{a}_1} = \begin{bmatrix} 1.00 \\ -.77 \end{bmatrix} \quad \text{and} \quad \hat{m}_2^* = -0.12$$

Since $\hat{\boldsymbol{a}}_2^* \boldsymbol{x}_0 = -0.18 < \hat{m}_2^* = -0.12$, classify $\boldsymbol{x}_0$ as π_2.

These results are consistent with the classification obtained for the case of equal prior probabilities in Example 11.3. These two classification results should be identical to those of Example 11.3.

11.15 $\frac{f_1(\underset{\sim}{x})}{f_2(\underset{\sim}{x})} \geq \left[\frac{c(1|2)}{c(2|1)}\frac{p_2}{p_1}\right]$ defines the same region as

$\ln f_1(\underset{\sim}{x}) - \ln f_2(\underset{\sim}{x}) \geq \ln\left[\frac{c(1|2)}{c(2|1)}\frac{p_2}{p_1}\right]$. For a multivariate normal distribution

$$\ln f_i(\underset{\sim}{x}) = -\frac{1}{2}\ln|\Sigma_i| - \frac{p}{2}\ln 2\pi - \frac{1}{2}(\underset{\sim}{x}-\underset{\sim}{\mu}_i)'\Sigma_i^{-1}(\underset{\sim}{x}-\underset{\sim}{\mu}_i),\ i=1,2$$

so

$$\ln f_1(\underset{\sim}{x}) - \ln f_2(\underset{\sim}{x}) = -\frac{1}{2}(\underset{\sim}{x}-\underset{\sim}{\mu}_1)'\Sigma_1^{-1}(\underset{\sim}{x}-\underset{\sim}{\mu}_1)$$

$$+\frac{1}{2}(\underset{\sim}{x}-\underset{\sim}{\mu}_2)'\Sigma_2^{-1}(\underset{\sim}{x}-\underset{\sim}{\mu}_2) - \frac{1}{2}\ln\left(\frac{|\Sigma_1|}{|\Sigma_2|}\right)$$

$$= -\frac{1}{2}[\underset{\sim}{x}'\Sigma_1^{-1}\underset{\sim}{x} - 2\underset{\sim}{\mu}_1'\Sigma_1^{-1}\underset{\sim}{x} + \underset{\sim}{\mu}_1'\Sigma_1^{-1}\underset{\sim}{\mu}_1$$

$$-\underset{\sim}{x}'\Sigma_2^{-1}\underset{\sim}{x} + 2\underset{\sim}{\mu}_2'\Sigma_2^{-1}\underset{\sim}{x} - \underset{\sim}{\mu}_2'\Sigma_2^{-1}\underset{\sim}{\mu}_2] - \frac{1}{2}\ln\left(\frac{|\Sigma_1|}{|\Sigma_2|}\right)$$

$$= -\frac{1}{2}\underset{\sim}{x}'(\Sigma_1^{-1} - \Sigma_2^{-1})\underset{\sim}{x} + (\underset{\sim}{\mu}_1'\Sigma_1^{-1} - \underset{\sim}{\mu}_2'\Sigma_2^{-1})\underset{\sim}{x} - k$$

where $k = \frac{1}{2}\ln\left(\frac{|\Sigma_1|}{|\Sigma_2|}\right) + \frac{1}{2}(\underset{\sim}{\mu}_1'\Sigma_1^{-1}\underset{\sim}{\mu}_1 - \underset{\sim}{\mu}_2'\Sigma_2^{-1}\underset{\sim}{\mu}_2)$.

11.16

$$Q = \ln\left[\frac{f_1(\underset{\sim}{x})}{f_2(\underset{\sim}{x})}\right] = -\frac{1}{2}\ln|\Sigma_1| - \frac{1}{2}(\underset{\sim}{x}-\underset{\sim}{\mu}_1)'\Sigma_1^{-1}(\underset{\sim}{x}-\underset{\sim}{\mu}_1)$$

$$+\frac{1}{2}\ln|\Sigma_2| + \frac{1}{2}(\underset{\sim}{x}-\underset{\sim}{\mu}_2)'\Sigma_1^{-1}(\underset{\sim}{x}-\underset{\sim}{\mu}_2)$$

$$= -\frac{1}{2}\underset{\sim}{x}'(\Sigma_1^{-1}-\Sigma_2^{-1})\underset{\sim}{x} + \underset{\sim}{x}'\Sigma_1^{-1}\underset{\sim}{\mu}_1 - \underset{\sim}{x}'\Sigma_2^{-1}\underset{\sim}{\mu}_2 - k$$

where $k = \frac{1}{2}\left[\ln\left(\frac{|\Sigma_1|}{|\Sigma_2|}\right) + \underset{\sim}{\mu}_1'\Sigma_1^{-1}\underset{\sim}{\mu}_1 - \underset{\sim}{\mu}_2'\Sigma_2^{-1}\underset{\sim}{\mu}_2\right]$.

When $\Sigma_1 = \Sigma_2 = \Sigma$,

$$Q = \underset{\sim}{x}'\Sigma^{-1}\underset{\sim}{\mu}_1 - \underset{\sim}{x}'\Sigma^{-1}\underset{\sim}{\mu}_2 - \frac{1}{2}(\underset{\sim}{\mu}_1'\Sigma^{-1}\underset{\sim}{\mu}_1 - \underset{\sim}{\mu}_2'\Sigma^{-1}\underset{\sim}{\mu}_2)$$

$$= \underset{\sim}{x}'\Sigma^{-1}(\underset{\sim}{\mu}_1 - \underset{\sim}{\mu}_2) - \frac{1}{2}(\underset{\sim}{\mu}_1 - \underset{\sim}{\mu}_2)'\Sigma^{-1}(\underset{\sim}{\mu}_1 + \underset{\sim}{\mu}_2)$$

11.17 Assuming equal prior probabilities and misclassification costs $c(2|1) = \$10$ and $c(1|2) = \$73.89$. In the table below,

$$\begin{aligned} Q &= -\frac{1}{2}\boldsymbol{x}_0'(\Sigma_1^{-1} - \Sigma_2^{-1})\boldsymbol{x}_0 + (\boldsymbol{\mu}_1'\Sigma_1^{-1} - \boldsymbol{\mu}_2'\Sigma_2^{-1})\boldsymbol{x}_0 \\ &\quad -\frac{1}{2}\ln\left(\frac{|\Sigma_1|}{|\Sigma_2|}\right) - \frac{1}{2}(\boldsymbol{\mu}_1'\Sigma_1^{-1}\boldsymbol{\mu}_1 - \boldsymbol{\mu}_2'\Sigma_2^{-1}\boldsymbol{\mu}_2) \end{aligned}$$

x	$P(\pi_1\|x)$	$P(\pi_2\|x)$	Q	Classification
[10, 15]′	1.00000	0	18.54	π_1
[12, 17]′	0.99991	0.00009	9.36	π_1
[14, 19]′	0.95254	0.04745	3.00	π_1
[16, 21]′	0.36731	0.63269	-0.54	π_2
[18, 23]′	0.21947	0.78053	-1.27	π_2
[20, 25]′	0.69517	0.30483	0.87	π_2
[22, 27]′	0.99678	0.00322	5.74	π_1
[24, 29]′	1.00000	0.00000	13.46	π_1
[26, 31]′	1.00000	0.00000	24.01	π_1
[28, 33]′	1.00000	0.00000	37.38	π_1
[30, 35]′	1.00000	0.00000	53.56	π_1

The quadratic discriminator was used to classify the observations in the above table. An observation $\boldsymbol{x}$ is classified as π_1 if

$$Q \geq \ln\left[\left(\frac{c(1|2)}{c(2|1)}\right)\left(\frac{p_2}{p_1}\right)\right] = \ln\left(\frac{73.89}{10}\right) = 2.0$$

Otherwise, classify $\boldsymbol{x}$ as π_2.

For (a), (b), (c) and (d), see the following plot.

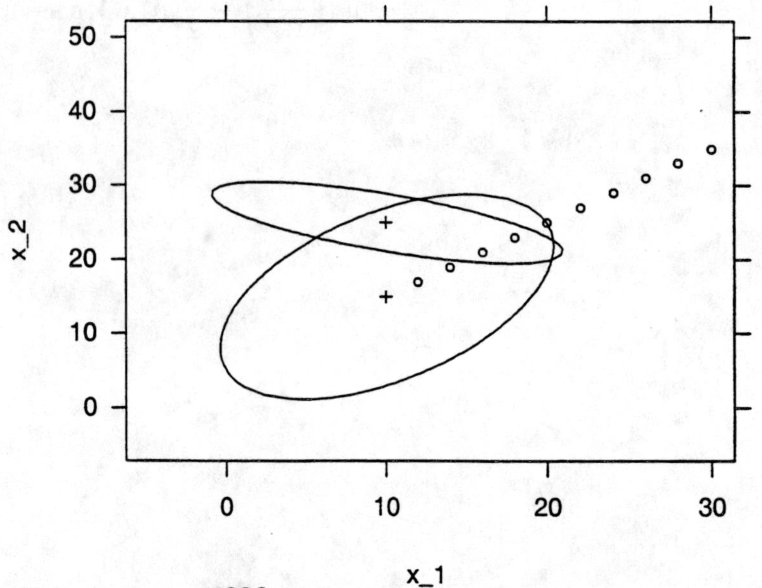

11.18 The vector $\underset{\sim}{e}$ is an (unscaled) eigenvector of $\Sigma^{-1}B$ since

$$\Sigma^{-1}B\underset{\sim}{e} = \Sigma^{-1}c(\underset{\sim}{\mu}_1-\underset{\sim}{\mu}_2)(\underset{\sim}{\mu}_1-\underset{\sim}{\mu}_2)'c\Sigma^{-1}(\underset{\sim}{\mu}_1-\underset{\sim}{\mu}_2)$$

$$= c^2\Sigma^{-1}(\underset{\sim}{\mu}_1-\underset{\sim}{\mu}_2)(\underset{\sim}{\mu}_1-\underset{\sim}{\mu}_2)'\Sigma^{-1}(\underset{\sim}{\mu}_1-\underset{\sim}{\mu}_2)$$

$$= \lambda\, \Sigma^{-1}(\underset{\sim}{\mu}_1-\underset{\sim}{\mu}_2) = \lambda\, \underset{\sim}{e}$$

where $\lambda = c^2\, (\underset{\sim}{\mu}_1-\underset{\sim}{\mu}_2)'\Sigma^{-1}(\underset{\sim}{\mu}_1-\underset{\sim}{\mu}_2)$.

11.19 (a) The calculated values agree with those in Example 11.6.

(b) Fisher's linear discriminant function is

$$\hat{y}_0 = \hat{\boldsymbol{a}}'\boldsymbol{x}_0 = -\frac{1}{3}x_1 + \frac{2}{3}x_2$$

where

$$\overline{y}_1 = \frac{17}{3};\ \overline{y}_2 = \frac{10}{3};\ \hat{m} = \frac{27}{6} = 4.5$$

Assign $\boldsymbol{x}_0'$ to π_1 if $-\frac{1}{3}x_1 + \frac{2}{3}x_2 - 4.5 \geq 0$

Otherwise assign $\boldsymbol{x}_0'$ to π_2.

π_1

Observation	$\hat{\boldsymbol{a}}'\boldsymbol{x}_0 - \hat{m}$	Classification
1	2.83	π_1
2	0.83	π_1
3	-0.17	π_2

π_2

Observation	$\hat{\boldsymbol{a}}'\boldsymbol{x}_0 - \hat{m}$	Classification
1	-1.50	π_2
2	0.50	π_1
3	-2.50	π_2

The results from this table verify the confusion matrix given in Example 11.6.

(c) This is the table of squared distances $\hat{D}_i^2(\boldsymbol{x})$ for the observations, where

$$D_i^2(\boldsymbol{x}) = (\boldsymbol{x} - \overline{\boldsymbol{x}}_i)' \boldsymbol{S}_{\text{pooled}}^{-1} (\boldsymbol{x} - \overline{\boldsymbol{x}}_i)$$

π_1				π_2			
Obs.	$\hat{D}_1^2(\boldsymbol{x})$	$\hat{D}_2^2(\boldsymbol{x})$	Classification	Obs.	$\hat{D}_1^2(\boldsymbol{x})$	$\hat{D}_2^2(\boldsymbol{x})$	Classification
1	$\frac{4}{3}$	$\frac{21}{3}$	π_1	1	$\frac{13}{3}$	$\frac{4}{3}$	π_2
2	$\frac{4}{3}$	$\frac{9}{3}$	π_1	2	$\frac{1}{3}$	$\frac{4}{3}$	π_1
3	$\frac{4}{3}$	$\frac{3}{3}$	π_2	3	$\frac{19}{3}$	$\frac{4}{3}$	π_2

The classification results are identical to those obtained in **(b)**

11.20 The result obtained from this matrix identity is identical to the result of Example 11.6

11.23 **(a)** Here are the normal probability plots for each of the variables x_1, x_2, x_3, x_4, x_5

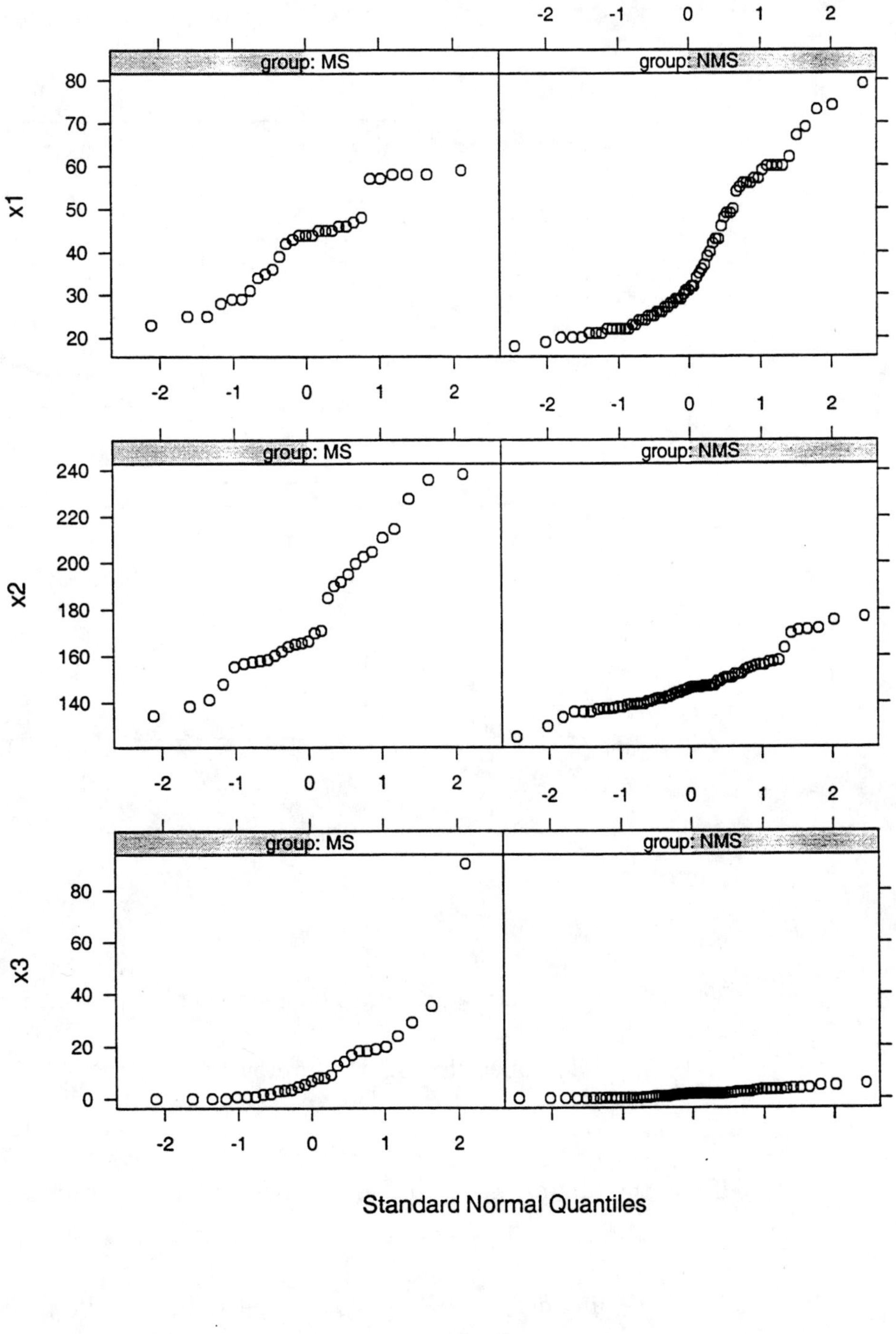
group: MS
group: NMS
x1
x2
x3
Standard Normal Quantiles

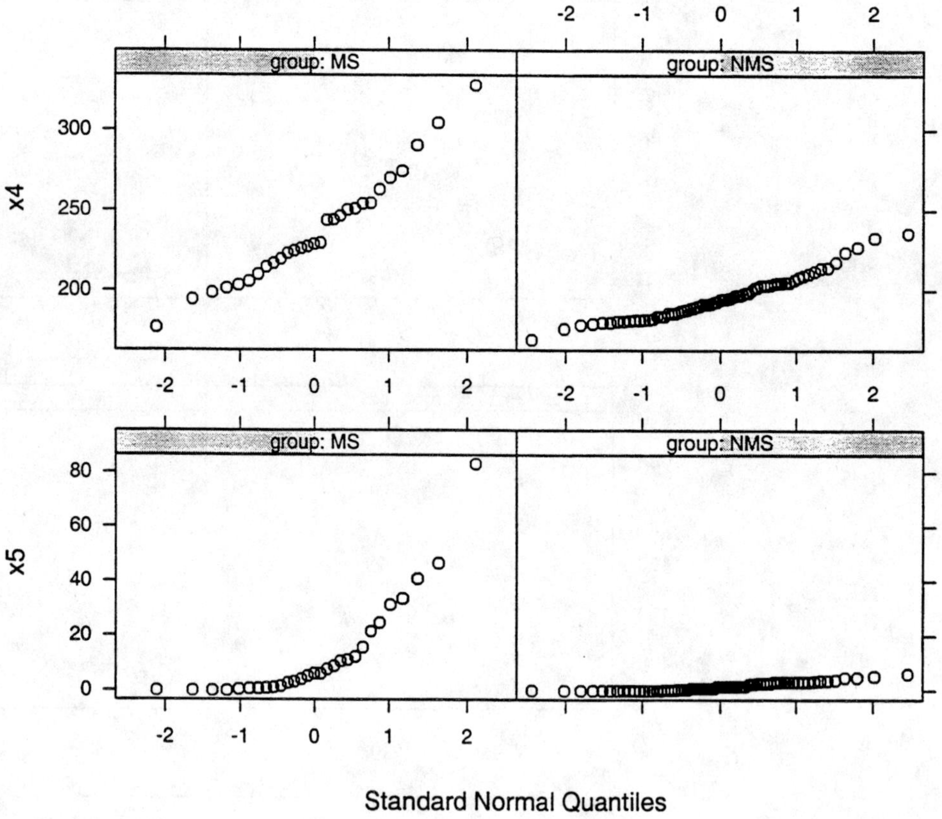

Variables x_1, x_3, and x_5 appear to be nonnormal. The transformations $\ln(x_1), \ln(x_3 + 1)$, and $\ln(x_5 + 1)$ appear to slightly improve normality.

(b) Using the original data, the linear discriminant function is:

$$\hat{y} = \hat{\boldsymbol{a}}'\boldsymbol{x} = 0.023x_1 - 0.034x_2 + 0.21x_3 - 0.08x_4 - 0.25x_5$$

where

$$\hat{m} = -23.23$$

Thus, we allocate $\boldsymbol{x}_0$ to π_1 (NMS group) if

$$\hat{\boldsymbol{a}}\boldsymbol{x}_0 - \hat{m} = 0.023x_1 - 0.034x_2 + 0.21x_3 - 0.08x_4 - 0.25x_5 + 23.23 \geq 0$$

Otherwise, allocate $\boldsymbol{x}_0$ to π_2 (MS group).

(c) Confusion matrix:

		Predicted Membership		
		π_1	π_2	Total
Actual membership	π_1	66	3	69
	π_2	7	22	29

APER= $\frac{3+7}{69+29} = .102$

This is the holdout confusion matrix:

		Predicted Membership		
		π_1	π_2	Total
Actual membership	π_1	64	5	69
	π_2	8	21	29

$\hat{E}(\text{AER}) = \frac{5+8}{69+29} = .133$

11.24 (a) Here are the scatterplots for the pairs of observations $(x_1, x_2), (x_1, x_3)$, and (x_1, x_4):

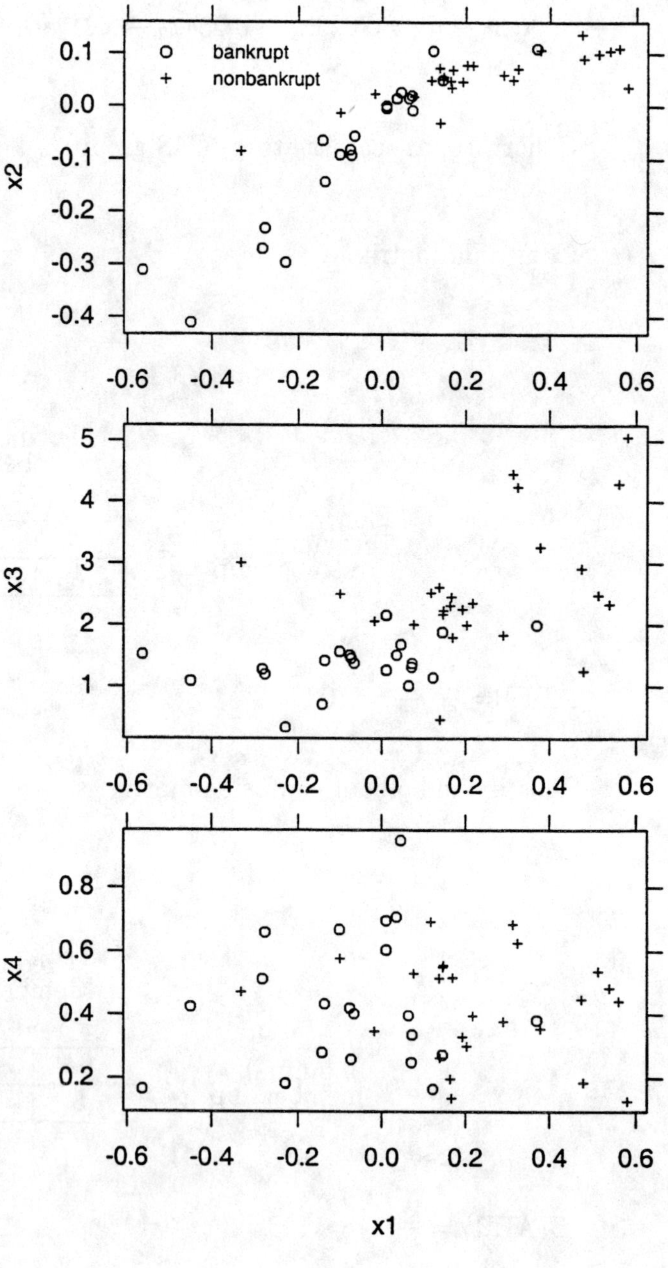

The data in the above plot appear to form fairly elliptical shapes, so bivariate normality does not seem like an unreasonable assumption.

(b) $\pi_1 \equiv$ bankrupt firms, $\pi_2 \equiv$ nonbankrupt firms. For (x_1, x_2):

$$\bar{\boldsymbol{x}}_1 = \begin{bmatrix} -0.0688 \\ -0.0819 \end{bmatrix}, \quad \boldsymbol{S}_1 = \begin{bmatrix} 0.04424 & 0.02847 \\ 0.02847 & 0.02092 \end{bmatrix}$$

$$\bar{\boldsymbol{x}}_2 = \begin{bmatrix} 0.2354 \\ 0.0551 \end{bmatrix}, \quad \boldsymbol{S}_2 = \begin{bmatrix} 0.04735 & 0.00837 \\ 0.00837 & 0.00231 \end{bmatrix}$$

(c), (d), (e) See the tables of part **(g)**

(f)

$$\boldsymbol{S}_{\text{pooled}} = \begin{bmatrix} 0.04594 & 0.01751 \\ 0.01751 & 0.01077 \end{bmatrix}$$

Fisher's linear discriminant function is

$$\hat{y} = \hat{\boldsymbol{a}}'\boldsymbol{x} = -4.67x_1 - 5.12x_2$$

where

$$\hat{m} = -.32$$

Thus, we allocate $\boldsymbol{x}_0$ to π_1 (Bankrupt group) if

$$\hat{\boldsymbol{a}}\boldsymbol{x}_0 - \hat{m} = -4.67x_1 - 5.12x_2 + .32 \geq 0$$

Otherwise, allocate $\boldsymbol{x}_0$ to π_2 (Nonbankrupt group).

APER= $\frac{9}{46} = .196$.

Since $\boldsymbol{S}_1$ and $\boldsymbol{S}_2$ look quite different, Fisher's linear discriminant function may not be appropriate. However the performance of this linear discriminant function is as good as that of the quadratic discriminant function, based on the APER criterion.

(g) For (x_1, x_3),

$$\overline{\boldsymbol{x}}_1 = \begin{bmatrix} -0.0688 \\ 1.3675 \end{bmatrix}, \quad \boldsymbol{S}_1 = \begin{bmatrix} 0.04424 & 0.03428 \\ 0.03428 & 0.16455 \end{bmatrix}$$

$$\overline{\boldsymbol{x}}_2 = \begin{bmatrix} 0.2354 \\ 2.5939 \end{bmatrix}, \quad \boldsymbol{S}_2 = \begin{bmatrix} 0.04735 & 0.07543 \\ 0.07543 & 1.04596 \end{bmatrix}$$

For (x_1, x_4),

$$\overline{\boldsymbol{x}}_1 = \begin{bmatrix} -0.0688 \\ 0.4368 \end{bmatrix}, \quad \boldsymbol{S}_1 = \begin{bmatrix} 0.04424 & 0.00431 \\ 0.00431 & 0.04441 \end{bmatrix}$$

$$\overline{\boldsymbol{x}}_2 = \begin{bmatrix} 0.2354 \\ 0.4264 \end{bmatrix}, \quad \boldsymbol{S}_2 = \begin{bmatrix} 0.04735 & -0.00662 \\ -0.00662 & 0.02618 \end{bmatrix}$$

For the various classification rules and error rates for these variable pairs, see the following tables.

This is the table of quadratic functions for the variable pairs (x_1, x_2), (x_1, x_3), and (x_1, x_5), both with $p_1 = 0.5$ and $p_1 = 0.05$. The classification rule for any of these functions is to classify a new observation into π_1 (bankrupt firms) if the quadratic function is ≥ 0, and to classify the new observation into

π_2 (nonbankrupt firms) otherwise. Notice in the table below that only the constant term changes when the prior probabilities change.

Variables	Prior	Quadratic function		
(x_1, x_2)	$p_1 = 0.5$	$-61.77x_1^2 + 35.84x_1x_2 + 407.20x_2^2 + 5.64x_1 - 30.60x_2$	$-$	0.17
	$p_1 = 0.05$		$-$	3.11
(x_1, x_3)	$p_1 = 0.5$	$-1.55x_1^2 + 3.89x_1x_3 - 3.08x_3^2 - 10.69x_1 + 7.90x_3$	$-$	3.14
	$p_1 = 0.05$		$-$	6.08
(x_1, x_4)	$p_1 = 0.5$	$-0.46x_1^2 + 7.75x_1x_4 + 8.43x_4^2 - 10.05x_1 - 8.11x_4$	$+$	2.23
	$p_1 = 0.05$		$-$	0.71

Here is a table of the APER and $\hat{E}$(AER) for the various variable pairs and prior probabilities.

	APER		$\hat{E}$(APR)	
Variables	$p_1 = 0.5$	$p_1 = 0.05$	$p_1 = 0.5$	$p_1 = 0.05$
(x_1, x_2)	0.20	0.26	0.22	0.26
(x_1, x_3)	0.11	0.37	0.13	0.39
(x_1, x_4)	0.17	0.39	0.22	0.46

For equal priors, it appears that the (x_1, x_3) variable pair is the best classifier, as it has the lowest APER. For unequal priors, $p_1 = 0.05$ and $p_2 = 0.95$, the variable pair (x_1, x_2) has the lowest APER.

(h) When using all four variables (X_1, X_2, X_3, X_4),

$$\overline{\boldsymbol{x}}_1 = \begin{bmatrix} -0.0688 \\ -0.0819 \\ 1.3675 \\ 0.4368 \end{bmatrix}, \quad \boldsymbol{S}_1 = \begin{bmatrix} 0.04424 & 0.02847 & 0.03428 & 0.00431 \\ 0.02847 & 0.02092 & 0.02580 & 0.00362 \\ 0.03428 & 0.02580 & 0.16455 & 0.03300 \\ 0.00431 & 0.00362 & 0.03300 & 0.04441 \end{bmatrix}$$

$$\overline{\boldsymbol{x}}_2 = \begin{bmatrix} 0.2354 \\ 0.0551 \\ 2.5939 \\ 0.4264 \end{bmatrix}, \quad \boldsymbol{S}_2 = \begin{bmatrix} 0.04735 & 0.00837 & 0.07543 & -0.00662 \\ 0.00837 & 0.00231 & 0.00873 & 0.00031 \\ 0.07543 & 0.00873 & 1.04596 & 0.03177 \\ -0.00662 & 0.00031 & 0.03177 & 0.02618 \end{bmatrix}$$

Assign a new observation $\boldsymbol{x}_0$ to π_1 if its quadratic function given below is less than 0:

Prior	Quadratic function
$p_1 = 0.5$	$\boldsymbol{x}_0' \begin{bmatrix} -49.232 & -20.657 & -2.623 & 14.050 \\ -20.657 & 526.336 & 11.412 & -52.493 \\ -2.623 & 11.412 & -3.748 & 1.4337 \\ 14.050 & -52.493 & 1.434 & 11.974 \end{bmatrix} \boldsymbol{x}_0 + \begin{bmatrix} 4.91 \\ -28.42 \\ 8.65 \\ -11.80 \end{bmatrix}' \boldsymbol{x}_0 - 2.69$
$p_1 = 0.05$	$- 5.64$

For $p_1 = 0.5$: APER $= \frac{3}{46} = .07$, $\quad \hat{E}(\text{AER}) = \frac{5}{46} = .11$

For $p_1 = 0.05$: APER $= \frac{9}{46} = .20$, $\quad \hat{E}(\text{AER}) = \frac{11}{46} = .24$

11.25 (a) Fisher's linear discriminant function is

$$\hat{y}_0 = \boldsymbol{a}'\boldsymbol{x}_0 - \hat{m} = -4.80x_1 - 1.48x_3 + 3.33$$

Classify $\boldsymbol{x}_0$ to π_1 (bankrupt firms) if

$$\boldsymbol{a}'\boldsymbol{x}_0 - \hat{m} \geq 0$$

Otherwise classify $\boldsymbol{x}_0$ to π_2 (nonbankrupt firms).

The APER is $\frac{2+4}{46} = .13$.

This is the scatterplot of the data in the (x_1, x_3) coordinate system, along with the discriminant line.

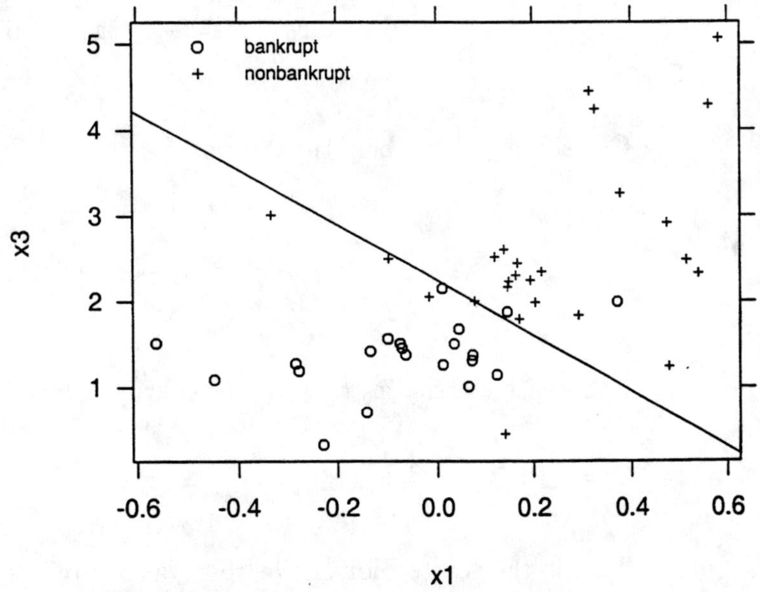

(b) With data point 16 for the bankrupt firms deleted, Fisher's linear discriminant function is given by

$$\hat{y}_0 = \boldsymbol{a}'\boldsymbol{x}_0 - \hat{m} = -5.93x_1 - 1.46x_3 + 3.31$$

Classify $\boldsymbol{x}_0$ to π_1 (bankrupt firms) if

$$\boldsymbol{a}'\boldsymbol{x}_0 - \hat{m} \geq 0$$

Otherwise classify $\boldsymbol{x}_0$ to π_2 (nonbankrupt firms).

The APER is $\frac{1+4}{45} = .11$.

With data point 13 for the nonbankrupt firms deleted, Fisher's linear discriminant function is given by

$$\hat{y}_0 = \boldsymbol{a}'\boldsymbol{x}_0 - \hat{m} = -4.35x_1 - 1.97x_3 + 4.36$$

Classify $\boldsymbol{x}_0$ to π_1 (bankrupt firms) if

$$\boldsymbol{a}'\boldsymbol{x}_0 - \hat{m} \geq 0$$

Otherwise classify $\boldsymbol{x}_0$ to π_2 (nonbankrupt firms).

The APER is $\frac{1+3}{45} = .089$.

This is the scatterplot of the observations in the (x_1, x_3), coordinate system with the discriminant lines for the three linear discriminant functions given above. Also labelled are observation 16 for bankrupt firms and observation

13 for nonbankrupt firms.

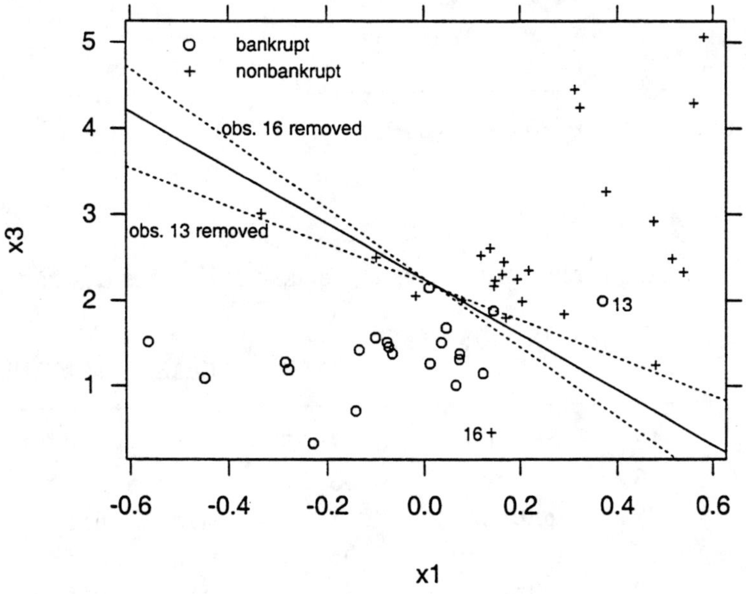

It appears that deleting these observations has changed the line significantly.

11.26 **(a)** The least squares regression results for the X, Z data are:

```
Parameter Estimates

                   Parameter       Standard      T for H0:
Variable  DF        Estimate          Error    Parameter=0     Prob > |T|

INTERCEP   1       -0.081412     0.13488497         -0.604         0.5492
X3         1        0.307221     0.05956685          5.158         0.0001
```

Here are the dot diagrams of the fitted values for the bankrupt firms and for the nonbankrupt firms:

```
          . :
 .   . ::::::. . ..
+---------+---------+---------+---------+---------+-------Bankrupt
                 .  .. .
  .      .  .  :  : ::.:. .. .      :  .    .
+---------+---------+---------+---------+---------+-------Nonbankrupt
0.00      0.30      0.60      0.90      1.20      1.50
```

This table summarizes the classification results using the fitted values:

OBS	GROUP	FITTED	CLASSIFICATION
13	bankrupt	0.57896	misclassify
16	bankrupt	0.53122	misclassify
31	nonbankr	0.47076	misclassify
34	nonbankr	0.06025	misclassify
38	nonbankr	0.48329	misclassify
41	nonbankr	0.30089	misclassify

The confusion matrix is:

		Predicted Membership		
		π_1	π_2	Total
Actual membership	π_1	19	2	21
	π_2	4	21	25

Thus, the APER is $\frac{2+4}{46} = .13$.

(b) The least squares regression results using all four variables X_1, X_2, X_3, X_4 are:

Parameter Estimates

Variable	DF	Parameter Estimate	Standard Error	T for H0: Parameter=0	Prob > |T|
INTERCEP	1	0.208915	0.18615284	1.122	0.2683
X1	1	0.156317	0.46653100	0.335	0.7393
X2	1	1.149093	0.90606395	1.268	0.2119
X3	1	0.225972	0.07030479	3.214	0.0026
X4	1	-0.305175	0.32336357	-0.944	0.3508

Here are the dot diagrams of the fitted values for the bankrupt firms and for the nonbankrupt firms:

```
                       .:
        . .  .. .  ... :: ... . .  .
   -+---------+---------+---------+---------+---------+---------+-----Bankrupt
                                .
                                :  . .
                    .        .::: :: :    ..  :   .    .
   -+---------+---------+---------+---------+---------+---------+-----Nonbankrupt
-0.35      0.00      0.35      0.70      1.05      1.40
```

This table summarizes the classification results using the fitted values:

OBS	GROUP	FITTED	CLASSIFICATION
15	bankrupt	0.62997	misclassify
16	bankrupt	0.72676	misclassify
20	bankrupt	0.55719	misclassify
34	nonbankr	0.21845	misclassify

The confusion matrix is:

		Predicted Membership		
		π_1	π_2	Total
Actual membership	π_1	18	3	21
	π_2	1	24	25

Thus, the APER is $\frac{3+1}{46} = .087$. Here is a scatterplot of the residuals against the fitted values, with points 16 of the bankrupt firms and 13 of the non-bankrupt firms labelled. It appears that point 16 of the bankrupt firms is an outlier.

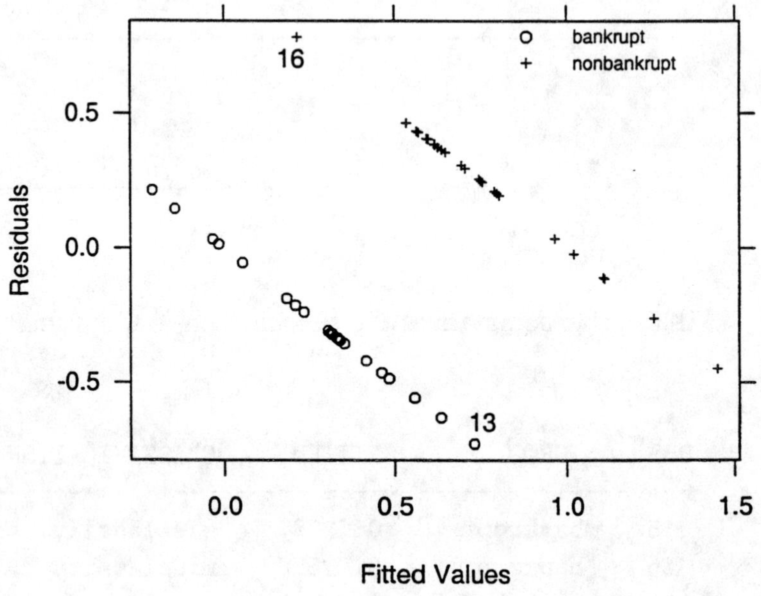

11.27 (a) Plot of the data in the (x_2, x_4) variable space:

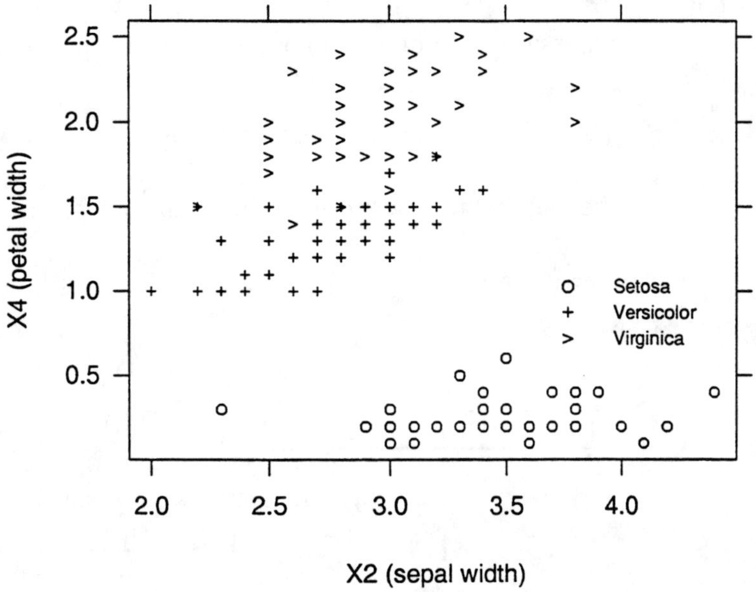

The points from all three groups appear to form an elliptical shape. However, it appears that the points of π_1 *(Iris setosa)* form an ellipse with a different orientation than those of π_2 *(Iris versicolor)* and π_3 *(Iris virginica)*. This indicates that the observations from π_1 may have a different covariance matrix from the observations from π_2 and π_3.

(b) Here are the results of a test of the null hypothesis $H_0 : \boldsymbol{\mu}_1 = \boldsymbol{\mu}_2 = \boldsymbol{\mu}_3$ versus H_1 : at least one of the $\boldsymbol{\mu}_i$'s is different from the others at the $\alpha = 0.05$ level of significance:

Statistic	Value	F	Num DF	Den DF	Pr > F
Wilks' Lambda	0.02343863	199.145	8	288	0.0001

Thus, the null hypothesis $H_0 : \boldsymbol{\mu}_1 = \boldsymbol{\mu}_2 = \boldsymbol{\mu}_3$ is rejected at the $\alpha = 0.05$ level of significance. As discussed earlier, the plots give us reason to doubt the assumption of equal covariance matrices for the three groups.

(c) $\pi_1 \equiv$ *Iris setosa*; $\pi_2 \equiv$ *Iris versicolor* $\pi_3 \equiv$ *Iris virginica*

The quadratic discriminant scores $\hat{d}_i^Q(\boldsymbol{x})$ given by (11-47) with $p_1 = p_2 = p_3 = \frac{1}{3}$ are:

population	$\hat{d}_i^Q(\boldsymbol{x}) = -\frac{1}{2}\ln\lvert\boldsymbol{S}_i\rvert - \frac{1}{2}(\boldsymbol{x}-\overline{\boldsymbol{x}}_i)'\boldsymbol{S}_i^{-1}(\boldsymbol{x}-\overline{\boldsymbol{x}}_i)$
π_1	$-3.68x_2^2 + 6.16x_2x_4 - 47.60x_4^2 + 23.71x_2 + 2.30x_4 - 37.67$
π_2	$-9.09x_2^2 + 19.57x_2x_4 - 22.87x_4^2 + 24.94x_2 + 7.63x_4 - 36.53$
π_3	$-6.76x_2^2 + 8.54x_2x_4 - 9.32x_4^2 + 22.92x_2 + 12.38x_4 - 44.04$

To classify the observation $\boldsymbol{x}_0' = [3.5 \quad 1.75]$, compute $\hat{d}_i^Q(\boldsymbol{x}_0)$ for $i = 1, 2, , 3$, and classify $\boldsymbol{x}_0$ to the population for which $\hat{d}_i^Q(\boldsymbol{x}_0)$ is the largest.

$$\hat{d}_1^Q(\boldsymbol{x}_0) = -103.77$$

$$\hat{d}_2^Q(\boldsymbol{x}_0) = 0.043$$

$$\hat{d}_3^Q(\boldsymbol{x}_0) = -1.23$$

So classify $\boldsymbol{x}_0$ to π_2 (*Iris versicolor*).

(d) The linear discriminant scores $\hat{d}_i(\boldsymbol{x})$ are:

population	$\hat{d}_i(\boldsymbol{x}) = \overline{\boldsymbol{x}}_i'\boldsymbol{S}_{\text{pooled}}\boldsymbol{x} - \frac{1}{2}\overline{\boldsymbol{x}}_i'\boldsymbol{S}_{\text{pooled}}\overline{\boldsymbol{x}}_i$	$\hat{d}_i(\boldsymbol{x}_0)$
π_1	$36.02x_2 - 22.26x_4 - 59.00$	28.12
π_2	$19.31x_2 + 16.58x_4 - 37.73$	58.86
π_3	$15.49x_2 + 36.28x_4 - 59.78$	57.92

Since $\hat{d}_i(\boldsymbol{x}_0)$ is the largest for $i = 2$, we classify the new observation $\boldsymbol{x}_0' = [3.5 \quad 1.75]$ to π_1 according to (11-52). The results are the same for **(c)** and **(d)**.

(e) To use rule (11-56), construct $\hat{d}_{ki}(\boldsymbol{x}) = \hat{d}_k(\boldsymbol{x}) - \hat{d}_i(\boldsymbol{x})$ for all $i \neq k$. Then classify $\boldsymbol{x}$ to π_k if $\hat{d}_{ki}(\boldsymbol{x}) \geq 0$ for all $i = 1, 2, 3$. Here is a table of $\hat{d}_{ki}(\boldsymbol{x}_0)$ for $i, k = 1, 2, 3$:

		i 1	2	3
	1	0	-30.74	-29.80
j	2	30.74	0	0.94
	1	29.80	-0.94	0

Since $\hat{d}_{ki}(\boldsymbol{x}_0) \geq 0$ for all $i \neq 2$, we allocate $\boldsymbol{x}_0$ to π_2, using (11-52)

Here is the scatterplot of the data in the (x_2, x_4) variable space, with the classification regions $\hat{R}_1$, $\hat{R}_2$, and $\hat{R}_3$ delineated.

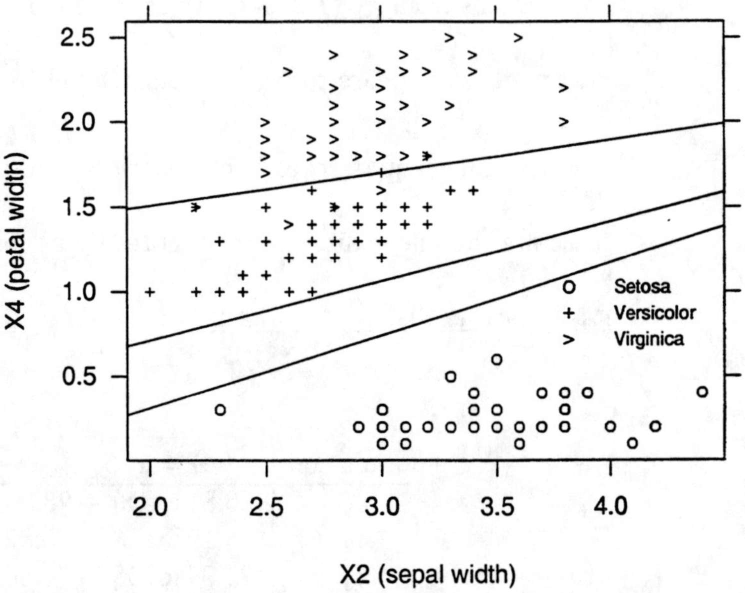

(f) The APER $= \frac{1+4}{150} = .033$. $\hat{E}(\text{AER}) = \frac{4+2}{150} = .04$

11.28 (a) This is the plot of the data in the $(\log Y_1, \log Y_2)$ variable space:

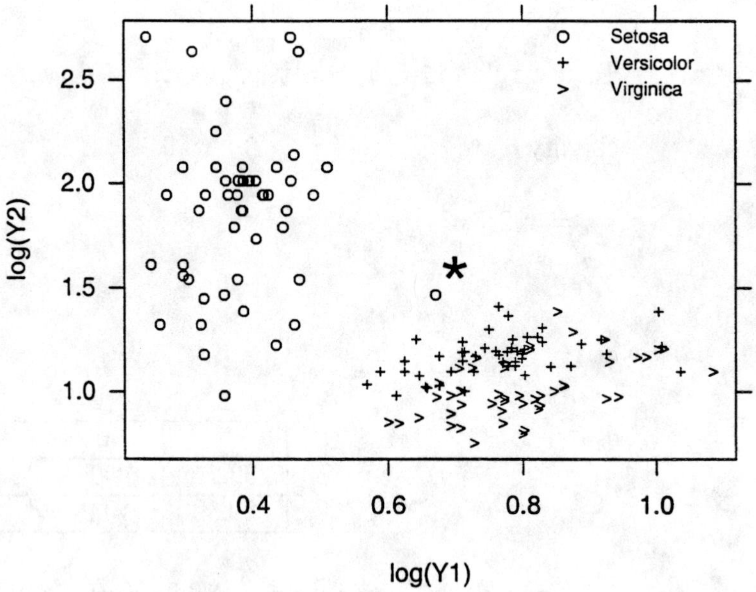

The points of all three groups appear to follow roughly an ellipse-like pattern. However, the orientation of the ellipse appears to be different for the observations from π_1 (*Iris setosa*), from the observations from π_2 and π_3. In π_1, there also appears to be an outlier, labelled with a "*".

(b), (c) Assuming equal covariance matrices and ivariate normal populations, these are the linear discriminant scores $\hat{d}_i(\boldsymbol{x})$ for $i = 1, 2, 3$.

For both variables $\log Y_1$, and $\log Y_2$:

population	$\hat{d}_i(\boldsymbol{x}) = \overline{\boldsymbol{x}}_i' \boldsymbol{S}_{\text{pooled}} \boldsymbol{x} - \frac{1}{2} \overline{\boldsymbol{x}}_i' \boldsymbol{S}_{\text{pooled}} \overline{\boldsymbol{x}}_i$
π_1	$26.81 \log Y_1 + 28.90 \log Y_2 - 31.97$
π_2	$75.10 \log Y_1 + 13.82 \log Y_2 - 36.83$
π_3	$79.94 \log Y_1 + 10.80 \log Y_2 - 37.30$

For variable $\log Y_1$ only:

population	$\hat{d}_i(\boldsymbol{x}) = \overline{\boldsymbol{x}}_i' \boldsymbol{S}_{\text{pooled}} \boldsymbol{x} - \frac{1}{2} \overline{\boldsymbol{x}}_i' \boldsymbol{S}_{\text{pooled}} \overline{\boldsymbol{x}}_i$
π_1	$40.90 \log Y_1 - 7.82$
π_2	$81.84 \log Y_1 - 31.30$
π_3	$85.20 \log Y_1 - 33.93$

For variable $\log Y_2$ only:

population	$\hat{d}_i(\boldsymbol{x}) = \overline{\boldsymbol{x}}_i' \boldsymbol{S}_{\text{pooled}} \boldsymbol{x} - \frac{1}{2} \overline{\boldsymbol{x}}_i' \boldsymbol{S}_{\text{pooled}} \overline{\boldsymbol{x}}_i$
π_1	$30.93 \log Y_2 - 28.73$
π_2	$19.52 \log Y_2 - 11.44$
π_3	$16.87 \log Y_2 + 8.54$

Variables	APER	E(AER)
$\log Y_1$, $\log Y_2$	$\frac{26}{150} = .17$	$\frac{27}{150} = .18$
$\log Y_1$	$\frac{49}{150} = .33$	$\frac{49}{150} = .33$
$\log Y_2$,	$\frac{34}{150} = .23$	$\frac{34}{150} = .23$

(d) The preceeding misclassification rates are not nearly as good as those in Example 11.12. Using "shape" is effective in discriminating π_1 (***iris versicolor***) from π_2 and π_3. It is not as good at discriminating π_2 from π_3, because of the overlap of π_1 and π_2 in both shape variables. Therefore, shape is not an effective discriminator of all three species of iris.

11.29 **(a)** The calculated values of $\overline{\boldsymbol{x}}_1, \overline{\boldsymbol{x}}_1, \overline{\boldsymbol{x}}_3, \overline{\boldsymbol{x}}$, and $\boldsymbol{S}_{\text{pooled}}$ agree with the results for these quantities given in Example 11.11

(b)

$$W^{-1} = \begin{bmatrix} 0.348899 & 0.000193 \\ 0.000193 & .000003 \end{bmatrix}, \quad B = \begin{bmatrix} 12.50 & 1518.74 \\ 1518.74 & 258471.12 \end{bmatrix}$$

The eigenvalues and scaled eigenvectors of $W^{-1}B$ are

$$\hat{\lambda}_1 = 5.646, \quad \hat{a}_1' = \begin{bmatrix} 5.009 \\ 0.009 \end{bmatrix}$$

$$\hat{\lambda}_2 = 0.191, \quad \hat{a}_2' = \begin{bmatrix} 0.207 \\ -0.014 \end{bmatrix}$$

To classify $x_0' = [3.21 \quad 497]$, use (11-67) and compute

$$\Sigma_{j=1}^{2}[\hat{a}_j'(x - \overline{x}_i)]^2 \qquad i = 1, 2, 3$$

Allocate x_0' to π_k if

$$\Sigma_{j=1}^{2}[\hat{a}_j'(x - \overline{x}_k)]^2 \leq \Sigma_{j=1}^{2}[\hat{a}_j'(x - \overline{x}_i)]^2 \qquad \text{for all} i \neq k$$

For x_0,

k	$\Sigma_{j=1}^{2}[\hat{a}_j'(x - \overline{x}_k)]^2$
1	2.63
2	16.99
3	2.43

Thus, classify x_0 to π_3 This result agrees with the classification given in Example 11.11. Any time there are three populations with only two discriminants, classification results using Fisher's Discriminants will be identical to those using the sample distance method of Example 11.11.

11.30 (a) Assuming normality and equal covariance matrices for the three populations π_1, π_2, and π_3, the minimum TPM rule is given by:

Allocate $\boldsymbol{x}$to π_k if the linear discriminant score $\hat{d}_k(\boldsymbol{x})$ = the largest of $\hat{d}_1(\boldsymbol{x}), \hat{d}_2(\boldsymbol{x}), \hat{d}_3(\boldsymbol{x}$

where $\hat{d}_i(\boldsymbol{x})$ is given in the following table for $i = 1, 2, 3$.

population	$\hat{d}_i(\boldsymbol{x}) = \overline{\boldsymbol{x}}_i' \boldsymbol{S}_{\text{pooled}} \boldsymbol{x} - \frac{1}{2} \overline{\boldsymbol{x}}_i' \boldsymbol{S}_{\text{pooled}} \overline{\boldsymbol{x}}_i$
π_1	$0.70x_1 + 0.58x_2 - 13.52x_3 + 6.93x_4 + 1.44x_5 - 44.78$
π_2	$1.85x_1 + 0.32x_2 - 12.78x_3 + 8.33x_4 - 0.14x_5 - 35.20$
π_3	$2.64x_1 + 0.20x_2 - 2.16x_3 + 5.39x_4 - 0.08x_5 - 23.61$

(b) Confusion matrix is:

		Predicted Membership			
		π_1	π_2	π_3	Total
Actual membership	π_1	7	0	0	7
	π_2	1	10	0	11
	π_3	0	3	35	38

And the APER $\frac{0+1+3}{56} = .071$

The holdout confusion matrix is:

		Predicted Membership			
		π_1	π_2	π_3	Total
Actual membership	π_1	7	0	0	7
	π_2	2	7	2	11
	π_3	0	3	35	38

$E(\text{AER}) = \frac{2+2+3}{56} = .125$

(c) One choice of transformations, $x_1, \log x_2, \sqrt{x_3}, \log x_4, \sqrt{x_5}$ appears to improve the normality of the data but the classification rule from these data has slightly higher error rates than the rule derived from the original data. The error rates (APER, $\hat{E}$(AER)) for the linear discriminants in Example 11.14 are also slightly higher than those for the original data.

11.31 (a) The data look fairly normal.

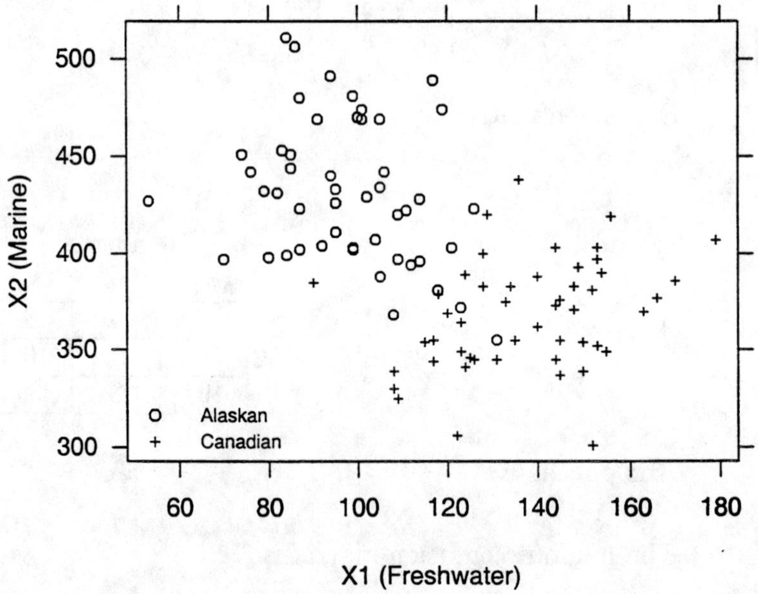

Although the covariances have different signs for the two groups, the correlations are small. Thus the assumption of bivariate normal distributions with equal covariance matrices does not seem unreasonable.

(b) The linear discriminant function is

$$\hat{\boldsymbol{a}}'\boldsymbol{x} - \hat{m} = -0.13x_1 + 0.052x_2 - 5.54$$

Classify an observation $\boldsymbol{x}_0$ to π_1 (Alaskan salmon) if $\hat{\boldsymbol{a}}'\boldsymbol{x}_0 - \hat{m} \geq 0$ and classify $\boldsymbol{x}_0$ to π_2 (Canadian salmon) otherwise.

Dot diagrams of the discriminant scores:

```
                                    .
                            .       :  .. : :
                 .    :  ...: : ::: ::::: :.::..:   ...
-------+---------+---------+---------+---------+---------Alaskan
         :      .
         :    ::: .:. . .
   :     ::...:::.::::::: :       .
-------+---------+---------+---------+---------+---------Canadian
    -8.0      -4.0       0.0       4.0       8.0      12.0
```

It does appear that growth ring diameters separate the two groups reasonably well, as APER= $\frac{6+1}{100} = .07$ and E(AER)= $\frac{6+1}{100} = .07$

(c) Here are the bivariate plots of the data for male and female salmon separately.

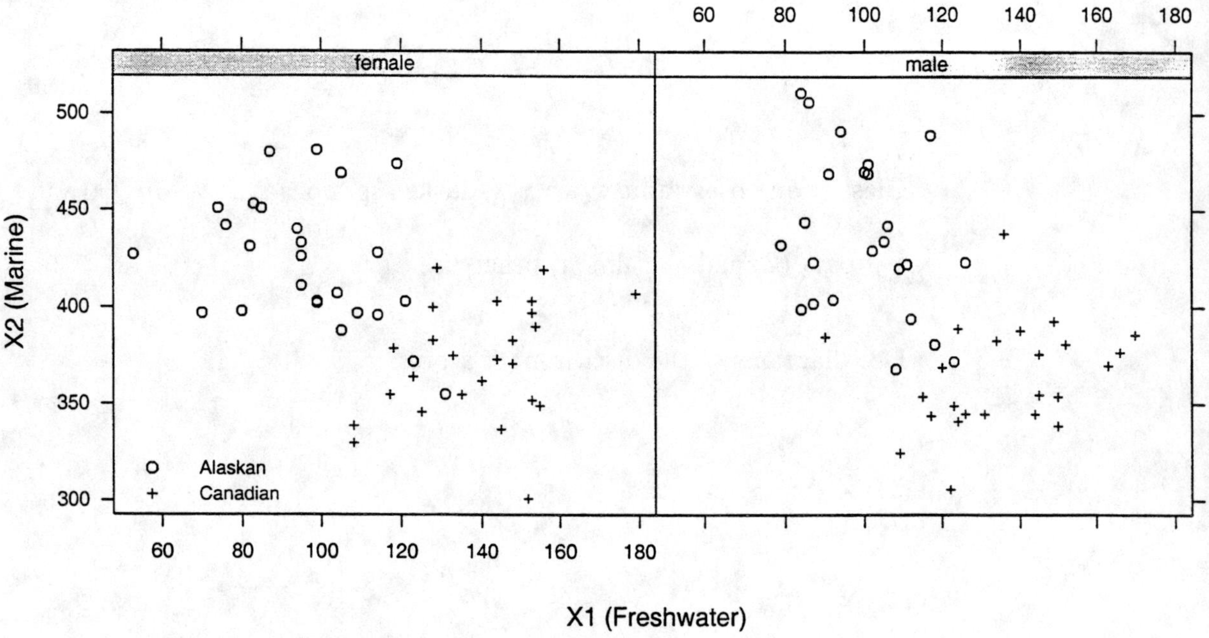

For the male salmon, these are some summary statistics

$$\overline{\boldsymbol{x}}_1 = \begin{bmatrix} 100.3333 \\ 436.1667 \end{bmatrix}, \quad \boldsymbol{S}_1 = \begin{bmatrix} 181.97101 & -197.71015 \\ -197.71015 & 1702.31884 \end{bmatrix}$$

$$\overline{\boldsymbol{x}}_2 = \begin{bmatrix} 135.2083 \\ 364.0417 \end{bmatrix}, \quad \boldsymbol{S}_2 = \begin{bmatrix} 370.17210 & 141.64312 \\ 141.64312 & 760.65036 \end{bmatrix}$$

The linear discriminant function for the male salmon only is

$$\hat{\boldsymbol{a}}'\boldsymbol{x} - \hat{m} = -0.12x_1 + 0.056x_2 - 8.12$$

Classify an observation $\boldsymbol{x}_0$ to π_1 (Alaskan salmon) if $\hat{\boldsymbol{a}}'\boldsymbol{x}_0 - \hat{m} \geq 0$ and classify $\boldsymbol{x}_0$ to π_2 (Canadian salmon) otherwise.

Using this classification rule, APER= $\frac{3+1}{48}$ = .08 and E(AER)= $\frac{3+2}{48}$ = .10.

For the female salmon, these are some summary statistics

$$\bar{\boldsymbol{x}}_1 = \begin{bmatrix} 96.5769 \\ 423.6539 \end{bmatrix}, \quad \boldsymbol{S}_1 = \begin{bmatrix} 336.33385 & -210.23231 \\ -210.23231 & 1097.91539 \end{bmatrix}$$

$$\bar{\boldsymbol{x}}_2 = \begin{bmatrix} 139.5385 \\ 369.0000 \end{bmatrix}, \quad \boldsymbol{S}_2 = \begin{bmatrix} 289.21846 & 120.64000 \\ 120.64000 & 1038.72000 \end{bmatrix}$$

The linear discriminant function for the female salmon only is

$$\hat{\boldsymbol{a}}'\boldsymbol{x} - \hat{m} = -0.13x_1 + 0.05x_2 - 2.66$$

Classify an observation $\boldsymbol{x}_0$ to π_1 (Alaskan salmon) if $\hat{\boldsymbol{a}}'\boldsymbol{x}_0 - \hat{m} \geq 0$ and classify $\boldsymbol{x}_0$ to π_2 (Canadian salmon) otherwise.

Using this classification rule, APER= $\frac{3+0}{52}$ = .06 and E(AER)= $\frac{3+0}{52}$ = .06.

It is unlikely that gender is a useful discriminatory variable, as splitting the data into female and male salmon did not improve the classification results greatly.

11.32 (a) Here is the bivariate plot of the data for the two groups:

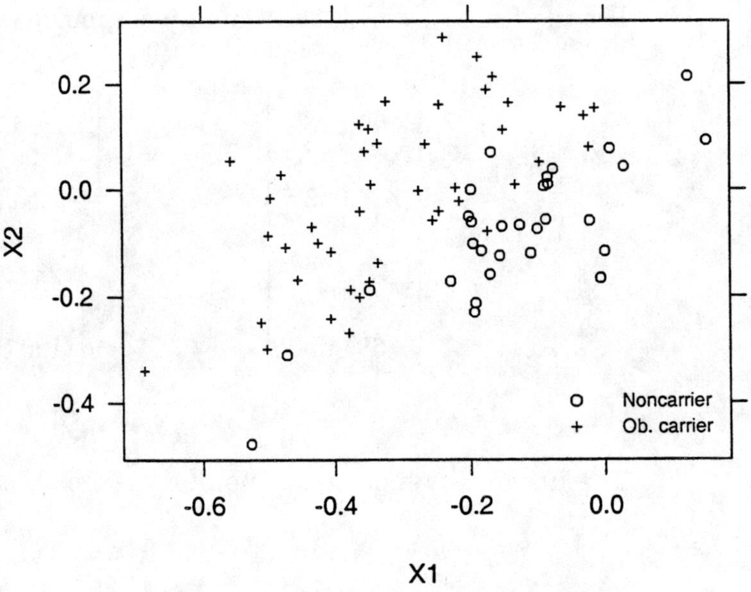

Because the points for both groups form fairly elliptical shapes, the bivariate normal assumption appears to be a reasonable one. Normal score plots for each group confirm this.

(b) Assuming equal prior probabilities, the sample linear discriminant function is

$$\hat{\boldsymbol{a}}'\boldsymbol{x} - \hat{m} = 19.32x_1 - 17.12x_2 + 3.56$$

Classify an observation $\boldsymbol{x}_0$ to π_1 (Noncarriers) if $\hat{\boldsymbol{a}}'\boldsymbol{x}_0 - \hat{m} \geq 0$ and classify $\boldsymbol{x}_0$ to π_2 (Obligatory carriers) otherwise.

The holdout confusion matrix is

		Predicted Membership		
		π_1	π_2	Total
Actual membership	π_1	26	4	30
	π_2	8	37	45

$\hat{E}(\text{AER}) = \frac{4+8}{75} = .16$

(c) The classification results for the 10 new cases using the discriminant function in part **(b)**:

Case	x_1	x_2	$\hat{\boldsymbol{a}}'\boldsymbol{x} - \hat{m}$	Classification
1	-0.112	-0.279	6.17	π_1
2	-0.059	-0.068	3.58	π_1
3	0.064	0.012	4.59	π_1
4	-0.043	-0.052	3.62	π_1
5	-0.050	-0.098	4.27	π_1
6	-0.094	-0.113	3.68	π_1
7	-0.123	-0.143	3.63	π_1
8	-0.011	-0.037	3.98	π_1
9	-0.210	-0.090	1.04	π_1
10	-0.126	-0.019	1.45	π_1

(d) Assuming that the prior probability of obligatory carriers is $\frac{1}{4}$ and that of noncarriers is $\frac{3}{4}$, the sample linear discriminant function is

$$\hat{\boldsymbol{a}}'\boldsymbol{x} - \hat{m} = 19.32x_1 - 17.12x_2 + 4.66$$

Classify an observation $\boldsymbol{x}_0$ to π_1 (Noncarriers) if $\hat{\boldsymbol{a}}'\boldsymbol{x}_0 - \hat{m} \geq 0$ and classify $\boldsymbol{x}_0$ to π_2 (Obligatory carriers) otherwise.

The holdout confusion matrix is

		Predicted Membership		
		π_1	π_2	Total
Actual membership	π_1	30	0	30
	π_2	18	27	45

$\hat{E}(\text{AER}) = \frac{18+0}{75} = 0.24$

The classification results for the 10 new cases using the discriminant function in part (b):

Case	x_1	x_2	$\hat{\boldsymbol{a}}'\boldsymbol{x} - \hat{m}$	Classification
1	-0.112	-0.279	7.27	π_1
2	-0.059	-0.068	4.68	π_1
3	0.064	0.012	5.69	π_1
4	-0.043	-0.052	4.72	π_1
5	-0.050	-0.098	5.37	π_1
6	-0.094	-0.113	4.78	π_1
7	-0.123	-0.143	4.73	π_1
8	-0.011	-0.037	5.08	π_1
9	-0.210	-0.090	2.14	π_1
10	-0.126	-0.019	2.55	π_1

11.33 Let $x_3 \equiv$ YrHgt, $x_4 \equiv$ FtFrBody, $x_6 \equiv$ Frame, $x_7 \equiv$ BkFat, $x_8 \equiv$ SaleHt, and $x_9 \equiv$ SaleWt.

(a) For $\pi_1 \equiv$ Angus, $\pi_2 \equiv$ Hereford, and $\pi_3 \equiv$ Simental, here are Fisher's linear discriminants

$$
\begin{aligned}
\hat{d}_1 &= -3737 + 126.88x_3 - 0.48x_4 + 19.08x_5 - 205.22x_6 \\
&\quad +275.84x_7 + 28.15x_8 - 0.03x_9 \\
\hat{d}_2 &= -3686 + 127.70x_3 - 0.47x_4 + 18.65x_5 - 206.18x_6 \\
&\quad +265.33x_7 + 26.80x_8 - 0.03x_9 \\
\hat{d}_1 &= -3881 + 128.08x_3 - 0.48x_4 + 19.39x_5 - 206.36x_6 \\
&\quad +245.50x_7 + 29.47x_8 - 0.03x_9
\end{aligned}
$$

When $\boldsymbol{x}_0' = [50, 1000, 73, 7, .17, 54, 1525]$ we obtain $\hat{d}_1 = 3596.31, \hat{d}_2 = 3593.32$, and $\hat{d}_3 = 3594.13$, so assign the new observation to π_2, Hereford.

This is the plot of the discriminant scores in the two-dimensional discriminant space:

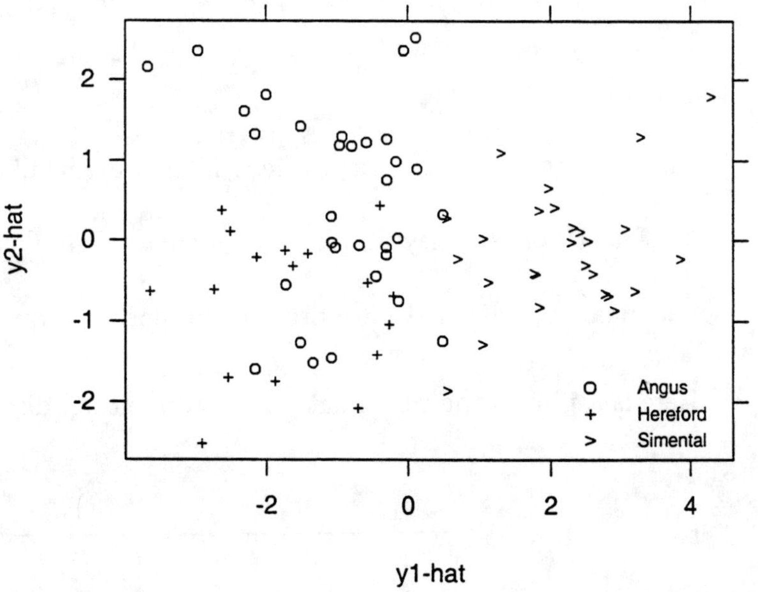

(b) Here is the APER and $\hat{E}$(AER) for different subsets of the variables:

Subset	APER	$\hat{E}$(AER)
$x_3, x_4, x_5, x_6, x_7, x_8, x_9$	.13	.25
x_4, x_5, x_7, x_8	.14	.20
x_5, x_7, x_8	.21	.24
x_4, x_5	.43	.46
x_4, x_7	.36	.39
x_4, x_8	.32	.36
x_7, x_8	.22	.22
x_5, x_7	.25	.29
x_5, x_8	.28	.32

11.34 For $\pi_1 \equiv$ General Mills, $\pi_2 \equiv$ Kellogg, and $\pi_3 \equiv$ Quaker and assuming multivariate normal data with a common covariance matrix, equal costs, and equal priors, these are Fisher's linear discriminant functions:

$$
\begin{aligned}
\hat{d}_1 &= .23x_3 + 3.79x_4 - 1.69x_5 - .01x_6 5.53x_7 \\
&\quad 1.90x_8 + 1.36x_9 - 0.12x_{10} - 33.14 \\
\hat{d}_2 &= .32x_3 + 4.15x_4 - 3.62x_5 - .02x_6 9.20x_7 \\
&\quad 2.07x_8 + 1.50x_9 - 0.20x_{10} - 43.07 \\
\hat{d}_3 &= .29x_3 + 2.64x_4 - 1.20x_5 - .02x_6 5.43x_7 \\
&\quad 1.22x_8 + .65x_9 - 0.13x_{10}
\end{aligned}
$$

The Kellogg cereals appear to have high protein, fiber, and carbohydrates, and low fat. However, they also have high sugar. The Quaker cereals appear to have low sugar, but also have low protein and carbohydrates.

Here is a plot of the cereal data in two-dimension discriminant space:

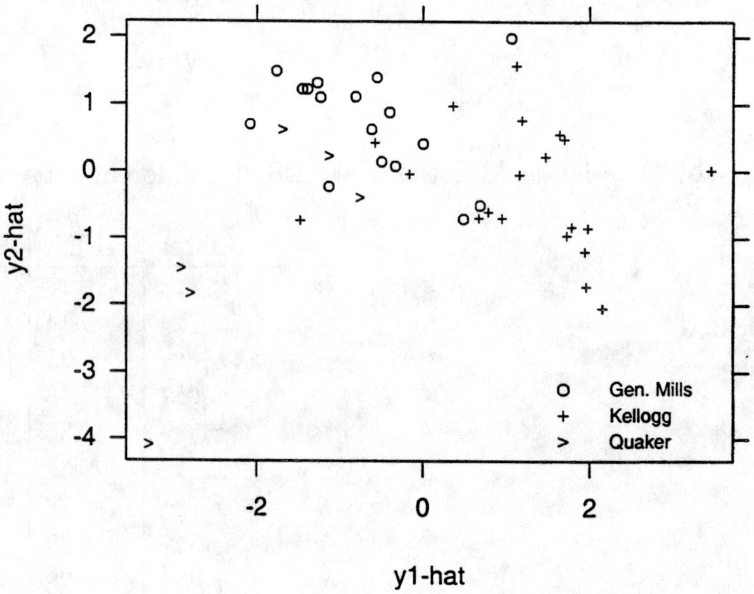

Chapter 12

12.1 a) Codes:

1 →	South	Yes	Democrat	Yes	Yes
0 →	non-South	No	Republican	No	No

e.g. Reagan - Carter:

	1	0
1	1	0
0	2	2

, $\frac{a+d}{p} = 3/5 = .60$

Pair	Coefficient (a+d)/p
R-C	.6
R-F	.4
R-N	.6
R-J	0
R-K	.6
C-F	0
C-N	.2
C-J	.4
C-K	.6
F-N	.8
F-J	.6
F-K	.4
N-J	.4
N-K	.6
J-K	.4

12.1 b)

Pair	Coefficient 1	2	3	Rank Order 1	2	3
R-C	.6	.75	.429	4.5	4.5	4.5
R-F	.4	.571	.25	10	10	10
R-N	.6	.75	.429	4.5	4.5	4.5
R-J	0	0	0	14.5	14.5	14.5
R-K	.6	.75	.429	4.5	4.5	4.5
C-F	0	0	0	14.5	14.5	14.5
C-N	.2	.333	.111	13	13	13
C-J	.4	.571	.25	10	10	10
C-K	.6	.75	.429	4.5	4.5	4.5
F-N	.8	.889	.667	1	1	1
F-J	.6	.75	.429	4.5	4.5	4.5
F-K	.4	.571	.25	10	10	10
N-J	.4	.571	.25	10	10	10
N-K	.6	.75	.429	4.5	4.5	4.5
J-K	.4	.571	.25	10	10	10

12.2

Pair	Coefficient 5	6	7	Rank Order 5	6	7
R-C	.333	.5	.2	9	9	9
R-F	0	0	0	14	14	14
R-N	.333	.5	.2	9	9	9
R-J	0	0	0	14	14	14
R-K	.333	.5	.2	9	9	9
C-F	0	0	0	14	14	14
C-N	.2	.333	.111	12	12	12
C-J	.4	.571	.25	6	6	6
C-K	.5	.667	.333	3	3	3
F-N	.667	.8	.5	1	1	1
F-J	.5	.667	.333	3	3	3
F-K	.25	.4	.143	11	11	11
N-J	.4	.571	.25	6	6	6
N-K	.5	.667	.333	3	3	3
J-K	.4	.571	.25	6	6	6

12.3

x\y	1	0	Total
1	a	b	a+b
0	c	d	c+d
Total	a+c	b+d	p=a+b+c+d

$$r = \frac{\sum_{i=1}^{p}(x_i-\bar{x})(y_i-\bar{y})}{\sqrt{\sum_{i=1}^{p}(x_i-\bar{x})^2 \sum_{i=1}^{p}(y_i-\bar{y})^2}}$$

$\bar{x} = (a+b)/p; \quad \bar{y} = (a+c)/p$

$$\Sigma(x_i-\bar{x})^2 = (a+b)(1-(a+b)/p)^2 + (c+d)(0-(a+b)/p)^2 = \frac{(c+d)(a+b)}{p}$$

$$\Sigma(y_i-\bar{y})^2 = (a+c)(1-(a+c)/p)^2 + (b+d)(0-(a+c)/p)^2 = \frac{(a+c)(b+d)}{p}$$

$$\Sigma(x_i-\bar{x})(y_i-\bar{y}) = \Sigma(x_iy_i-y_i\bar{x}-x_i\bar{y}+\bar{x}\bar{y})$$

$$= a - \frac{(a+c)(a+b)}{p} - \frac{(a+b)(a+c)}{p} + p\,\frac{(a+b)(a+c)}{p^2}$$

$$= \frac{a(a+b+c+d)-(a+c)(a+b)}{p} = \frac{ad-bc}{p}$$

Therefore

$$r = \frac{(ad-bc)/p}{\left[\frac{(c+d)(a+b)(a+c)(b+d)}{p^2}\right]^{\frac{1}{2}}} = \frac{ad-bc}{[(a+b)(c+d)(a+c)(b+d)]^{\frac{1}{2}}}$$

12.4 Let $c_1 = \frac{a+d}{p}$, $c_2 = \frac{2(a+d)}{2(a+d)+(b+c)}$ and $c_3 = \frac{a+d}{(a+d)+2(b+c)}$

then $c_3 = \frac{1}{1+2(c_1^{-1}-1)}$ so c_3 increases as c_1 increases

Also, $c_2 = \frac{2}{c_1^{-1}+1}$ so c_2 increases as c_1 increases

Finally, $c_2 = \frac{4}{c_3^{-1}+3}$ so c_2 increases as c_3 increases

12.5 a) Single linkage

$$\begin{array}{c} \\ 1 \\ 2 \\ 3 \\ 4 \end{array}\begin{array}{c} \begin{array}{cccc} 1 & 2 & 3 & 4 \end{array} \\ \begin{bmatrix} 0 & & & \\ \textcircled{1} & 0 & & \\ 11 & 2 & 0 & \\ 5 & 3 & 4 & 0 \end{bmatrix} \end{array} \rightarrow \begin{array}{c} \\ (12) \\ 3 \\ 4 \end{array}\begin{array}{c} \begin{array}{ccc} (12) & 3 & 4 \end{array} \\ \begin{bmatrix} 0 & & \\ \textcircled{2} & 0 & \\ 3 & 4 & 0 \end{bmatrix} \end{array} \rightarrow \begin{array}{c} \\ (123) \\ 4 \end{array}\begin{array}{c} \begin{array}{cc} (123) & 4 \end{array} \\ \begin{bmatrix} 0 & \\ 3 & 0 \end{bmatrix} \end{array}$$

Dendogram

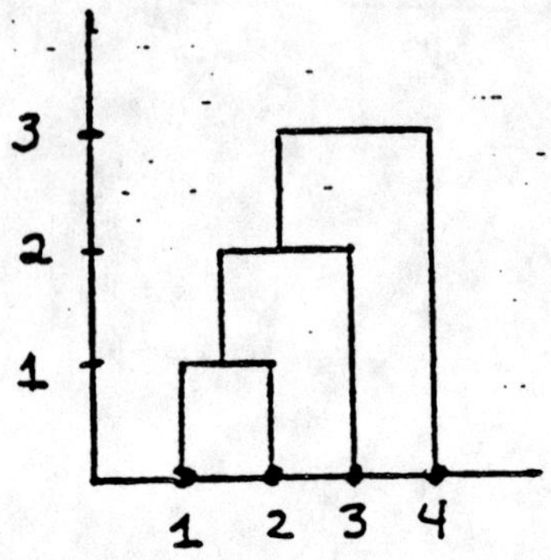

12.5 b) Complete Linkage

Dendogram

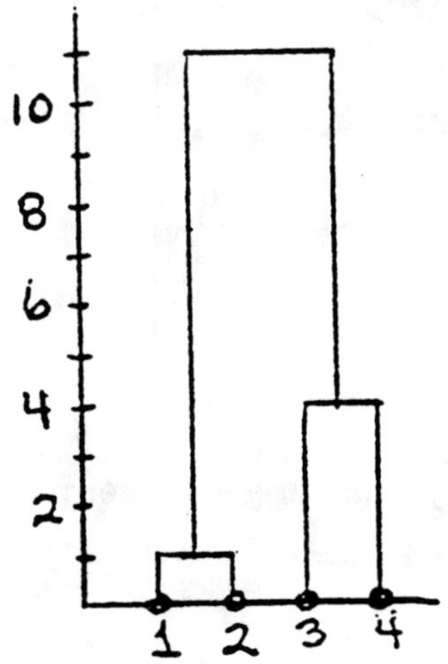

c) Average Linkage

Dendogram

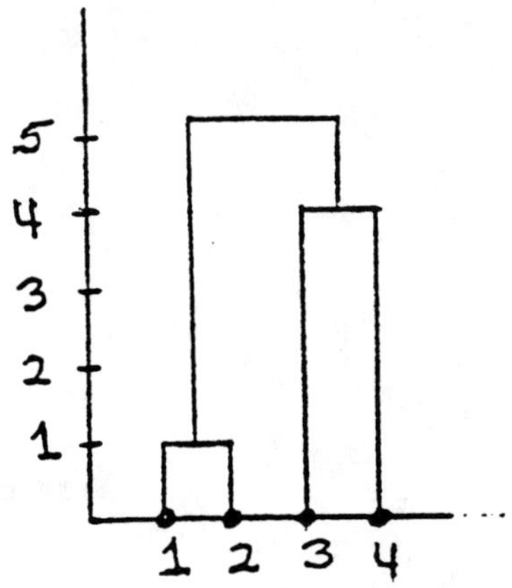

12.6 Dendograms

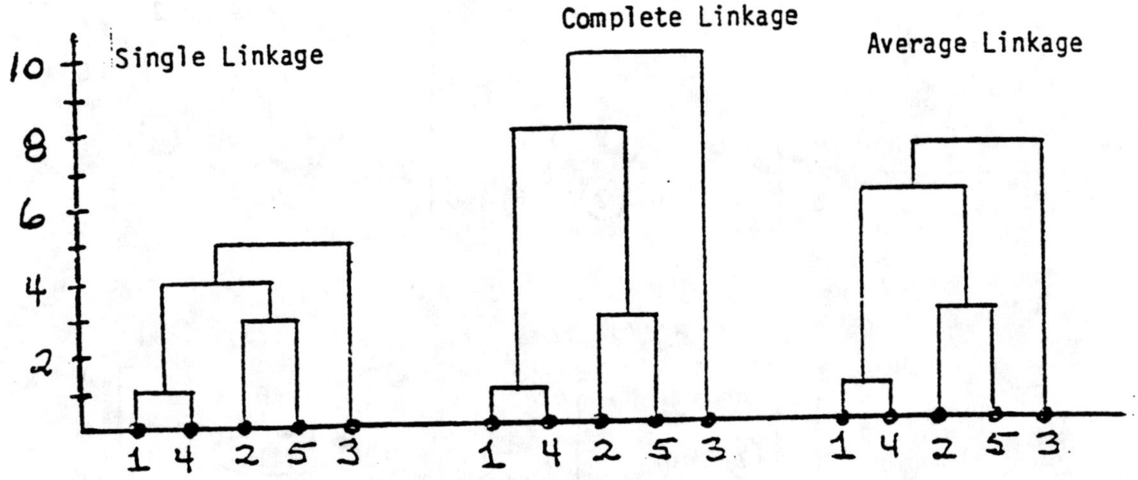

All three methods produce the same hierarchical arrangements. Item 3 is somewhat different from the other items.

12.7 Treating correlations as <u>similarity</u> coefficients we have:

<u>Single linkage</u>

$S_{23} = .6$

$S_{(23)1} = \max\,(S_{21}, S_{31}) = .58$

$S_{(23)4} = .44$, $S_{(23)5} = .43$, etc.

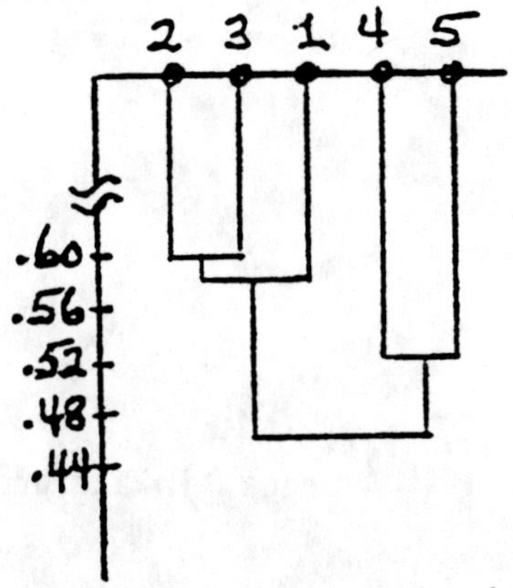

Complete Linkage

$S_{23} = .6$
$S_{(23)1} = \min(S_{21}, S_{31}) = .51$
$S_{(23)4} = .39,\ S_{(23)5} = .32$, etc.

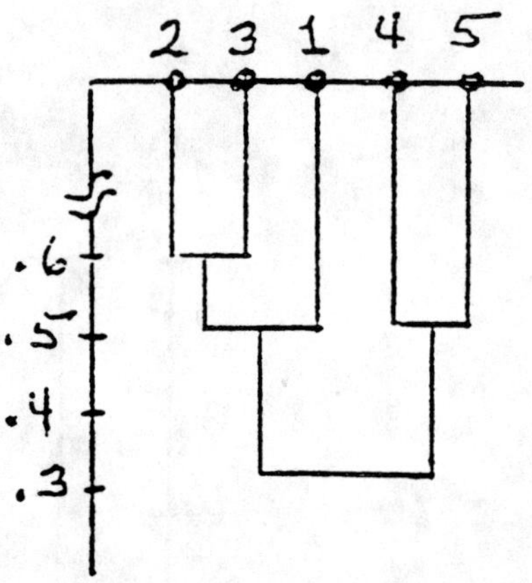

Both methods arrive at nearly the same clustering.

12.8

$$\begin{array}{c} \\ 1 \\ 2 \\ 3 \\ 4 \\ 5 \end{array}\begin{array}{c} \begin{array}{ccccc} 1 & 2 & 3 & 4 & 5 \end{array} \\ \left[\begin{array}{ccccc} 0 & & & & \\ 9 & 0 & & & \\ 3 & 7 & 0 & & \\ 6 & 5 & 9 & 0 & \\ 11 & 10 & \textcircled{2} & 8 & 0 \end{array}\right]\end{array} \rightarrow \begin{array}{c} \\ 1 \\ 2 \\ (35) \\ 4 \end{array}\begin{array}{c} \begin{array}{cccc} 1 & 2 & (35) & 4 \end{array} \\ \left[\begin{array}{cccc} 0 & & & \\ 9 & 0 & & \\ 7 & 8.5 & 0 & \\ 6 & \textcircled{5} & 8.5 & 0 \end{array}\right]\end{array} \rightarrow$$

$$\begin{array}{c} \\ 1 \\ (35) \\ (24) \end{array}\begin{array}{c} \begin{array}{ccc} 1 & (35) & (24) \end{array} \\ \left[\begin{array}{ccc} 0 & & \\ \textcircled{7} & 0 & \\ 7.5 & 8.5 & 0 \end{array}\right]\end{array} \rightarrow \begin{array}{c} \\ (135) \\ (24) \end{array}\begin{array}{c} \begin{array}{cc} (135) & (24) \end{array} \\ \left[\begin{array}{cc} 0 & \\ \textcircled{8.167} & 0 \end{array}\right]\end{array}$$

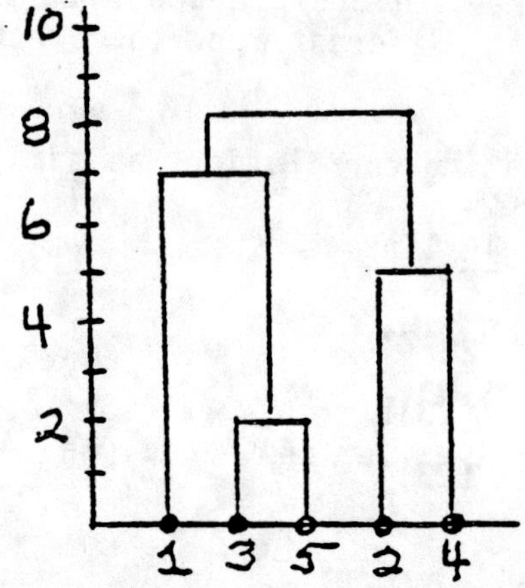

Average linkage produces results similar to single linkage.

12.9 Dendograms

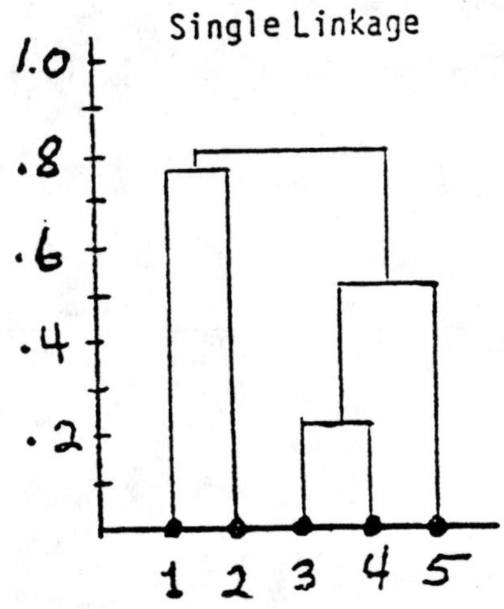

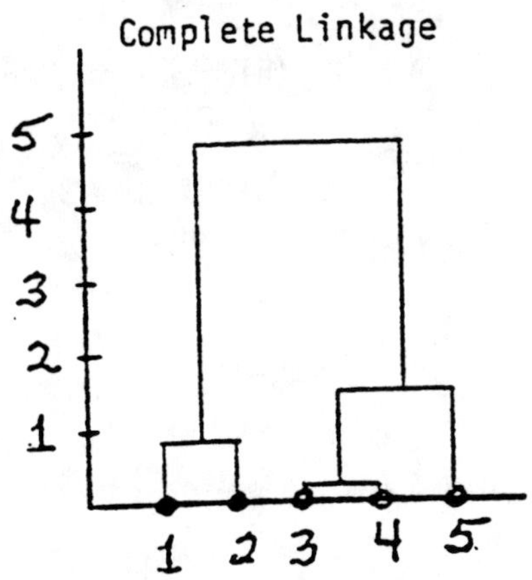

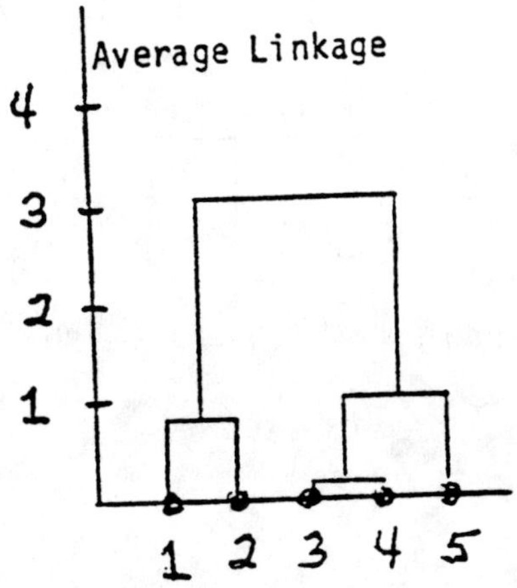

Although the vertical scales are different, all three linkage methods produce the same groupings. (Note different vertical scales.)

12.10 (a) $ESS_1 = (2-2)^2 = 0$, $ESS_2 = (1-1)^2 = 0$, $ESS_3 = (5-5)^2 = 0$, and $ESS_4 = (8-8)^2 = 0$.

(b) At step 2

Clusters			Increase in ESS
{12}	{3}	{4}	.5
{13}	{2}	{4}	4.5
{14}	{2}	{3}	18.0
{1}	{23}	{4}	8.0
{1}	{24}	{3}	24.5
{1}	{2}	{34}	4.5

(c) At step 3

Clusters		Increase in ESS
{12}	{34}	5.0
{123}	{4}	8.7

Finally all four together have
$ESS = (2-4)^2 + (1-4)^2 + (5-4)^2 + (8-4)^2 = 30$

12.11 K = 2 initial clusters (AB) and (CD)

	$\bar{x}_1$	$\bar{x}_2$
(AB)	3	1
(CD)	1	1

$d^2(A,(AB)) = 13$, $d^2(A,(CD)) = 25$ } do not reassign A

$d^2(B,(AB)) = 13$, $d^2(B,(CD)) = 9$ ✓ } reassign B to (CD)

	$\bar{x}_1$	$\bar{x}_2$
A	5	4
(BCD)	1	0

	Squared distance to group centroids			
Cluster	A	B	C	D
A	0	52	45	13
(BCD)	32✓	4✓	5✓	5✓

Therefore, no more reassignments take place and final clusters are A and (BCD).

12.12 K = 2 initial clusters (AC) and (BD)

	$\bar{x}_1$	$\bar{x}_2$
(AC)	3	.5
(BD)	-2	-.5

	Squared distance to group centroids			
Cluster	A	B	C	D
(AC)	10.25✓	16.25	10.25✓	42.25
(BD)	61.25	3.25✓	11.25	3.25✓

Therefore, no reassignments take place and final clusters are (AC) and (BC). This result is different from the result in Example 12.11. In Example 12.11 the initial clusters (AB) and (CD) became the final clusters A and (BCD). A graph of the items lends support to the A and (BCD) groupings.

12.13 K = 2 initial clusters (AB) and (CD)

	$\bar{x}_1$	$\bar{x}_2$
(AB)	2	2
(CD)	-1	-2

$d^2(D,(AB)) = 41$, $d^2(D,(CD)) = 4$ } do not reassign D

$d^2(C,(AB)) = 17$, $d^2(C,(CD)) = 4$ } do not reassign C

$d^2(B,(AB)) = 10$, $d^2(B,(CD)) = 9$ } reassign B to (CD)

	$\bar{x}_1$	$\bar{x}_2$
A	5	3
(BCD)	-1	-1

	Squared distance to group centroids			
Cluster	A	B	C	D
A	0✓	40	41	89
(BCD)	52	4✓	5✓	5✓

No more reassignments take place and the final clusters, A and (BCD), are the same as they are in Example 12.11. In this case we start with the same initial groups and the first, and only, reassignment is the same. Thus it makes no difference if you start at the bottom or top of the list of items.

12.14 We first use (12-31) to express $\tilde{\boldsymbol{P}}$ in terms of its singular decomposition $\tilde{\boldsymbol{P}} = \sum_{j=1}^{J-1} \lambda_j \boldsymbol{u}_j \boldsymbol{v}_j'$ so $\tilde{\boldsymbol{P}} - \hat{\boldsymbol{P}} = \sum_{j=2}^{J-1} \lambda_j \boldsymbol{u}_j \boldsymbol{v}_j'$, where $\hat{\boldsymbol{P}} = \lambda_1 \boldsymbol{u}_1 \boldsymbol{v}_1'$, and

$$(\tilde{\boldsymbol{P}} - \hat{\boldsymbol{P}})'\mathrm{D}_c^{-1}(\tilde{\boldsymbol{P}} - \hat{\boldsymbol{P}}) = (\sum_{j=2}^{J-1} \lambda_j \boldsymbol{u}_j \boldsymbol{v}_j')'\mathrm{D}_c^{-1}(\sum_{k=2}^{J-1} \lambda_k \boldsymbol{u}_k \boldsymbol{v}_k')$$

$$= (\sum_{j=2}^{J-1} \lambda_j \boldsymbol{v}_j \boldsymbol{u}_j')\mathrm{D}_c^{-1}(\sum_{k=2}^{J-1} \lambda_k \boldsymbol{u}_k \boldsymbol{v}_k') = \sum_{j=2}^{J-1} \lambda_j^2 \boldsymbol{v}_j \boldsymbol{v}_j'$$

since, by (12-32), the $\boldsymbol{u}_j$ satisfy $\boldsymbol{u}_j'\mathrm{D}_c^{-1}\boldsymbol{u}_k = 1$ if $j = k$ and $= 0$, otherwise. Consequently,

$$tr[\mathrm{D}_r^{-1}(\tilde{\boldsymbol{P}} - \hat{\boldsymbol{P}})'\mathrm{D}_c^{-1}(\tilde{\boldsymbol{P}} - \hat{\boldsymbol{P}})] = tr[\mathrm{D}_r^{-1}(\sum_{j=2}^{J-1} \lambda_j^2 \boldsymbol{v}_j \boldsymbol{v}_j')]$$

$$= \sum_{j=2}^{J-1} \lambda_j^2 \, tr[\mathrm{D}_r^{-1} \boldsymbol{v}_j \boldsymbol{v}_j'] = \sum_{j=2}^{J-1} \lambda_j^2 \, tr[\boldsymbol{v}_j' \mathrm{D}_r^{-1} \boldsymbol{v}_j] = \sum_{j=2}^{J-1} \lambda_j^2$$

since, by (12-32), $\boldsymbol{v}_j'\mathrm{D}_r^{-1}\boldsymbol{v}_j = 1$.

12.15 Recall four equations from the text

$$\tilde{\mathbf{P}} = \tilde{\mathbf{U}}\boldsymbol{\Lambda}\tilde{\mathbf{V}} \quad (12\text{-}31) \qquad \text{where } \mathbf{I} = \tilde{\mathbf{U}}\mathbf{D}_r^{-1}\tilde{\mathbf{U}} = \tilde{\mathbf{V}}\mathbf{D}_c^{-1}\tilde{\mathbf{V}} \quad (12\text{-}32)$$

$$\mathbf{Y} = \mathbf{D}_r^{-1}\tilde{\mathbf{U}}\boldsymbol{\Lambda} \quad (12\text{-}33) \qquad \mathbf{Z} = \mathbf{D}_c^{-1}\tilde{\mathbf{V}}\boldsymbol{\Lambda} \quad (12\text{-}34)$$

First, multipying both sides of (12-33), $\mathbf{Y}\tilde{\mathbf{V}}' = \mathbf{D}_r^{-1}\tilde{\mathbf{U}}\boldsymbol{\Lambda}\tilde{\mathbf{V}}' = \mathbf{D}_r^{-1}\tilde{\mathbf{P}}$ by (12-31). Next, (12-34) gives $\mathbf{Z}\boldsymbol{\Lambda}^{-1} = \mathbf{D}_c^{-1}\tilde{\mathbf{V}}$ so post-multiplying the equation above we obtain

$$\mathbf{Y}\tilde{\mathbf{V}}'\mathbf{D}_c^{-1}\tilde{\mathbf{V}} = \mathbf{D}_r^{-1}\tilde{\mathbf{P}}\mathbf{D}_c^{-1}\tilde{\mathbf{V}} = \mathbf{D}_r^{-1}\tilde{\mathbf{P}}\mathbf{Z}\boldsymbol{\Lambda}^{-1}$$

By (12-32), $\tilde{\mathbf{V}}'\mathbf{D}_c^{-1}\tilde{\mathbf{V}} = \mathbf{I}$, and the last equation reduces to

$$\mathbf{Y} = \mathbf{D}_r^{-1}\tilde{\mathbf{P}}\mathbf{Z}\boldsymbol{\Lambda}^{-1}$$

Starting with (12-33) and $\mathbf{Y}\boldsymbol{\Lambda}^{-1} = \mathbf{D}_r^{-1}\tilde{\mathbf{U}}$, we use (12-34) and (12-31) to give $\mathbf{Z}\tilde{\mathbf{U}}' = \mathbf{D}_c^{-1}\tilde{\mathbf{P}}'$. Then

$$\mathbf{Z}\tilde{\mathbf{U}}'\mathbf{D}_r^{-1}\tilde{\mathbf{U}} = \mathbf{D}_c^{-1}\tilde{\mathbf{P}}'\mathbf{D}_r^{-1}\tilde{\mathbf{U}} = \mathbf{D}_c^{-1}\tilde{\mathbf{P}}'\mathbf{Y}\boldsymbol{\Lambda}^{-1}$$

Finally, by (12-32), $\tilde{\mathbf{U}}'\mathbf{D}_r^{-1}\tilde{\mathbf{U}} = \mathbf{I}$ so the last equation reduces to

$$Z = \mathbf{D}_r^{-1}\tilde{\mathbf{P}}'\mathbf{Y}\boldsymbol{\Lambda}^{-1}$$

12.16. (a) The Euclidean distances between pairs of cereal brands

	C1	C2	C3	C4	C5	C6	C7	C8	C9	C10	C11	C12
C1	0.0											
C2	116.0	0.0										
C3	15.5	121.7	0.0									
C4	6.4	117.9	10.0	0.0								
C5	103.2	61.6	100.6	102.1	0.0							
C6	72.8	44.1	78.4	74.4	54.3	0.0						
C7	86.4	71.9	82.5	84.9	22.3	52.4	0.0					
C8	15.3	121.5	1.4	10.1	100.6	78.3	82.4	0.0				
C9	46.2	72.6	54.7	48.9	75.8	32.1	65.2	54.5	0.0			
C10	54.9	123.0	68.9	59.5	134.7	87.8	122.5	68.8	65.7	0.0		
C11	81.3	154.7	94.7	85.8	169.6	121.3	157.0	94.6	94.5	47.1	0.0	
C12	42.3	114.2	31.3	38.5	81.1	75.3	60.2	31.0	59.8	92.9	121.9	0.0
C13	163.2	163.4	177.9	168.1	208.0	155.4	205.1	177.9	148.9	112.4	110.7	198.0
C14	46.7	90.8	60.4	51.5	103.8	55.4	92.9	60.3	28.5	44.3	67.5	75.9
C15	60.3	170.5	50.0	56.6	141.5	127.8	121.5	50.0	103.8	101.7	115.6	62.0
C16	46.9	90.8	60.5	51.6	103.8	55.5	92.9	60.3	28.5	44.3	67.6	75.8
C17	23.1	101.0	21.6	21.6	81.4	58.5	63.6	21.4	37.5	70.1	100.7	26.0
C18	265.7	221.1	280.0	270.6	278.9	233.9	283.3	280.0	235.6	227.7	218.6	294.5
C19	68.2	181.9	60.5	65.2	155.9	138.7	136.2	60.5	113.2	102.7	111.7	76.6
C20	116.6	71.0	113.2	115.3	19.7	69.9	32.1	113.1	89.3	150.5	183.5	90.6
C21	103.0	217.7	96.6	100.6	191.7	174.7	171.6	96.6	148.1	129.7	130.5	111.7
C22	98.6	160.1	112.6	103.4	181.3	130.5	170.2	112.6	106.9	54.1	22.5	139.2
C23	58.0	102.8	49.1	54.9	62.4	68.1	41.3	48.9	61.2	105.4	136.9	20.7
C24	68.1	181.8	60.4	65.2	155.8	138.7	136.1	60.4	113.1	102.7	111.6	76.5
C25	49.4	121.0	36.2	44.8	82.5	82.1	62.8	36.2	68.9	101.7	130.2	14.7
C26	182.8	290.3	186.0	183.8	285.6	250.4	267.2	185.9	220.2	173.8	145.7	210.7
C27	134.7	99.9	148.2	139.1	150.9	101.1	152.2	148.2	104.2	99.6	113.7	160.9
C28	16.1	128.3	14.2	14.2	111.1	85.7	92.3	13.7	59.2	63.5	86.3	39.4
C29	107.5	159.0	120.3	111.6	180.7	132.1	170.7	120.3	116.0	54.1	64.6	144.1
C30	33.5	120.1	21.2	29.2	90.7	78.8	71.2	21.0	61.7	83.1	113.7	17.2
C31	78.9	80.5	90.9	82.8	108.5	59.2	103.1	90.8	56.9	52.6	90.6	101.7
C32	32.1	122.6	43.5	36.0	120.8	83.1	105.0	43.3	51.3	50.9	60.0	65.9
C33	143.1	68.0	141.3	142.4	42.0	84.5	61.1	141.2	109.8	170.6	203.8	120.8
C34	173.0	157.7	187.8	177.9	207.5	155.6	206.8	187.8	151.8	127.0	123.8	205.9
C35	116.2	70.4	112.7	114.9	16.9	69.2	30.4	112.6	89.9	148.8	183.8	90.0
C36	114.1	230.0	111.1	112.9	210.2	186.9	190.8	111.1	158.8	129.8	122.7	131.2
C37	53.1	78.2	51.4	52.4	51.6	41.3	34.2	51.1	38.1	91.1	124.5	36.6
C38	54.2	100.4	45.8	51.0	61.8	63.5	43.5	45.8	59.0	99.2	133.6	25.8
C39	48.3	93.5	42.5	45.9	61.0	55.1	43.3	42.5	49.6	90.7	125.9	27.3
C40	40.6	140.9	51.6	44.3	139.8	100.7	123.8	51.4	70.3	44.1	46.2	79.4
C41	197.8	309.6	194.3	196.6	288.1	268.0	268.1	194.3	237.8	215.5	194.4	209.9
C42	191.1	301.3	190.3	190.8	286.6	260.4	267.3	190.2	229.3	200.8	174.0	209.7
C43	185.2	290.7	189.2	186.6	288.1	251.4	270.2	189.2	221.4	173.6	143.7	214.8

	C13	C14	C15	C16	C17	C18	C19	C20	C21	C22	C23	C24
C13	0.0											
C14	127.4	0.0										
C15	213.2	105.0	0.0									
C16	127.4	1.0	105.0	0.0								
C17	173.1	51.3	69.7	51.3	0.0							
C18	134.4	220.7	321.2	220.8	270.1	0.0						
C19	212.5	110.8	16.2	110.9	81.2	322.6	0.0					

C20	223.2	117.3	151.2	117.3	94.3	288.6	166.1	0.0				
C21	234.6	142.8	50.3	142.8	117.2	347.4	36.5	201.2	0.0			
C22	91.5	79.1	135.2	79.2	116.8	204.1	131.1	195.9	148.8	0.0		
C23	204.9	83.3	81.1	83.2	36.8	295.9	96.2	70.9	130.9	153.2	0.0	
C24	212.5	110.7	16.0	110.8	81.1	322.6	1.4	166.0	36.5	131.1	96.1	0.0
C25	207.5	86.0	60.0	86.1	35.2	303.9	75.3	91.8	110.1	147.9	23.2	75.3
C26	233.8	200.3	159.3	200.3	204.2	342.0	143.8	297.3	121.0	152.7	231.2	143.8
C27	67.1	92.1	193.3	92.2	136.5	141.1	197.4	164.6	227.0	105.1	162.0	197.4
C28	174.0	59.3	46.7	59.3	30.1	278.3	55.0	123.1	89.7	104.7	58.5	54.9
C29	83.1	93.3	144.4	93.3	122.6	214.5	141.7	197.4	160.4	51.8	156.3	141.7
C30	191.2	73.8	53.3	73.8	24.6	293.2	66.8	102.5	102.5	130.6	34.3	66.8
C31	104.8	49.4	135.7	49.3	78.9	207.0	141.7	124.7	173.2	91.2	104.5	141.7
C32	150.5	37.5	75.3	37.5	47.4	248.1	78.9	132.4	108.8	79.4	80.7	78.7
C33	230.0	136.6	181.8	136.5	121.5	283.5	196.3	31.7	231.9	214.1	101.6	196.3
C34	30.1	132.2	226.4	132.3	180.7	107.3	226.8	221.3	250.8	107.0	210.8	226.8
C35	221.6	117.8	150.9	117.7	93.7	289.9	165.8	10.1	201.0	195.7	70.2	165.7
C36	226.8	148.7	71.8	148.7	131.9	341.0	56.0	221.0	28.8	139.2	151.3	56.0
C37	182.4	63.6	95.5	63.6	31.1	270.0	108.7	64.4	144.7	138.6	27.7	108.6
C38	198.4	80.8	81.3	80.9	34.1	292.4	95.7	74.1	131.3	148.9	17.1	95.7
C39	188.6	71.5	83.1	71.6	27.4	282.6	96.8	74.6	132.8	140.6	21.8	96.7
C40	146.6	52.5	71.8	52.6	62.1	252.4	70.9	152.7	96.8	66.6	96.6	70.8
C41	301.1	227.1	153.1	227.1	213.8	401.5	140.2	295.1	108.9	210.5	228.7	140.1
C42	277.2	214.8	154.9	214.9	209.3	375.5	140.8	294.9	112.9	188.1	229.2	140.7
C43	229.1	200.6	165.0	200.7	207.1	335.7	149.7	300.2	128.8	149.4	235.2	149.6

	C25	C26	C27	C28	C29	C30	C31	C32	C33	C34	C35	C36
C25	0.0											
C26	213.9	0.0										
C27	170.1	257.2	0.0									
C28	46.5	175.0	148.2	0.0								
C29	152.5	172.5	103.0	113.8	0.0							
C30	20.8	200.3	158.2	30.2	132.8	0.0						
C31	111.4	225.7	66.9	91.2	79.1	97.2	0.0					
C32	75.0	170.7	126.2	36.4	101.6	62.2	81.5	0.0				
C33	122.5	324.8	167.2	151.1	214.1	131.9	137.3	157.0	0.0			
C34	215.5	253.2	58.3	184.8	107.8	201.1	112.6	158.5	225.1	0.0		
C35	91.3	297.5	163.7	122.7	194.6	101.0	121.9	133.6	33.3	220.7	0.0	
C36	131.0	93.2	227.1	102.7	152.9	120.7	178.1	114.7	250.8	244.4	220.8	0.0
C37	43.5	234.6	136.1	60.4	141.6	44.5	81.7	72.4	91.2	186.6	63.7	161.4
C38	24.7	230.4	156.4	57.3	148.9	30.7	97.7	81.1	103.2	205.3	72.0	150.5
C39	30.1	227.7	146.5	53.6	140.6	30.7	87.9	74.5	102.6	195.3	72.6	150.5
C40	86.9	150.1	132.6	41.9	88.9	71.1	88.4	24.1	177.4	158.4	153.0	98.1
C41	209.3	98.9	305.4	186.0	236.3	204.2	264.3	190.2	325.4	315.9	297.0	96.8
C42	210.6	71.2	286.8	180.8	216.6	203.0	251.2	179.4	324.1	292.0	296.8	94.0
C43	218.2	17.7	254.4	178.3	170.3	204.2	225.5	172.3	327.1	248.4	300.5	100.9

	C37	C38	C39	C40	C41	C42	C43
C37	0.0						
C38	27.0	0.0					
C39	20.2	10.1	0.0				
C40	90.2	94.6	88.5	0.0			
C41	241.1	232.1	233.1	177.4	0.0		
C42	237.9	231.7	231.2	164.5	35.2	0.0	
C43	237.2	233.9	230.8	151.2	108.2	78.7	0

(b) Complete linkage produces results similar to single linkage.

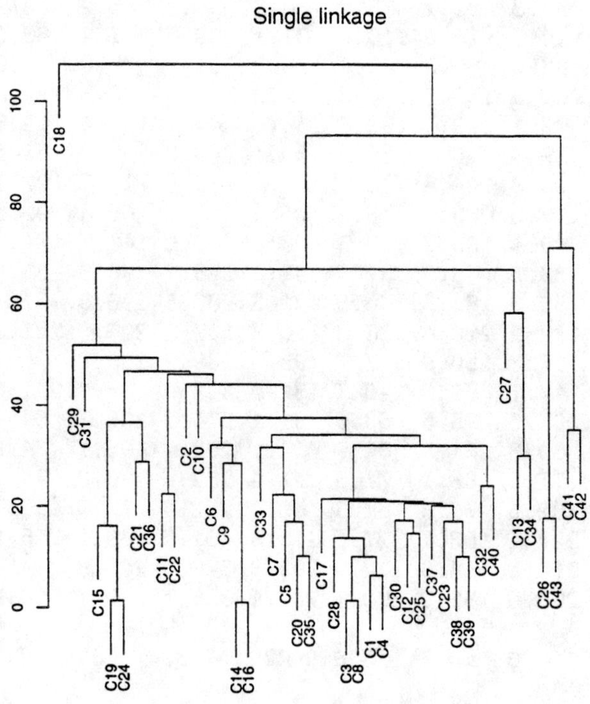

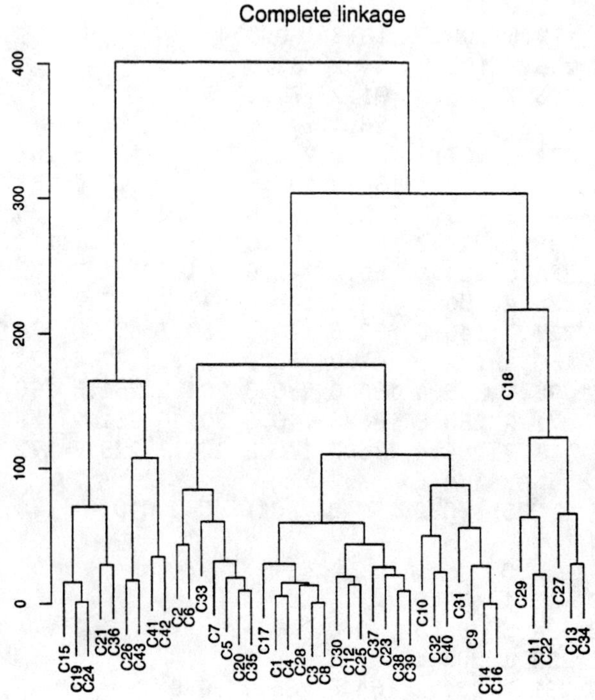

12.17. In K-means method, we use the means of the clusters identified by average linkage as the initial cluster centers.

Final cluster centers for K = 4

	1	2	3	4	5	6	7	8
1	110.0	2.1	0.9	215.0	0.7	15.3	7.9	50.0
2	114.4	3.1	1.7	171.1	2.8	15.0	6.6	123.9
3	86.7	2.3	0.5	26.7	1.4	10.0	5.8	55.8
4	112.5	3.2	0.8	225.0	5.8	12.5	10.8	245.0

Distances between centers

	1	2	3	4
1	0.0			
2	86.1	0.0		
3	190.0	162.2	0.0	
4	195.4	132.7	275.4	0.0

	K-means						4 clusters			
	K = 2		K = 3		K = 4		Single		Complete	
1	C1	1	C1	1	C1	1	C1	1	C1	1
2	C2	1	C2	1	C2	1	C2	1	C2	1
3	C3	1	C3	1	C3	1	C3	1	C3	1
4	C4	1	C4	1	C4	1	C4	1	C4	1
5	C5	1	C5	1	C5	1	C5	1	C5	1
6	C6	1	C6	1	C6	1	C6	1	C6	1
7	C7	1	C7	1	C7	1	C7	1	C7	1
8	C8	1	C8	1	C8	1	C8	1	C8	1
9	C9	1	C9	1	C9	1	C9	1	C9	1
10	C10	1	C10	1	C12	1	C10	1	C10	1
11	C12	1	C12	1	C15	1	C11	1	C12	1
12	C14	1	C14	1	C17	1	C12	1	C14	1
13	C15	1	C15	1	C19	1	C13	1	C16	1
14	C16	1	C16	1	C20	1	C14	1	C17	1
15	C17	1	C17	1	C23	1	C15	1	C20	1
16	C19	1	C19	1	C24	1	C16	1	C23	1
17	C20	1	C20	1	C25	1	C17	1	C25	1
18	C21	1	C23	1	C28	1	C19	1	C28	1
19	C23	1	C24	1	C30	1	C20	1	C30	1
20	C24	1	C25	1	C33	1	C21	1	C31	1
21	C25	1	C28	1	C35	1	C22	1	C32	1
22	C26	1	C30	1	C37	1	C23	1	C33	1
23	C28	1	C31	1	C38	1	C24	1	C35	1
24	C30	1	C32	1	C39	1	C25	1	C37	1
25	C32	1	C33	1	C10	2	C27	1	C38	1
26	C33	1	C35	1	C11	2	C28	1	C39	1
27	C35	1	C37	1	C14	2	C29	1	C40	1
28	C36	1	C38	1	C16	2	C30	1	C11	11
29	C37	1	C39	1	C22	2	C31	1	C13	11
30	C38	1	C40	1	C29	2	C32	1	C22	11
31	C39	1	C21	2	C31	2	C33	1	C27	11
32	C40	1	C26	2	C32	2	C34	1	C29	11
33	C41	1	C36	2	C40	2	C35	1	C34	11
34	C42	1	C41	2	C21	3	C36	1	C15	15
35	C43	1	C42	2	C26	3	C37	1	C19	15
36	C11	2	C43	2	C36	3	C38	1	C21	15
37	C13	2	C11	3	C41	3	C39	1	C24	15
38	C18	2	C13	3	C42	3	C40	1	C26	15
39	C22	2	C18	3	C43	3	C18	18	C36	15
40	C27	2	C22	3	C13	4	C26	26	C41	15
41	C29	2	C27	3	C18	4	C43	26	C42	15
42	C31	2	C29	3	C27	4	C41	41	C43	15
43	C34	2	C34	3	C34	4	C42	41	C18	18

12.18

(a)

NATION	1	2	3	4	5	6	7	8	9	10
1	.	26.394	19.547	20.844	8.627	9.924	12.618	29.441	7.349	10.176
2	26.394	.	7.063	5.607	17.787	16.510	38.922	3.114	19.530	16.720
3	19.547	7.063	.	2.106	10.961	9.645	31.991	9.996	12.843	10.196
4	20.844	5.607	2.106	.	12.244	10.973	33.352	8.668	13.938	11.145
5	8.627	17.787	10.961	12.244	.	1.347	21.170	20.836	2.702	2.729
6	9.924	16.510	9.645	10.973	1.347	.	22.433	19.546	3.722	2.639
7	12.618	38.922	31.991	33.352	21.170	22.433	.	41.945	19.643	22.542
8	29.441	3.114	9.996	8.668	20.836	19.546	41.945	.	22.602	19.803
9	7.349	19.530	12.843	13.938	2.702	3.722	19.643	22.602	.	2.906
10	10.176	16.720	10.196	11.145	2.729	2.639	22.542	19.803	2.906	.
11	13.206	13.347	6.669	7.762	4.679	3.491	25.669	16.408	6.218	3.550
12	55.207	81.570	74.641	75.996	63.796	65.070	42.664	84.601	62.172	65.057
13	7.910	20.899	14.722	15.464	5.556	6.477	19.517	24.002	3.572	4.818
14	20.770	7.192	2.958	4.283	12.427	11.114	33.044	9.631	14.571	12.181
15	26.815	2.880	8.218	6.327	18.277	17.059	39.322	4.431	19.718	16.829
16	25.434	51.791	44.859	46.236	34.028	35.296	12.923	54.809	32.533	35.426
17	24.707	2.093	5.237	4.131	16.113	14.813	37.188	4.786	17.953	15.210
18	23.438	2.984	4.302	2.648	14.835	13.564	35.968	6.069	16.559	13.752
19	21.863	6.118	3.408	4.118	13.472	12.158	34.185	8.467	15.585	13.118
20	30.402	4.099	10.933	9.623	21.799	20.506	42.891	1.007	23.555	20.757
21	29.126	2.778	9.737	8.339	20.527	19.244	41.642	0.534	22.275	19.467
22	3.883	30.144	23.264	24.572	12.378	13.658	8.844	33.188	10.843	13.736
23	36.641	62.973	56.015	57.426	45.219	46.478	24.112	65.980	43.737	46.633
24	22.370	4.093	3.148	1.566	13.761	12.480	34.868	7.131	15.474	12.685
25	9.655	35.821	28.851	30.273	18.104	19.341	3.349	38.825	16.714	19.610
26	22.816	49.148	42.201	43.595	31.392	32.652	10.286	52.161	29.915	32.812
27	29.189	3.879	10.347	8.573	20.631	19.402	41.700	3.450	22.099	19.212
28	18.062	9.057	4.449	3.984	9.664	8.556	30.562	12.140	10.955	8.060
29	26.847	1.350	7.715	6.039	18.256	16.996	39.364	3.179	19.859	17.010
30	28.057	3.688	9.451	7.628	19.521	18.307	40.546	4.291	20.922	18.029
31	3.315	28.772	21.797	23.230	11.147	12.346	10.268	31.763	9.943	12.788
32	14.013	13.331	7.488	7.936	6.046	5.236	26.380	16.419	6.786	3.914
33	4.477	27.041	20.007	21.532	9.864	10.913	12.518	29.982	8.720	11.423
34	4.647	23.040	16.383	17.458	5.829	7.057	16.386	26.120	3.553	6.335
35	4.023	30.181	23.307	24.615	12.417	13.704	8.909	33.233	10.864	13.748
36	82.725	109.040	102.068	103.503	91.306	92.560	70.197	112.030	89.824	92.720
37	20.019	6.892	3.335	2.155	11.489	10.296	32.527	9.973	12.934	10.059
38	26.144	1.388	7.128	5.393	17.554	16.303	38.681	3.870	19.163	16.307
.										
.										
.										

(b)

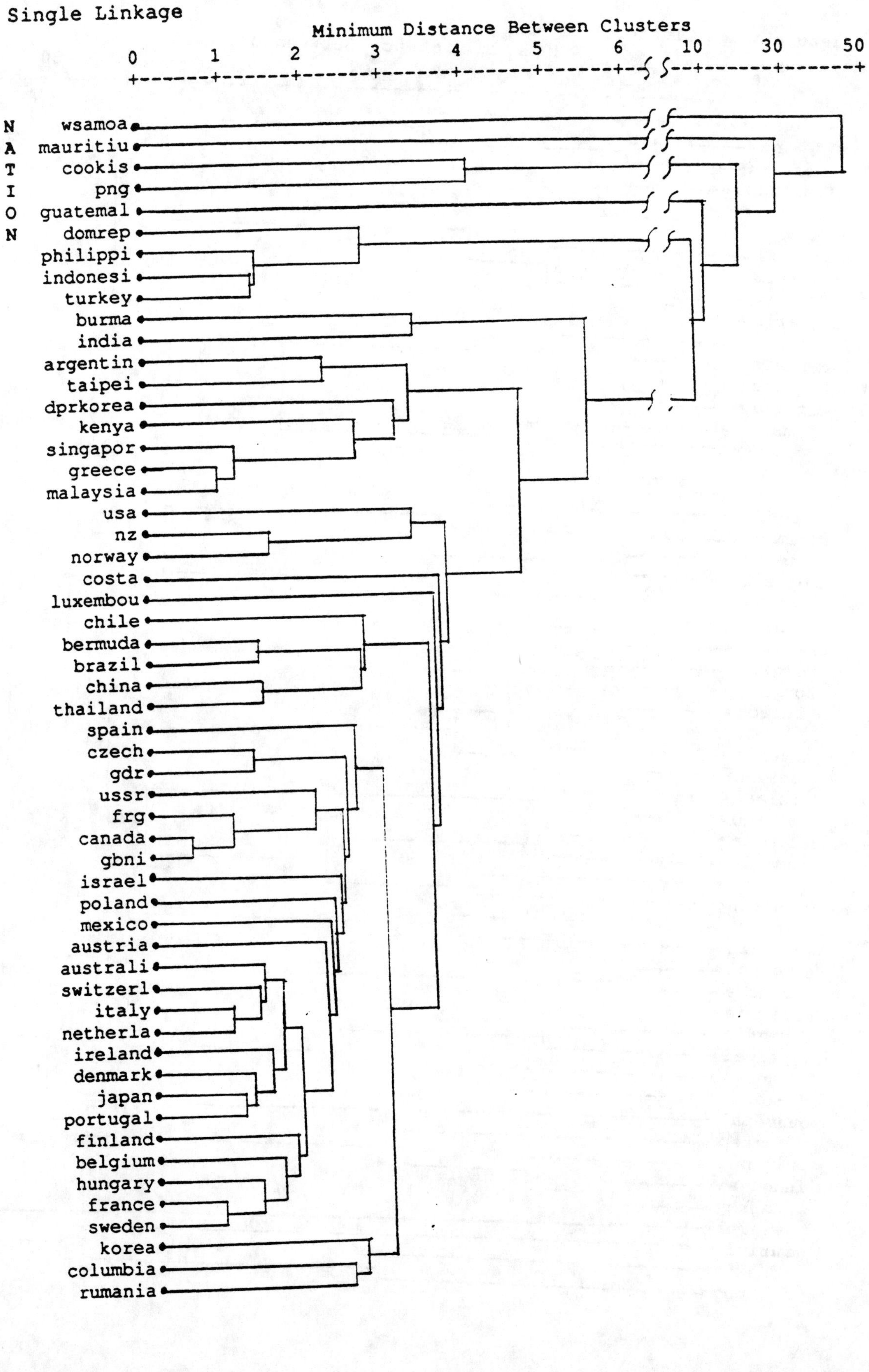
Single Linkage
Minimum Distance Between Clusters
0
1
2
3
4
5
6
10
30
50
NATION
wsamoa
mauritiu
cookis
png
guatemal
domrep
philippi
indonesi
turkey
burma
india
argentin
taipei
dprkorea
kenya
singapor
greece
malaysia
usa
nz
norway
costa
luxembou
chile
bermuda
brazil
china
thailand
spain
czech
gdr
ussr
frg
canada
gbni
israel
poland
mexico
austria
australi
switzerl
italy
netherla
ireland
denmark
japan
portugal
finland
belgium
hungary
france
sweden
korea
columbia
rumania

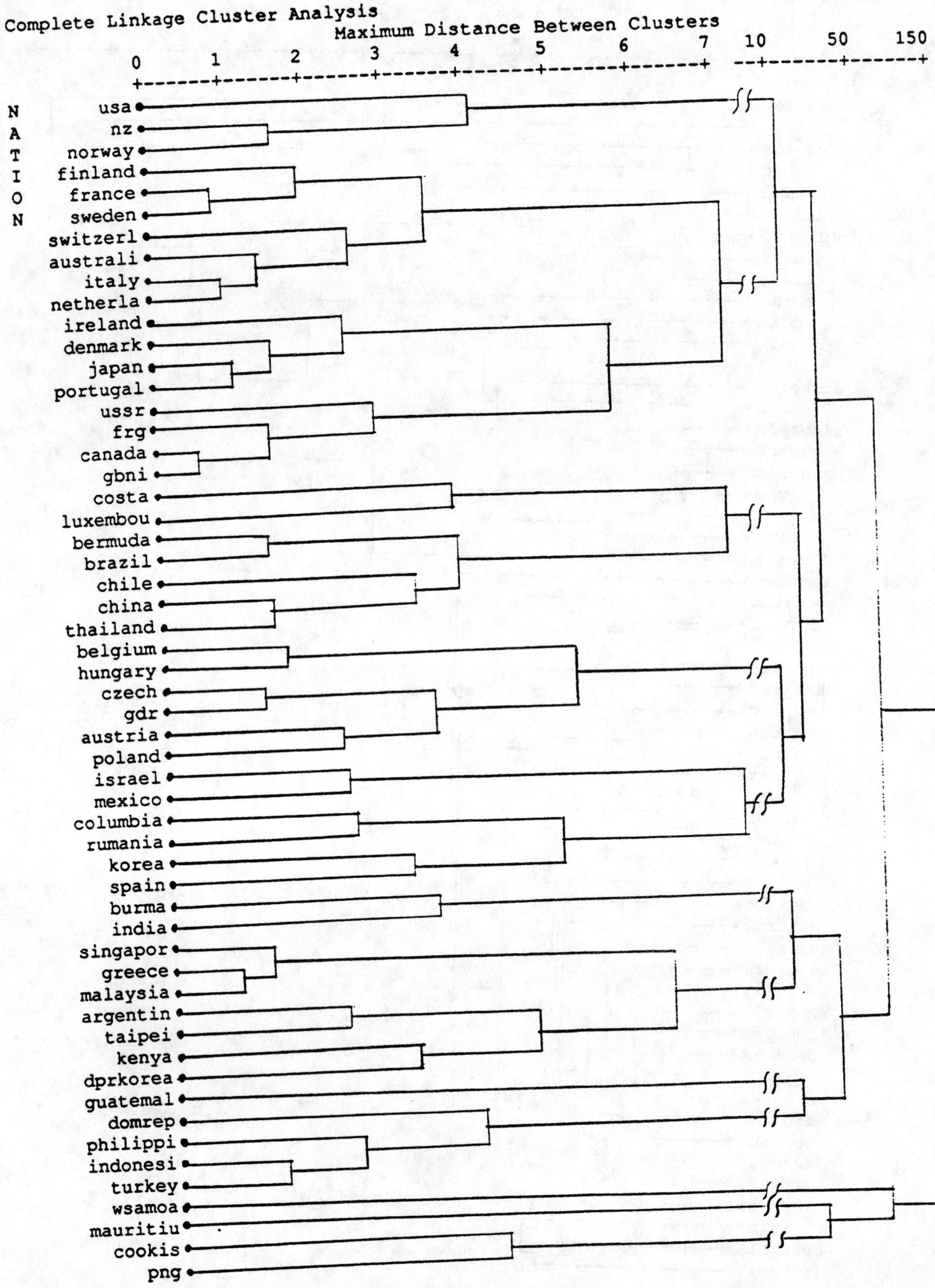

Complete Linkage Cluster Analysis
Maximum Distance Between Clusters
0 1 2 3 4 5 6 7 10 50 150
NATION
usa
nz
norway
finland
france
sweden
switzerl
australi
italy
netherla
ireland
denmark
japan
portugal
ussr
frg
canada
gbni
costa
luxembou
bermuda
brazil
chile
china
thailand
belgium
hungary
czech
gdr
austria
poland
israel
mexico
columbia
rumania
korea
spain
burma
india
singapor
greece
malaysia
argentin
taipei
kenya
dprkorea
guatemal
domrep
philippi
indonesi
turkey
wsamoa
mauritiu
cookis
png

(c)

	CLUSTER=1	CLUSTER=2	CLUSTER=3	CLUSTER=4
Single Linkage	argentin australi austria belgium bermuda brazil burma canada chile china columbia costa czech denmark domrep dprkorea finland france frg gbni gdr greece guatemal hungary india indonesi ireland israel italy japan kenya korea luxembou malaysia mexico netherla norway nz philippi poland portugal rumania singapor spain sweden switzerl taipei thailand turkey usa ussr	cookis png	mauritiu	wsamoa
Complete Linkage	australi austria belgium bermuda brazil canada chile china columbia costa czech denmark finland france frg gbni gdr hungary ireland israel italy japan korea luxembou mexico netherla norway nz poland portugal rumania spain sweden switzerl thailand usa ussr	argentin burma domrep dprkorea greece guatemal india indonesi kenya malaysia philippi singapor taipei turkey	cookis mauritiu png	wsamoa
K-means	australi austria belgium canada columbia czech denmark finland france frg gbni gdr hungary ireland israel italy japan korea mexico netherla norway nz poland portugal rumania spain sweden switzerl usa ussr	argentin bermuda brazil burma chile china costa domrep dprkorea greece india indonesi kenya luxembou malaysia philippi singapor taipei thailand turkey	cookis guatemal mauritiu png	wsamoa

12.19

(a)

	1	2	3	4	5	6	7	8	9	10
1	.	9.8384	2.5737	8.2443	9.0536	4.9629	2.9510	7.8043	3.7706	4.2318
2	9.8384	.	7.8910	1.7879	18.6383	4.9655	12.5797	2.1060	6.1686	6.0812
3	2.5737	7.8910	.	6.2190	11.2193	3.3771	5.2124	5.9166	2.5539	2.8845
4	8.2443	1.7879	6.2190	.	17.0420	3.4279	11.0006	1.1002	4.6472	4.6250
5	9.0536	18.6383	11.2193	17.0420	.	13.6879	7.1659	16.6973	12.7025	13.2729
6	4.9629	4.9655	3.3771	3.4279	13.6879	.	7.8045	3.0717	1.5524	2.3096
7	2.9510	12.5797	5.2124	11.0006	7.1659	7.8045	.	10.5105	6.4615	6.6410
8	7.8043	2.1060	5.9166	1.1002	16.6973	3.0717	10.5105	.	4.1431	4.0067
9	3.7706	6.1686	2.5539	4.6472	12.7025	1.5524	6.4615	4.1431	.	1.2885
10	4.2318	6.0812	2.8845	4.6250	13.2729	2.3096	6.6410	4.0067	1.2885	.
11	6.6090	3.4659	4.6236	1.8746	15.5563	2.2289	9.2606	1.5609	3.0489	2.8160
12	28.5954	38.4155	30.7670	36.8130	20.3588	33.5192	25.8952	36.3708	32.2811	32.5275
13	2.5354	9.4291	2.6704	7.8417	10.6773	5.1711	3.7501	7.3787	3.8163	3.4945
14	3.8467	6.1263	2.0421	4.4774	12.5921	1.3779	6.7169	4.1997	1.3063	2.1671
15	7.1444	2.7895	5.2333	1.4149	16.0734	2.5275	9.8271	0.8287	3.4962	3.3225
16	16.5587	26.2294	18.6754	24.6219	7.6115	21.2807	14.3397	24.2730	20.2572	20.7581
17	7.2631	2.7313	5.2084	1.0318	16.1079	2.5754	10.0073	1.1167	3.7157	3.6727
18	5.8753	4.0541	3.9622	2.4513	14.6293	1.1034	8.7106	2.2079	2.4359	2.8072
19	8.3409	1.6746	6.3245	0.4535	17.0932	3.4774	11.1337	1.1501	4.7628	4.8047
20	6.2967	3.9608	4.3757	2.3768	14.8316	1.5895	9.1850	2.4695	3.0658	3.5964
21	9.0778	0.9274	7.0708	1.0075	17.8641	4.2150	11.8448	1.5009	5.4439	5.4132
22	3.3146	6.7399	2.0297	5.1797	12.2210	2.2438	5.9913	4.6937	1.4714	1.6832
23	2.6433	12.0305	4.7059	10.4407	7.8642	7.3268	0.8506	9.9692	5.9462	6.0596
24	5.3200	4.5478	3.5076	2.9904	14.2486	1.0241	8.0518	2.5129	1.7135	1.8541
25	5.9139	4.2257	4.2916	2.7170	14.7996	1.6679	8.5309	2.3565	2.2677	2.2675
26	11.2883	21.0807	13.4284	19.4510	3.0775	16.1394	8.9136	19.0788	15.0220	15.4282
27	5.6692	4.4114	3.5969	2.7647	14.6093	1.7289	8.3389	2.4704	2.2929	2.1253
28	1.1918	9.8564	2.3304	8.2472	9.4797	5.1877	2.9580	7.8026	3.9073	4.0683
29	7.2095	2.8519	5.2042	1.5365	15.9386	2.4736	10.0364	1.3214	3.7669	3.9014
30	9.3103	1.3193	7.3354	1.5956	18.2574	4.6572	11.9319	1.6963	5.6407	5.3294
31	8.4893	1.6133	6.4500	0.3017	17.2686	3.6659	11.2501	1.3150	4.8956	4.8827
32	1.4839	8.4324	1.7159	6.8514	10.5299	3.6496	4.1816	6.3769	2.3574	2.7614
33	6.9851	4.6223	5.6233	3.6968	15.9161	3.6907	9.1749	3.0837	3.6379	2.8971
34	3.6275	13.3208	5.5842	11.6862	5.8423	8.4503	2.3082	11.3111	7.3424	7.7518
35	16.4833	26.1084	18.5551	24.4869	7.5132	21.1656	14.3252	24.1718	20.1700	20.6939
.										
.										
.										

12.20

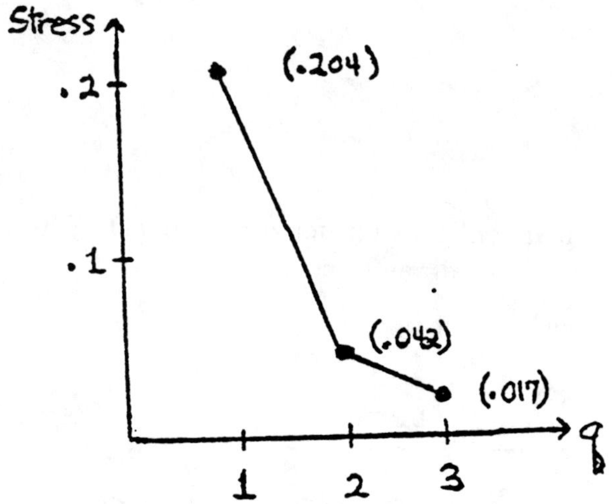

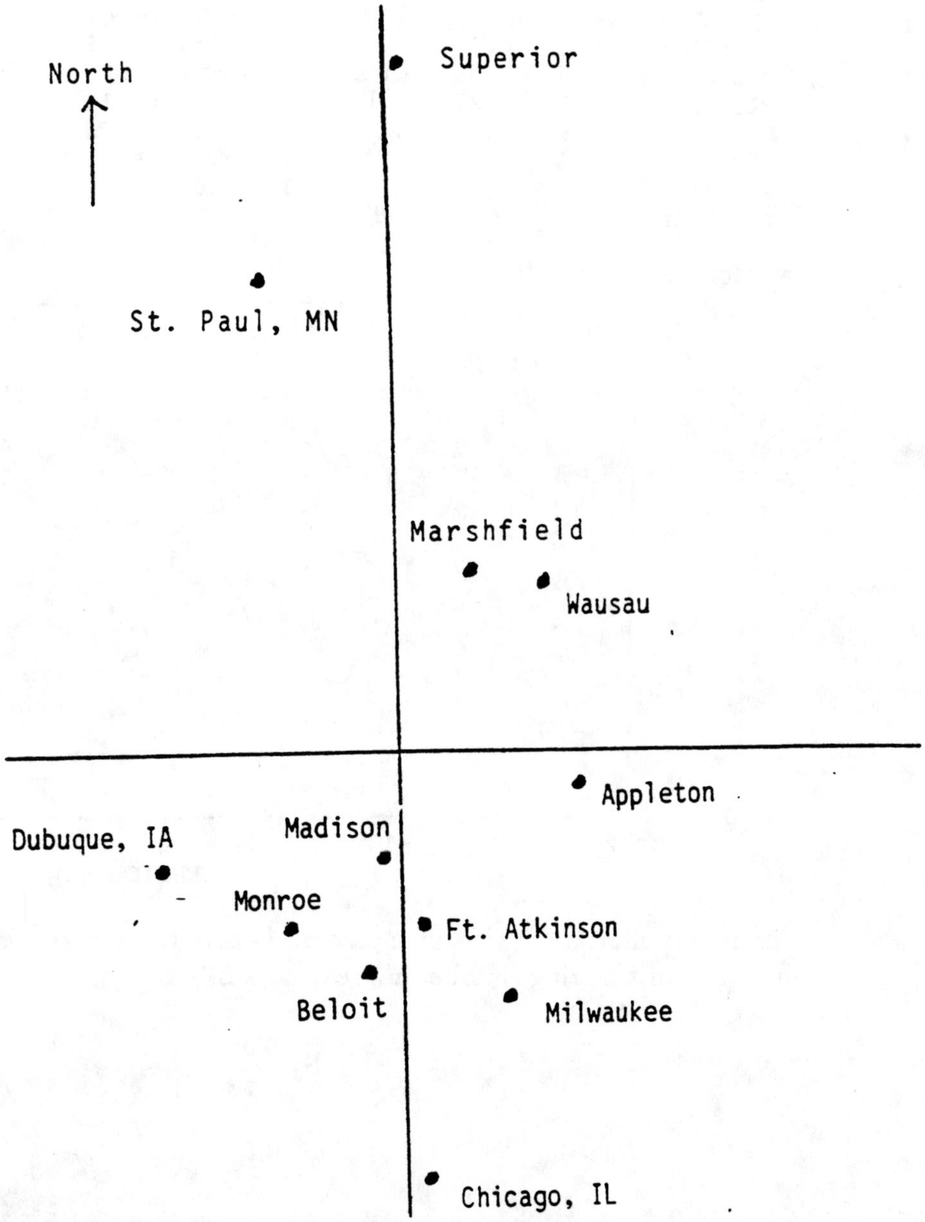

The configuration of these cities by multidimensional scaling (roughly) corresponds to the locations of the cities on a map of the region.

12.21

The stress of final configuration for q=5 is 0.000. The sites in 5 dimensions and the plot of the sites in two dimensions are

COORDINATES IN 5 DIMENSIONS

VARIABLE	PLOT	DIMENSION 1	2	3	4	5
P1980918	A	.51	-.28	.24	-.68	.12
P1931131	B	-1.32	.69	.62	-.05	-.02
P1550960	C	.47	-.07	.19	.30	.06
P1530987	D	.39	.09	.05	.34	.10
P1361024	E	.23	.30	-.32	.05	.12
P1351005	F	.47	.14	-.22	-.14	-.28
P1340945	G	.58	-.35	.46	.18	-.10
P1311137	H	-1.12	-1.12	-.31	.05	-.01
P1301062	I	-.22	.61	-.70	-.06	.01

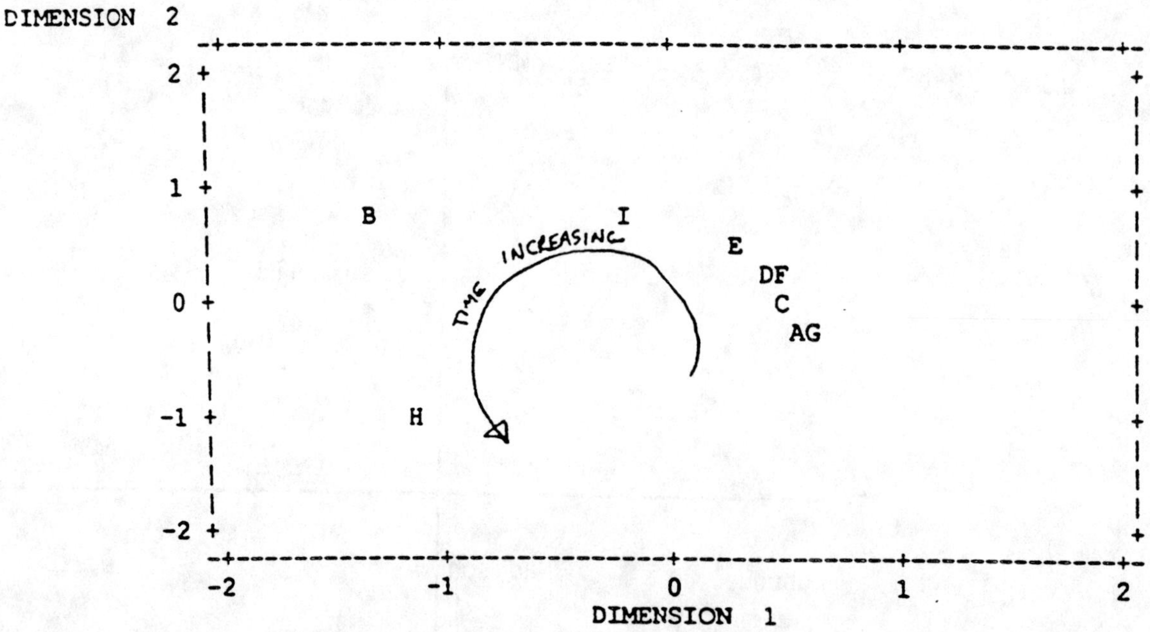

The results show a definite time pattern (where time of site is frequently determined by C-14 and tree ring (lumber in great houses) dating).

12.22 A correspondence analysis of the mental health-socioeconomic data

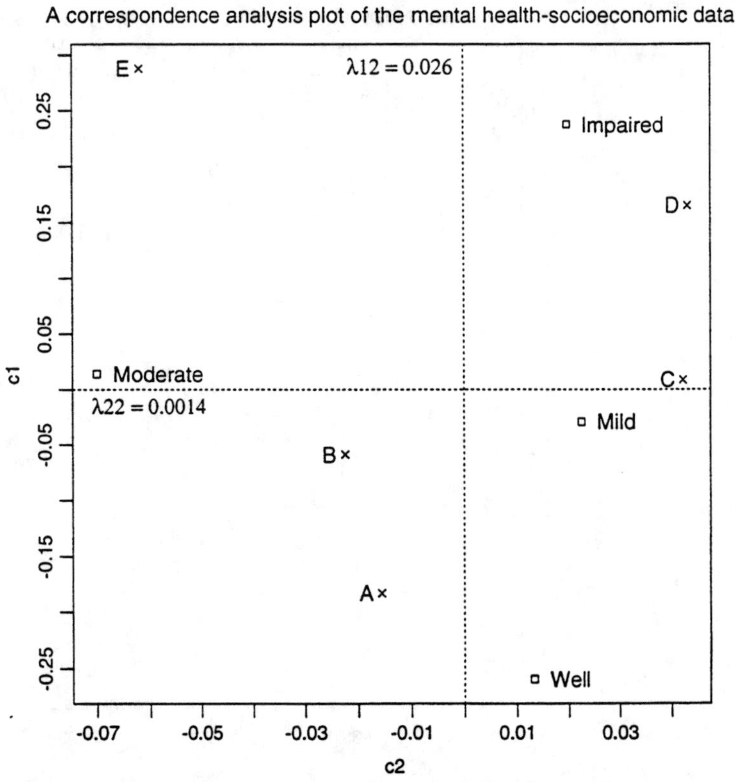


```
U                                    V
-0.6922  0.1539  0.5588 0.4300      -0.6266 -0.2313  0.0843 -0.3341
-0.1100  0.3665 -0.7007 0.6022      -0.1521 -0.2516 -0.5109 -0.6407
 0.0411 -0.8809 -0.0659 0.4670       0.0265  0.5490  0.5869 -0.5756
 0.7121  0.2570  0.4388 0.4841       0.4097  0.4668 -0.5519 -0.2297
                                     0.6448 -0.6032  0.2879 -0.3062

lambda
0.1613 0.0371 0.0082 0.0000

Cumulative inertia
0.0260 0.0274 0.0275

Cumulative proportion
0.9475 0.9976 1.0000
```

The lowest economic class is located between moderate and impaired. The next lowest class is closest to impaired.

12.23. A correspondence analysis of the income and job satisfaction data

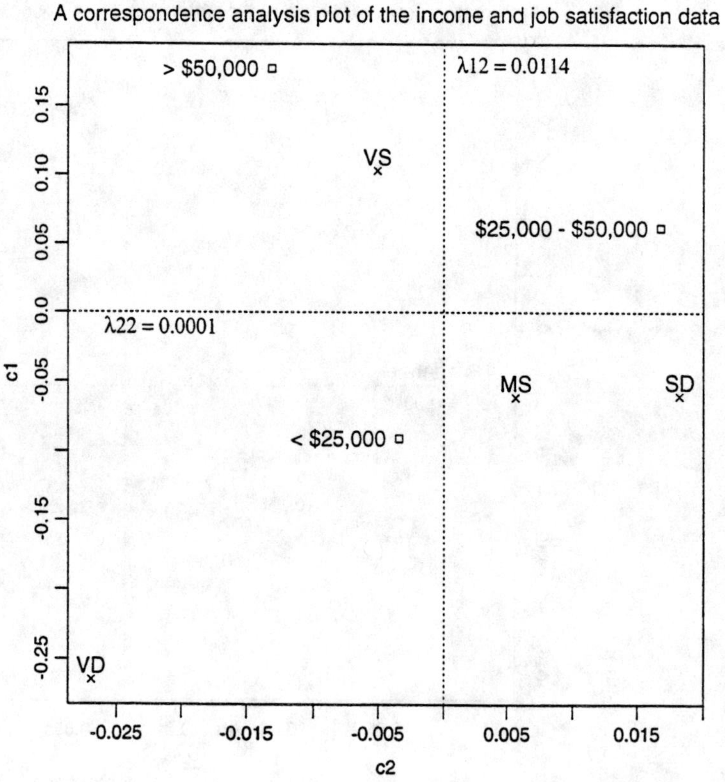


```
U                                V
-0.6272 -0.2392 0.7412           -0.6503 -0.6661 -0.3561
 0.2956  0.8073 0.5107           -0.1944  0.5933 -0.7758
 0.7206 -0.5394 0.4356           -0.3400  0.3159  0.2253
                                  0.6510 -0.3233 -0.4696

lambda
0.1069 0.0106 0.0000

Cumulative inertia
0.0114 0.0116

Cumulative proportion
0.9902 1.0000
```

Very satisfied is closest to the highest income group, and very dissatisfied is below the lowest income group. Satisfaction appears to increase with income.

12.24. A correspondence analysis of the Wisconsin forest data

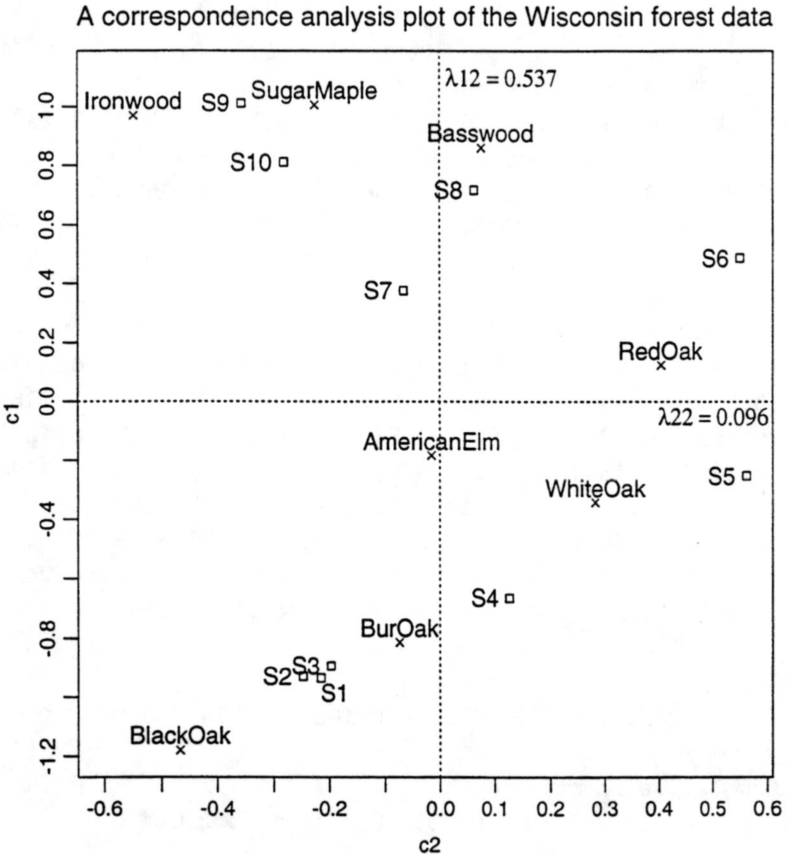

U

-0.3877	-0.2108	-0.0616	0.4029	-0.0582	0.3269	0.4247	-0.1590
-0.3856	-0.2428	-0.0106	0.4345	-0.1950	-0.1968	-0.2635	-0.3835
-0.3495	-0.1821	0.4079	-0.5718	0.2343	-0.1167	0.3294	-0.1272
-0.3006	0.1355	0.0540	-0.2646	0.0006	-0.0826	-0.6644	-0.3192
-0.1108	0.5817	-0.4856	-0.1598	-0.2333	0.1607	0.0772	-0.0518
0.2022	0.5400	0.4626	0.2687	-0.0978	-0.3943	0.2668	-0.3606
0.1852	-0.0756	-0.5090	-0.0291	0.6026	-0.1955	0.1520	-0.5154
0.3140	0.0644	0.3394	0.1567	0.3366	0.6573	-0.2507	-0.2267
0.4200	-0.3484	-0.0394	0.1165	-0.0625	-0.3772	-0.1456	0.1381
0.3549	-0.2897	-0.0345	-0.3393	-0.5994	0.2002	0.1262	-0.4907

V

-0.3904	-0.0831	-0.4781	0.4562	-0.0377	0.3369	0.4071	-0.3511
-0.5327	-0.4985	0.4080	0.0925	-0.0738	-0.3420	-0.2464	-0.3310
-0.1999	0.3889	0.4089	-0.3622	0.4391	0.3217	0.1808	-0.4260
0.0698	0.5382	-0.1726	0.3181	-0.0544	-0.1596	-0.6122	-0.4138
-0.0820	-0.0151	-0.4271	-0.7086	-0.4160	-0.1685	0.0307	-0.3258
0.4005	0.0831	0.1478	0.1866	-0.0042	-0.5895	0.5587	-0.3412
0.3634	-0.4850	-0.3232	-0.0937	0.6298	0.0164	-0.2172	-0.2745
0.4689	-0.2476	0.3150	0.0726	-0.4771	0.5142	-0.0763	-0.3412

lambda
0.7326 0.3101 0.2685 0.2134 0.1052 0.0674 0.0623 0.0000

Cumulative inertia
0.5367 0.6329 0.7050 0.7506 0.7616 0.7662 0.7700

Cumulative proportion
0.6970 0.8219 0.9155 0.9747 0.9891 0.9950 1.0000

12.25. We construct biplot of the pottery type-site data, with row proportions as variables.

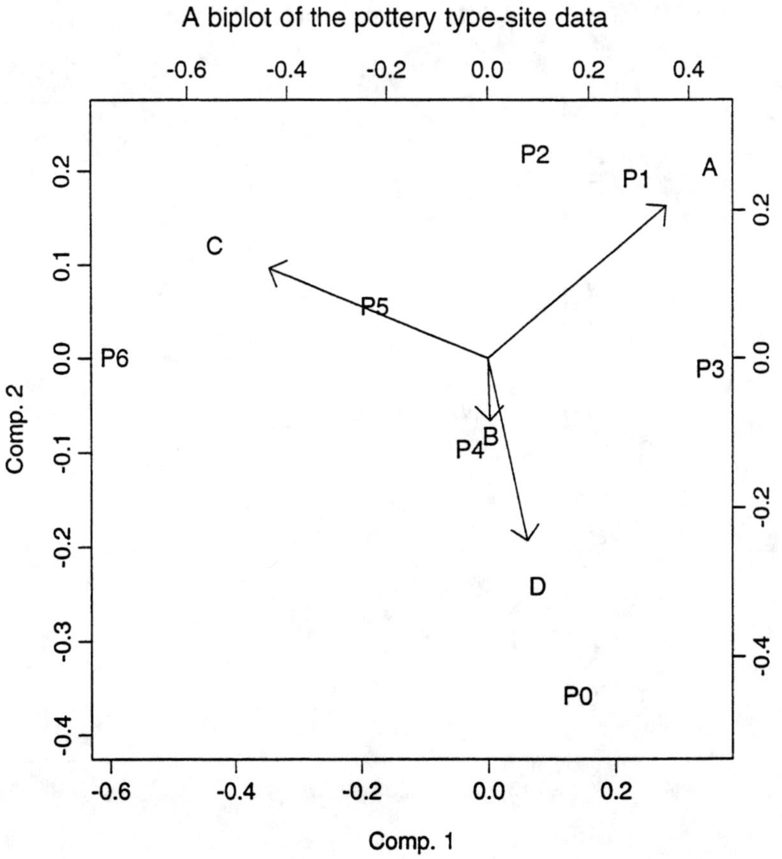

S				Eigenvectors of S			
0.0511	-0.0059	-0.0390	-0.0061	0.6233	0.5853	0.1374	-0.5
-0.0059	0.0084	-0.0051	0.0025	0.0064	-0.2385	-0.8325	-0.5
-0.0390	-0.0051	0.0628	-0.0187	-0.7694	0.3464	0.1951	-0.5
-0.0061	0.0025	-0.0187	0.0223	0.1396	-0.6932	0.5000	-0.5

Eigenvalues of S
0.0978 0.0376 0.0091 0.0000

	pc1	pc2	pc3	pc4
St. Dev.	0.3128	0.1940	0.0952	0
Prop. of Var.	0.6769	0.2604	0.0627	0
Cumulative Prop.	0.6769	0.9373	1.0000	1

As in the correspondence analysis.

12.26. We construct biplot of the mental health-socioeconomic data, with column proportions as variables.

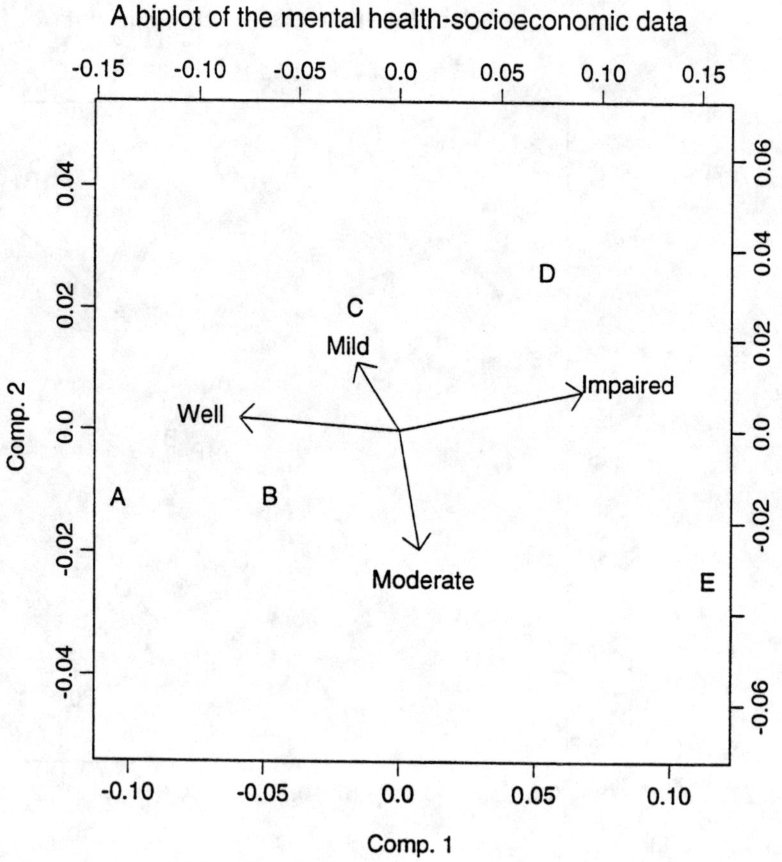

S

0.003089	0.000809	-0.000413	-0.003485
0.000809	0.000329	-0.000284	-0.000853
-0.000413	-0.000284	0.000379	0.000318
-0.003485	-0.000853	0.000318	0.004021

Eigenvectors of S

-0.6487	0.0837	-0.5676	0.5
-0.1685	0.4764	0.7033	0.5
0.0794	-0.8320	0.2270	0.5
0.7379	0.2719	-0.3628	0.5

Eigenvalues of S

0.007314 0.000480 0.000024 0.000000

	pc1	pc2	pc3	pc4
St. Dev.	0.0855	0.0219	0.0049	0
Prop. of Var.	0.9355	0.0614	0.0031	0
Cumulative Prop.	0.9355	0.9969	1.0000	1

The biplot gives similar locations for health and socioeconomic status. A reflection about the 45 degree line would make them appear more alike.

12.27. A Procrustes analysis of archaeological data

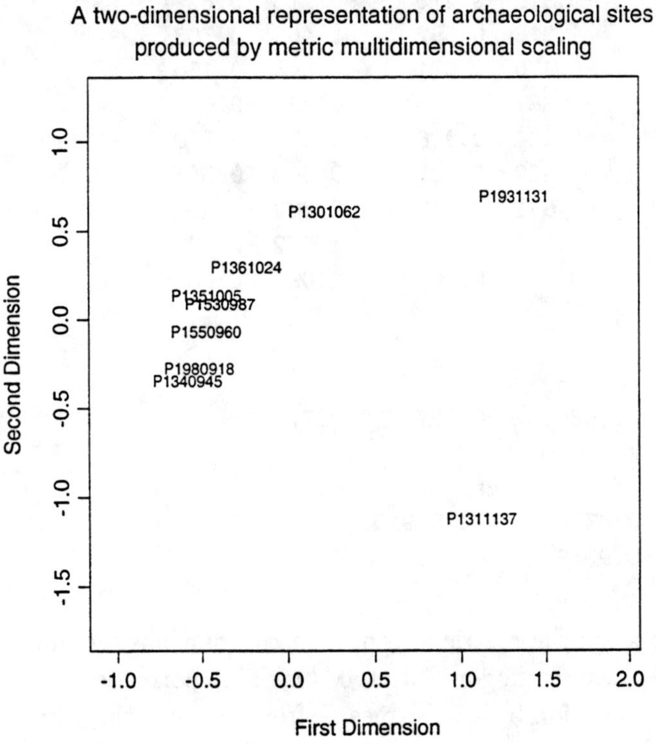

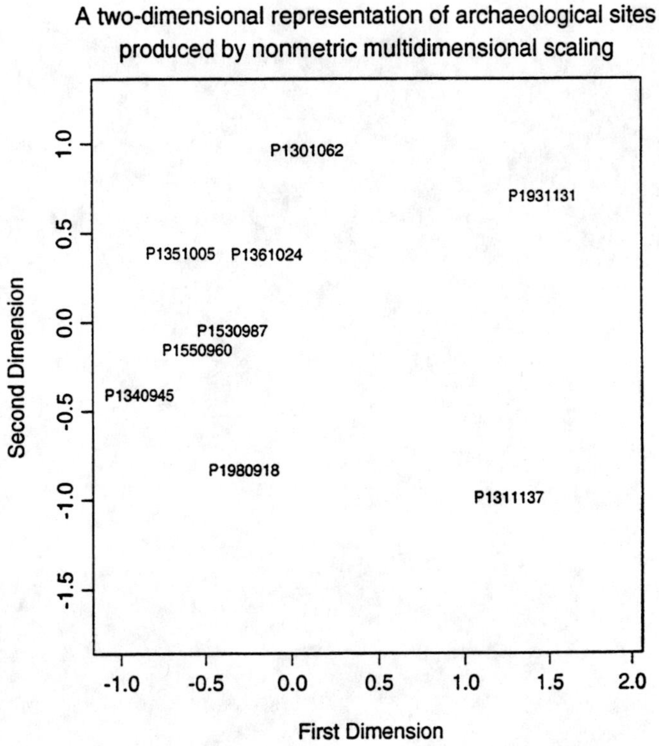

Site	Metric MDS		Nonmetric MDS	
P1980918	-0.512	-0.278	-0.276	-0.829
P1931131	1.318	0.692	1.469	0.703
P1550960	-0.470	-0.071	-0.545	-0.156
P1530987	-0.387	0.088	-0.338	-0.048
P1361024	-0.234	0.296	-0.137	0.379
P1351005	-0.469	0.137	-0.642	0.387
P1340945	-0.581	-0.349	-0.889	-0.409
P1311137	1.118	-1.122	1.262	-0.989
P1301062	0.216	0.608	0.096	0.963

U		V	
-0.9893	-0.1459	-0.9977	-0.0679
-0.1459	0.9893	-0.0679	0.9977

Q		Lambda	
0.9969	0.0784	4.7819	0.000
-0.0784	0.9969	0.0000	2.715

To better align the metric and nonmetric solutions, we multiply the nonmetric scaling solution by the orthogonal matrix $\hat{\mathbf{Q}}$. This corresponds to clockwise rotation of the nonmetric solution by 4.5 degrees. After rotation, the sum of squared distances, 0.803, is reduced to the Procrustes measure of fit $PR^2 = 0.756$.